Advances in
INORGANIC CHEMISTRY

Volume 58

Advances in
INORGANIC CHEMISTRY
Including Bioinorganic Studies

EDITED BY

Rudi van Eldik

Institute for Inorganic Chemistry
University of Erlangen-Nürnberg
91058 Erlangen
Germany

Jan Reedijk

Leiden Institute of Chemistry
Gorlaeus Laboratories
Leiden University
PO Box 9502, 2300 RA Leiden
The Netherlands

VOLUME 58: Homogeneous Biomimetic Oxidation Catalysis

ELSEVIER

AMSTERDAM • BOSTON • HEIDELBERG • LONDON • NEW YORK • OXFORD
PARIS • SAN DIEGO • SAN FRANCISCO • SINGAPORE • SYDNEY • TOKYO
Academic Press is an imprint of Elsevier

ACADEMIC PRESS

Academic Press is an imprint of Elsevier
84 Theobald's Road, London WC1X 8RR, UK
Radarweg 29, PO Box 211, 1000 AE, Amsterdam, The Netherlands
30 Corporate Drive, Suite 400, Burlington, MA 01803, USA
525 B Street, Suite 1900, San Diego, California 92101-4495, USA

This book is printed on acid-free paper ♾

Library of Congress Cataloging in Publication Data

British Library Cataloguing in Publication Data

ISBN–13: 978-0-12-023658-9
ISBN–10: 0-12-023658-3
ISSN: 0898-8838

For information on all Academic Press publications
visit our web site at http://books.elsevier.com

Printed and bound in The Netherlands

06 07 08 09 10 11   10 9 8 7 6 5 4 3 2 1

CONTENTS

Ligand Influences in Copper-Dioxygen Complex-Formation and Substrate Oxidations

LANYING Q. HATCHER and KENNETH D. KARLIN

Biomimetic Oxidations by Dinuclear and Trinuclear Copper Complexes

GIUSEPPE BATTAINI, ALESSANDRO GRANATA, ENRICO MONZANI,
MICHELE GULLOTTI and LUIGI CASELLA

Green Oxidation of Alcohols using Biomimetic Cu Complexes and Cu Enzymes as Catalysts

ISABEL W.C.E ARENDS, PATRICK GAMEZ and ROGER A. SHELDON

PREFACE

This thematic issue focuses on homogeneous biomimetic oxidation catalysis. I have invited Professor Jan Reedijk (Leiden University, The Netherlands) to be the co-editor of this volume, since he has an active interest in this area and edited a monograph on Bioinorganic Catalysis in 1999. Jan Reedijk is a well-known coordination chemist for his wide interest in transition-metal chemistry and its application in biological and environmental processes.

The present volume consists of six contributions from groups that are all actively involved in oxidation catalysis. The first contribution comes from Albrecht Berkessel (University of Cologne, Germany) and focuses on diversity-based approaches to selective biomimetic oxidation catalysis. It covers such general themes as combinatorial methods in catalysis research, generation of libraries and screening methods, and applies these to biomimetic oxidation catalysis, both based on metal complexes and on purely organic catalysts. In the second contribution, Stefania Tanase and Elisabeth Bouwman (Leiden University, The Netherlands) present the selective conversion of hydrocarbons with H_2O_2 using biomimetic non-heme iron and manganese oxidation catalysis. The catalytic processes covered include alkane hydroxylation, *cis*-dihydroxylation, aromatic hydroxylation, and olefin epoxidation. The third contribution on DNA oxidation by copper and manganese complexes comes from the CNRS in Toulouse and is written by Marguerite Pitié, Christophe Boldron and Geneviève Pratviel. It presents a systematic treatment of oxidative DNA damage by copper and manganese complexes. The next contribution comes from the Johns Hopkins University (Baltimore, USA) and presents the influences of the ligand in copper-dioxygen complex-formation and substrate oxidations. The authors, Lanying Hatcher and Kenneth Karlin, present tetradentate ligand influences, tridentate ligand influences, and ligand effects on the exogenous substrate reactivity of Cu_2O_2 compounds. The fifth contribution comes from the Universities of Pavia and Milano in Italy, is written by Giuseppe Battaini, Alessandro Granata, Enrico Monzani, Michele Gullotti, and Luigi Casella, and presents biomimetic oxidations by dinuclear and trinuclear copper complexes. The authors concentrate on dinuclear complexes as tyrosinase and catechol oxidase models, phenolase activity, and catecholase activity. In the final contribution, Isabel Arends, Patrick Gamez, and Roger Sheldon (Delft University of Technology, The Netherlands) focus on green oxidation of alcohols using biomimetic copper complexes and enzymes as catalysts. They discuss enzymatic systems using copper oxidases and present a detailed account of various possible copper catalyst systems used in biomimetic oxidation processes.

The mentioned authors are gratefully acknowledged for their efficient collaboration and remarkable efforts made to meet all set deadlines for the production of this volume.

I thoroughly believe that these contributions cover important advances in inorganic and bioinorganic chemistry, since information on biomimetic oxidation catalysis is of fundamental interest to so many environmental, biological, and industrial processes. I trust that the inorganic and bioinorganic chemistry community will specially benefit from this thematic volume.

Rudi van Eldik
University of Erlangen-Nürnberg
Germany
August 2005

DIVERSITY-BASED APPROACHES TO SELECTIVE BIOMIMETIC OXIDATION CATALYSIS

ALBRECHT BERKESSEL

Institut für Organische Chemie der Universität zu Köln, Greinstr. 4, D-50939 Köln, Germany

I. Introduction

The development of methods for the selective and atom-economic oxidation/oxygenation of organic substrates belongs to the most challenging fields of contemporary chemical research, both in the academic and the industrial environment (*1*). Whereas breath-taking turnover numbers and selectivities have been achieved for numerous reductive processes (such as catalytic asymmetric hydrogenation) (*2*), oxidative transformations such as the epoxidation of olefins still lag behind. Biomimetic oxidation catalysis aims at achieving the efficiencies and selectivities of enzymes, such as monooxygenases or peroxidases, with low-molecular weight compounds (*3*). The design of biomimetic oxidation catalysts is based on the active-site structure and function of oxidative enzymes. For the sake of atom-economy and sustainability, terminal oxidants such as molecular oxygen or dihydrogen peroxide are employed preferentially (*3*). It is an intrinsic feature of biomimetic oxidation catalysts that they aim at imitating *soluble* enzymes, which means

ADVANCES IN INORGANIC CHEMISTRY
VOLUME 58 ISSN 0898-8838 / DOI: 10.1016/S0898-8838(05)58001-2

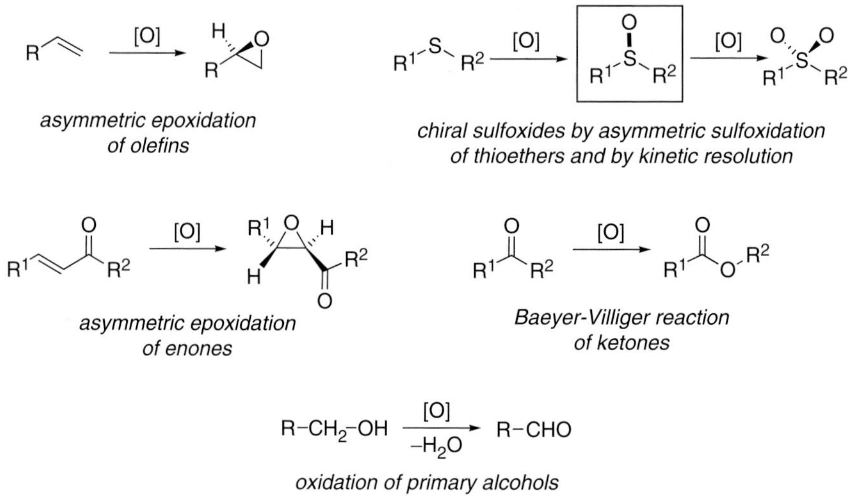

SCHEME 1. Oxygen transfer and dehydrogenation reactions treated in this chapter.

that they are generally *homogeneous* in nature. In some cases, immobilized variants have been prepared and will be discussed as well. On the other hand, purely heterogeneous (and non-biomimetic) oxidation catalysts such as transition metal mixed oxides or silicalites are not covered by this chapter.

The majority of the homogeneous and biomimetic oxidation catalysts being investigated to date fall into two classes: (i) Metal–ligand (peptidic or non-peptidic) combinations, (ii) metal-free organocatalysts. Typical reactions effected by the former are electrophilic oxygen transfer to e.g., C=C double bonds or other nucleophilic substrates (such as thioethers), whereas the latter have been used both for this purpose and to effect the addition of nucleophilic oxidants to acceptors such as enones. The biomimetic Baeyer-Villiger reaction of ketones, using dioxygen or dihydrogen peroxide, is another important oxidative transformation. Dehydrogenations, such as the oxidation of alcohols, form a fourth important class of reactions (Scheme 1).

The identification of an active and selective catalyst for a given transformation is a multi-dimensional problem, involving *inter alia* optimization of the ligand structure, the nature of the metal ion, etc. (Scheme 2). The classical approach to catalyst discovery and optimization involves one-by-one synthesis and evaluation of catalyst candidates. However, both of the above classes of biomimetic oxidation catalysts lend themselves excellently for combinatorial variation and optimization: In metal complexes, alterations of the ligand structure, the type of metal, counterions, etc. allow rapid diversification. The same holds for purely organic catalysts. In a subsequent step, appropriate screening methods identify the library member(s) having the desired properties (Scheme 3). This chapter will summarize achievements and challenges in diversity-based biomimetic oxidation, with respect

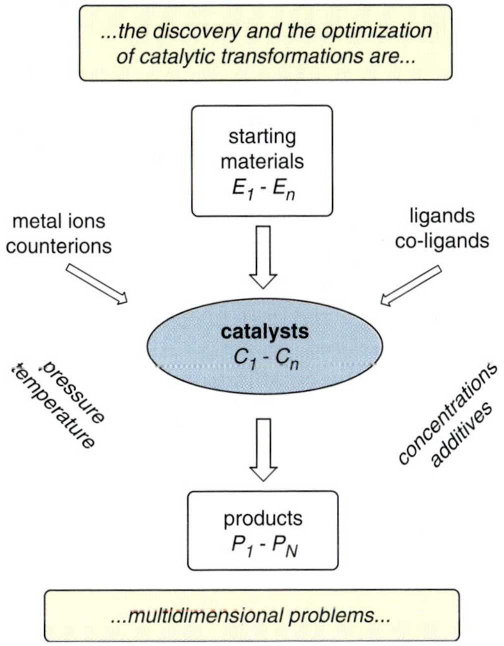

SCHEME 2. Parameter space in catalyst discovery and optimization.

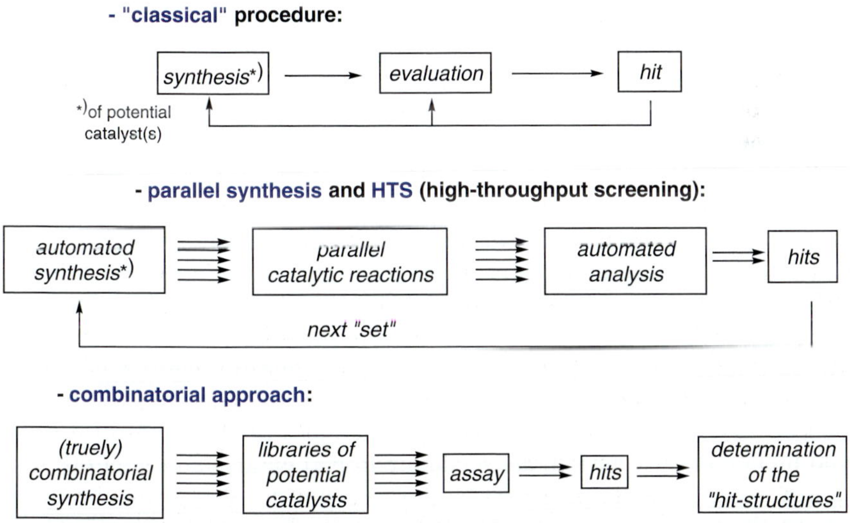

SCHEME 3. Classical and diversity-based catalyst discovery and optimization.

to both the *discovery* of novel and the *optimization* of existing catalyst systems. Besides work on low-molecular weight compounds, recent progress in the fields of directed evolution of oxidative enzymes will be addressed as well.

A. Combinatorial Methods in Catalysis Research

The search for catalysts has many features in common with the search for biologically active compounds, most importantly pharmaceuticals: In both cases, the aim is the discovery of a property, an effect. In classical pharmaceutical (or catalysis) research (Scheme 3, top), compound isolation and characterization precedes testing. As a consequence, a lot of effort may be invested on compounds that eventually do not show the desired properties (*4,5*). In the combinatorial approach, (Scheme 3, bottom) characterization comes *after* the discovery of a desired property, and it is *limited to those compounds that show the desired property.* As summarized in Scheme 3 (bottom), the key features of the combinatorial approach to catalyst discovery (and optimization) are (i) the generation of libraries of catalyst candidates, and (ii) the screening of the libraries for catalytic activity. Both aspects are dealt with in a general sense in Sections I.B and I.C below. Section I.D summarizes the features specific for library generation and screening in the field of biomimetic oxidation catalysis.

B. Library Generation

As with other applications of combinatorial chemistry, libraries of catalyst candidates are mainly generated either by parallel synthesis (Scheme 3, middle) (*6*) or by the split-and-combine approach (Scheme 4) (*7*). Parallel (in many instances automated) synthesis has the advantage that the identity of the synthesis products (e.g., ligands) is known, and that existing methods of solution-phase synthesis (and analysis) can be applied with only little modification.

By the split-and-combine ("one bead – one compound") approach, which is usually practiced using polymer beads as supports, large product numbers can be generated with only few synthetic steps. However, only a hand full of reactions proceed sufficiently clean and effective such that they can be applied to the split-and-combine protocol – the latter does not allow the purification of the solid-phase-bound intermediates or final products. Furthermore, the identity of the bead-bound materials is lost in the course of the split-and-combine synthesis. This intrinsic feature requires either post-screening identification of "hit"-compounds, or the "tagging" of the solid supports (*8*). Thus, it is not surprising that most of the work in diversity-based biomimetic oxidation, as in combinatorial catalysis in general, is carried out using libraries prepared by parallel synthesis, typical library sizes being in the tens to hundreds. Much larger numbers of (bio)catalyst candidates are generated in the directed-evolution approach which is treated in Section IV of this chapter. In this approach, the use of biological reproduction in combination with mutagenesis allows the rapid generation of thousands of enzyme variants (*9*). Another exception is the use of split-and-combine peptide libraries, mainly as hydrolase or aldolase models. Due to the extremely high standard of solid-phase peptide synthesis, purely peptidic catalyst libraries can be synthesized reliably by the split-and-combine

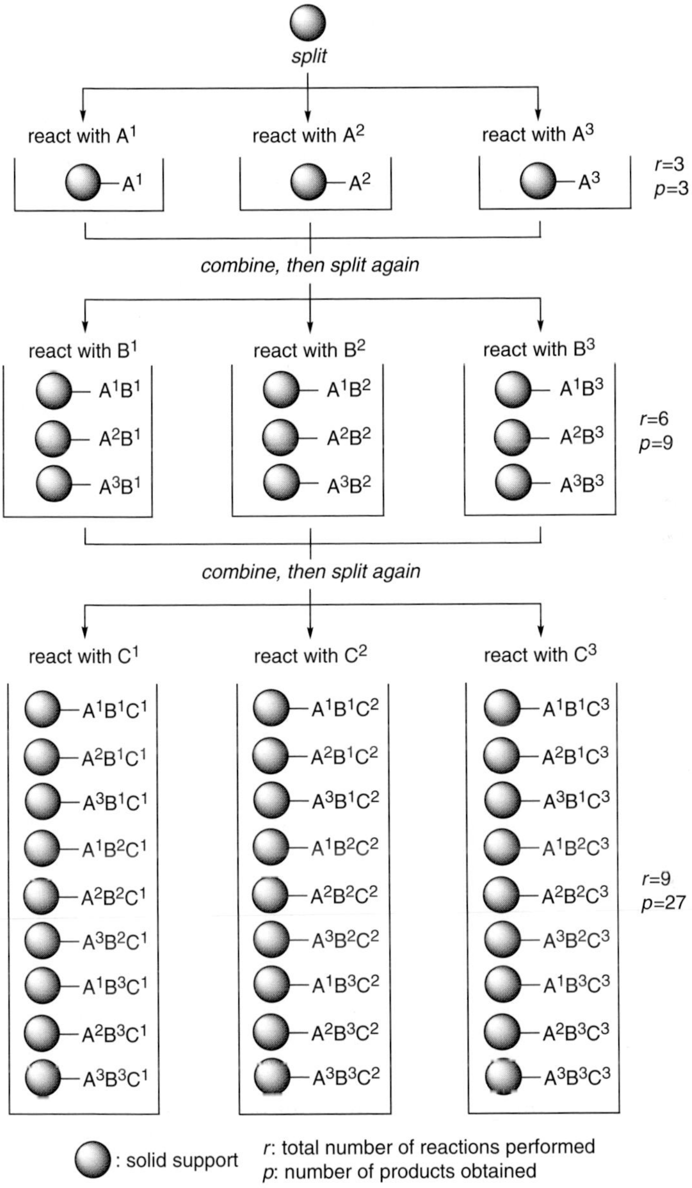

SCHEME 4. Split-and-combine synthesis.

protocol on solid phase, and libraries comprising hundreds to thousands of members have been reported (*10*).

C. Screening Methodology

In the screening process, a library of catalyst candidates is exposed to a test reaction. The methods applied for the identification of catalytically active library members, i.e., for the detection of substrate conversion, largely depend on library size and type. In many cases, catalyst candidates are screened in separate homogeneous solutions (e.g., in titer plates). This case may well include the synthesis of catalyst candidates on solid phase, followed by detachment and screening in homogeneous solution. If the test reaction forms or destroys a chromophore (or fluorophore), simple optical inspection or automated titer plate readers can be used for catalyst identification. More importantly, analytical methods such as GC, capillary electrophoresis, NMR, ESI-MS, IR, and others have been adapted successfully to library screening over the last years (*11*). Automated versions now allow the analysis of thousands of samples per day. In other words, the rapid screening of catalyst libraries for both activity and selectivity (including enantioselectivity) has become technically feasible (*11*). The latter methods have the additional and quite significant advantage that no "artificial", e.g., chromogenic, test substrates need to be devised, but the "real" substrates of interest can be used directly and without modification.

The bulk screening of bead-bound libraries for catalytic activity mainly relies on optical inspection (microscopy) and thus requires site-specific chromogenic test reactions. Site-specificity means that the dye generated stays with the catalytically active bead and does not diffuse away. A number of solutions to this problem have been devised, (*12*) including the formation of insoluble dyes (*13*) or the use of gels as the medium in which the screening is performed (*14*).

Infrared thermography is a very attractive method for the detection of catalytic activity (*15*). As it is an "optical inspection", it can be applied to all kinds of libraries and screening reactions. The detection of catalytic activity is based on reaction enthalpy (negative or positive) (*16*). As most oxidation reactions are significantly exothermic, IR thermography lends itself very well as a sensitive tool for the discovery of oxidation catalysts (see Section II.C) (*17*).

D. Opportunities for Diversity-Based Methods in Biomimetic Oxidation Catalysis: Enzyme Models and Enzymes

Biomimetic oxidation catalysts aim at imitating the activity and selectivity of oxidative enzymes such as oxidases, oxygenases, and peroxidases. Studies in the field of diversity-based biomimetic oxidation catalysis aim primarily both at the discovery of new catalyst types, and in many cases at high enantioselectivity. For the latter endeavour, catalyst designs are preferentially employed that are in principle known to provide activity (but not necessarily enantioselectivity). Most of the enzymes mentioned above harbor transition metal ions. Thus, combinatorial methods can provide diversity in the form of ligand libraries, additional diversity is introduced by the choice of the metal, counter ions(s), co-ligands, etc. (Sections II, III). A second and highly successful approach applies the methodology of directed evolution to oxidative enzymes that are in principle known to convert a given substrate, but at low selectivity. As shown in Section IV quite remarkable enantioselectivities have been achieved by this method in both the sulfoxidation of thioethers and in the Baeyer-Villiger oxidation of cyclic ketones.

II. Biomimetic Oxidation Catalysis Based on Metal Complexes

A. Asymmetric Epoxidation of Olefins

The epoxidation of olefins is a typical electrophilic oxygen transfer reaction (see Scheme 1). In Nature, the epoxidation of non-functionalized olefins is mostly effected by monooxygenases, the most prominent class being the cytochromes P450 (*18*). Like other monooxygenases, cytochrome P450 naturally employs molecular oxygen as the terminal oxidant. In the presence of a co-reductant, one oxygen atom of the dioxygen molecule is transferred to the substrate, whereas the other one is reduced to water (Scheme 5). By the so-called "shunt pathway", substrate oxygenation can also be effected by single-oxygen atom donors (such as iodosoarenes, peracids, hydroperoxides, including dihydrogen peroxide) and in the *absence* of a co-reductant (Scheme 5). Obviously, most efforts have concentrated on the use of single-oxygen atom donors which allow much simpler reaction conditions.

Cytochrome P450, nowadays called P450, is a heme enzyme, and numerous functional low-molecular weight models for cytochrome P450 have been devised (*19*). In contrast to "classical" biomimetic approaches to functional P450 models, Francis and Jacobsen have reported a diversity-oriented approach to P450-like reactivity (*20*). The goal was to discover a catalyst that allows the asymmetric oxidation of *trans*-β-methylstyrene **1** with dihydrogen peroxide as the source of oxygen, affording the epoxide **2** in good chemical yield and enantioselectivity. The strategy employed by Francis and Jacobsen is illustrated in Scheme 6a. The first and discovery-oriented ligand library (192 members) was prepared by a combination of the 16 core structures **3**-**8** and the 12 terminal building blocks **9**. The core structures **3**-**7** were

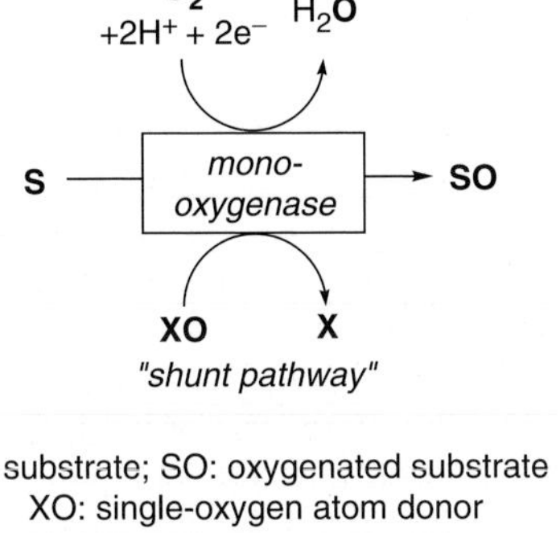

Scheme 5. Substrate oxygenation by monooxygenases.

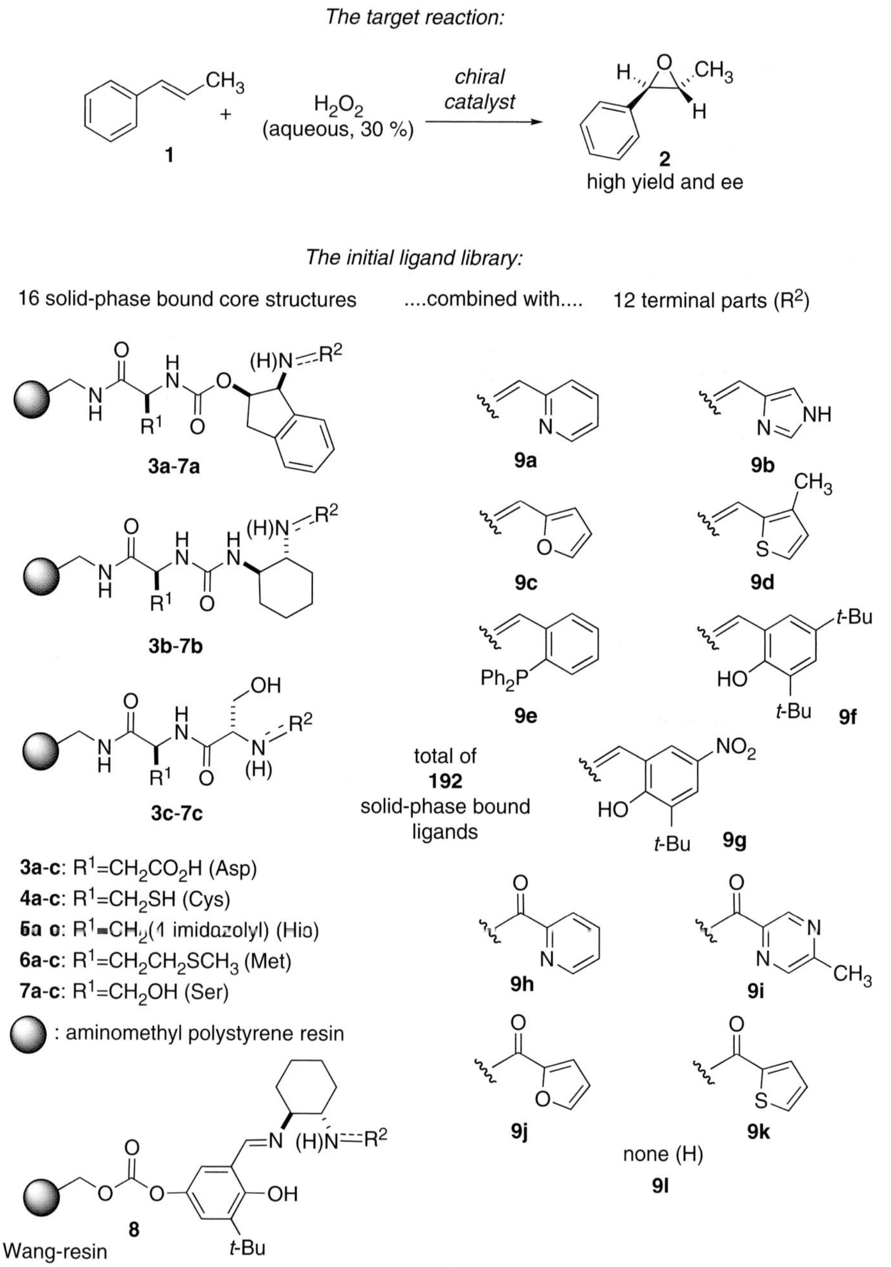

SCHEME 6a. Diversity-based approach to iron-based enantioselective catalysts – the ligand library used for catalyst discovery.

synthesized on solid phase, using aminomethyl polystyrene beads as the solid support. The first building block is always an amino acid, followed by either *cis*-aminoindanol (the "a"-series), *trans*-1,2-diaminocyclohexane (the "b"-series), or serine as a second amino acid (the "c"-series). The core structure **8**

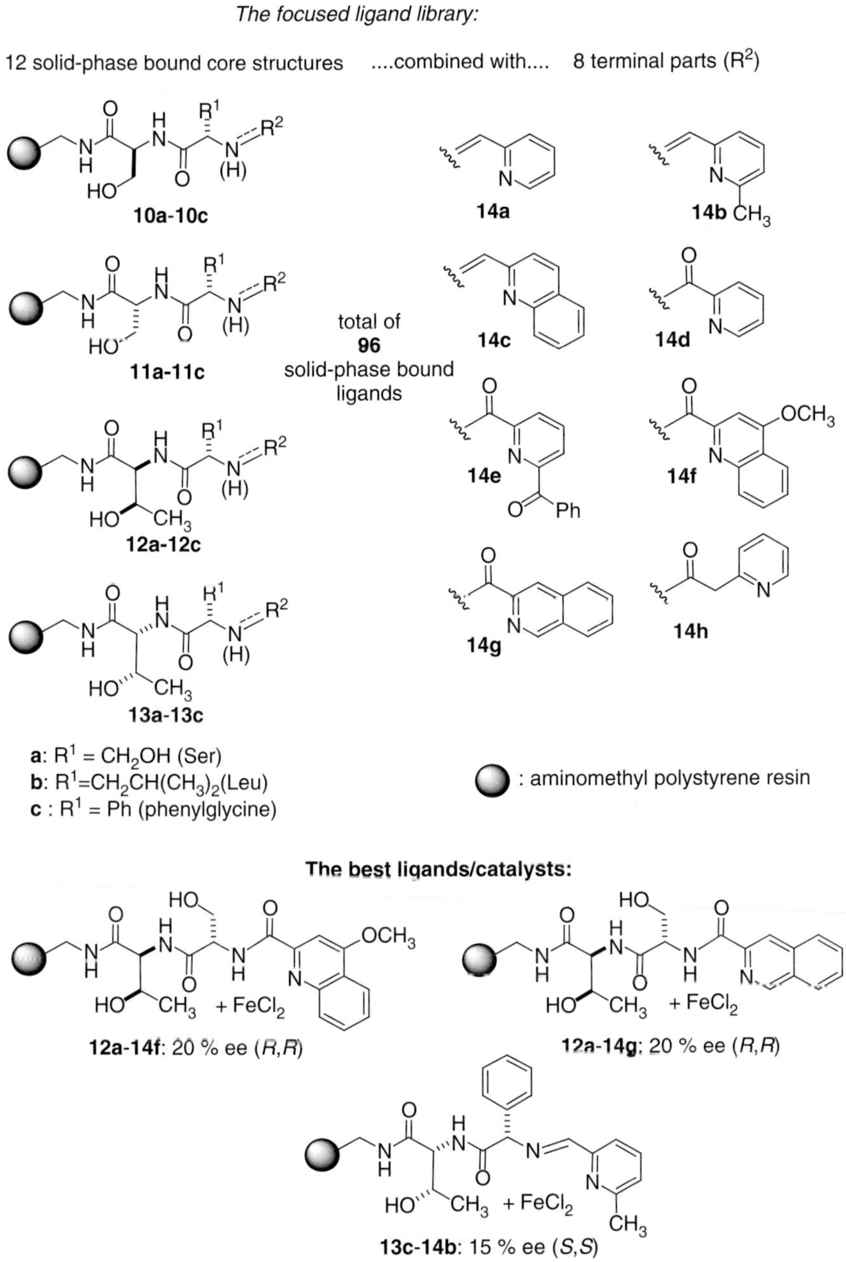

SCHEME 6b. Diversity-based approach to iron-based enantioselective catalysts – the focused ligand library used for ee optimization.

was prepared on Wang-resin, using the corresponding hydroxysalicylic aldehyde. The 16 core structures were coupled with the 12 terminal building blocks **9a-l**. Finally, the combined 192 ligands were incubated with 30 metal salts (including different oxidation states, counter anions).

For the identification of active catalysts for the epoxidation of the olefin **1**, a three-step protocol was applied, using GC as the analytical tool. In the first step, and using the whole combined library, aqueous dihydrogen peroxide in a 1:1 solvent mixture of dichloromethane/*tert*-butanol was identified as the optimal oxidant. In the next step, the metal sources were evaluated individually, still using the combined 192-mer ligand library. $FeCl_2$ proved to be the optimum metal source, providing both high catalytic activity in the presence of the ligands, and at the same time low background activity (i.e., strong ligand accelerated catalysis). The third step consisted in the parallel evaluation of each of the 192-mer ligand library members. The three most efficient ligands consisted of the following combinations: **4c**-**9h**; **7c**-**9a**; **7c**-**9h**, highlighting both the importance of serine (or cysteine) as parts of the solid-phase bound core structures, and the importance of pyridine-derived and metal-coordinating "end caps".

Unfortunately, the three hit-ligands **4c**-**9h**, **7c**-**9a**, and **7c**-**9h** afforded at best 7 % ee in the epoxidation of *trans*-β-methylstyrene (**1**). Therefore, Francis and Jacobsen synthesized and screened a second ligand library, using *L*-Ser, *L*-Leu, and *L*-phenylglycine in the solid-phase-bound core structures, and combined 12 of those (**10a-c**–**13a-c**) with 8 end caps (**14a-h**) derived from pyridine, quinoline, and *iso*-quinoline (total of 96 ligands). The screening of this second library identified the ligands **12a-14f**, **12a-14g**, and **13c-14b** (Scheme 6b), affording up to 20 % ee of the epoxide **2**, at epoxide yields of up to 78 % (*20*).

In summary, the goal of this combinatorial approach – the discovery of catalysts for the asymmetric epoxidation of olefins using dihydrogen peroxide as the terminal oxidant – was met: The Fe-complexes of the ligands **12a-14f**, **12a-14g**, and **13c-14b** indeed afforded up to 20 % ee, at epoxide yields of up to 78 %. In this study proof-of-principle was clearly achieved. For practical applications in asymmetric synthesis, further improvement of the catalyst enantioselectivity appears desirable and represents the next challenge to be met.

B. Asymmetric Sulfoxidation of Thioethers

Enantiomerically pure sulfoxides are important intermediates in organic synthesis (*21*) and quite a number of pharmaceuticals and other biologically active compounds harbor a chiral sulfoxide unit (*22*). With respect to oxidation catalysis, enantiomerically enriched sulfoxides can either be accessed by asymmetric sulfoxidation of prochiral thioethers (Scheme 7, path a), or by kinetic resolution of racemic sulfoxides (Scheme 7, path b). For the latter purpose, enantio-specific oxidation of one sulfoxide enantiomer to the sulfone, followed by separation, is the method of choice.

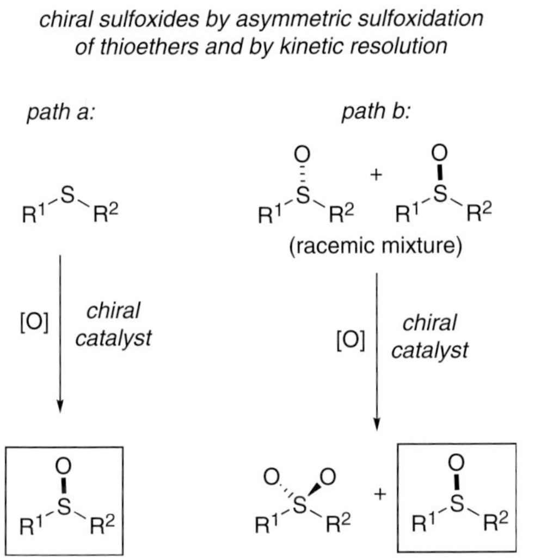

SCHEME 7. Chiral sulfoxides by enantioselective oxidation of prochiral thioethers, and by oxidative kinetic resolution of racemic sulfoxides.

Asymmetric sulfoxidation of thioethers has been achieved enzymatically, mostly using heme- or vanadium-based haloperoxidases as the bio-catalysts (*23*). In 1995, Bolm *et al.* reported that the vanadium complexes of the tridentate and amino-acid derived ligands **15** catalyze the enantioselective sulfoxidation of prochiral thioethers with dihydrogen peroxide as the terminal oxidant (Scheme 8) (*24*). The ligands are synthesized by a one-step condensation of the readily available salicylic aldehydes **16** with amino-acid derived amino alcohols, such as **17a**.

The highest enantioselectivity was achieved in the case of the thioacetal **18** as substrate (85 % ee), whereas thioanisole (**19**) gave the sulfoxide with up to 70 % ee (*25*). In 2003, Jackson *et al.* reported the optimization of this ligand type based on a parallel synthesis and screening approach (Scheme 8) (*26*). First, the amine/amino alcohol component was optimized using the solid-phase-bound salicylic aldehyde **20** as the second component. The aldehyde **20** was condensed with a total of 29 amine components, and the resulting solid-phase-bound ligands were evaluated in the V-catalyzed sulfoxidation of thioanisole (**19**) with dihydrogen peroxide. In this assays, the amines **17a** and **17b** (Scheme 8) gave the best enantioselectivities. In the next round, *cis*-aminoindanol (**17b**) was combined (by solution-phase condensation) with a total of 44 salicylic aldehydes. Interestingly, in 41 of these 44 combinations, the Schiff-base ligands could be "harvested" in sufficient purity by simple precipitation/filtration. In the subsequent assay, the combination of the amine **17b** with the aldehydes **21a** and **21b** proved best. In a final round of fine-tuning, the ligands **22a** and **22b** (Scheme 8) were identified as the most enentioselective ones: In the case of thioanisole (**19**) as substrate, up to 90 % ee were achieved, and even higher enantioselectivity (97 % ee) was reported for the sulfoxidation of methyl-2-naphthylsulfide as substrate. The latter

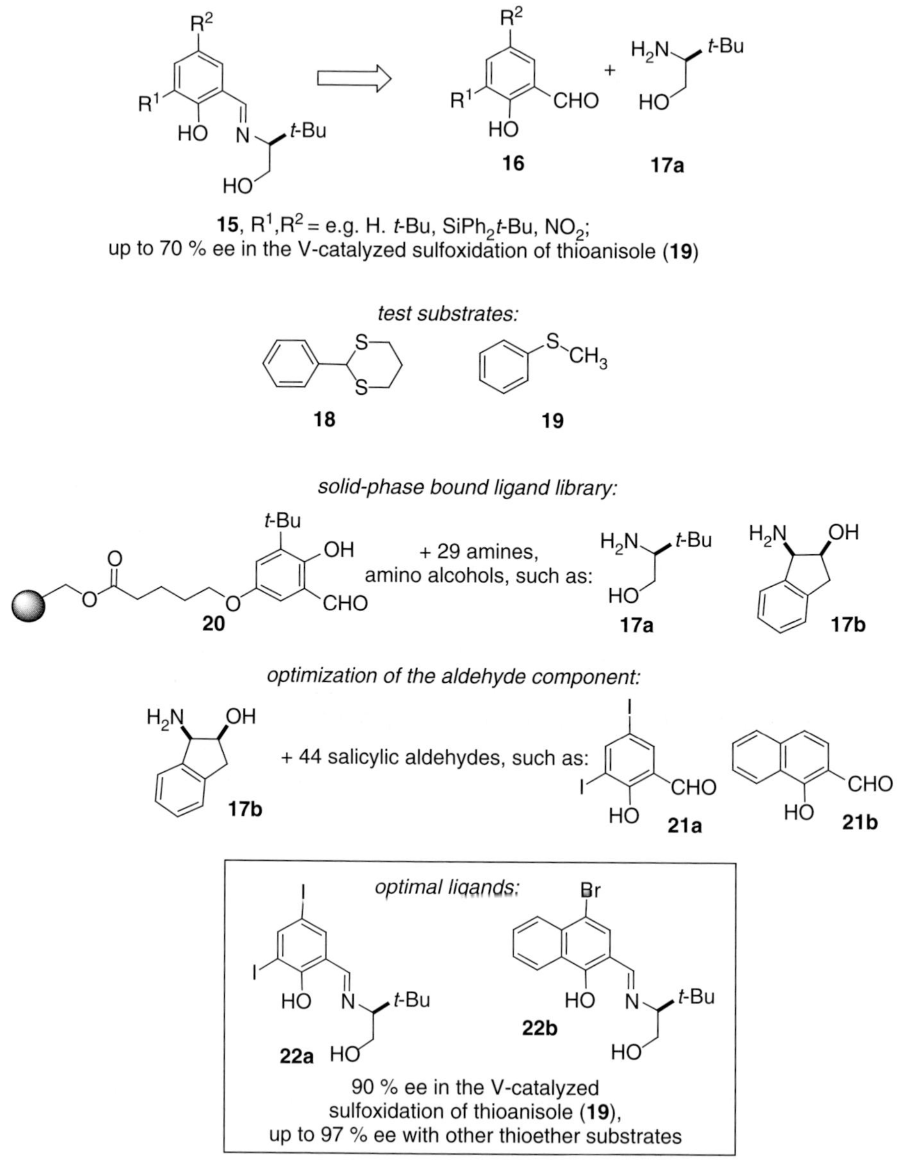

SCHEME 8. Highly efficient and selective modular ligands for the vanadium-catalyzed asymmetric sulfoxidation of prochiral thioethers.

values mark the highest enantioselectivity ever achieved in the V-catalyzed sulfoxidation of thioethers with dihydrogen peroxide.

Jackson *et al.* furthermore prepared the solid-phase-bound ligand libraries **23** shown in Scheme 9 (*27*). The ligands consist of a solid-phase-bound dipeptide which is capped by a salicylic aldehyde. The screening of a number of

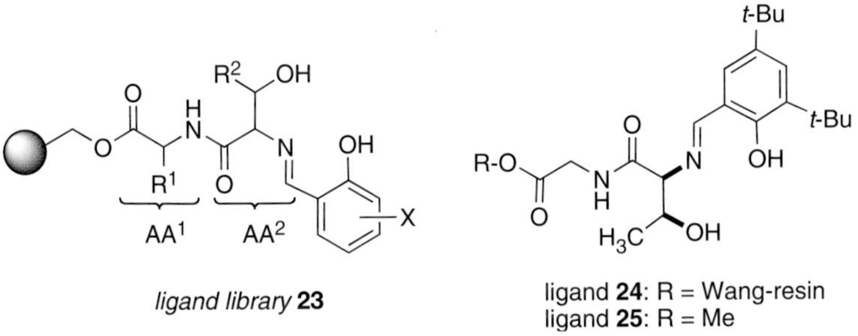

SCHEME 9. Solid-phase-bound ligands for the titanium-catalyzed asymmetric sulfoxidation of thioethers.

libraries in the vanadium-catalyzed sulfoxidation of thioanisole (**19**, Scheme 8) with dihydrogen peroxide identified the ligand **24** as the optimal one, providing 23 % ee and excellent conversion. Exchange of vanadium for titanium quite significantly raised the ee to 63–64 %. With methyl-2-naphthyl sulfide as the substrate, an ee of 72 % was achieved. Please note that the titanium complex of the solid-phase-bound ligand **24** is quite stable and can be reused a number of times. Alternatively, the ligand **25** can be used for sulfoxidations in homogeneous phase, with enantioselectivity identical to **23** (*27*).

Highly enantioselective sulfoxidations have recently been achieved using directed evolution of flavin monooxygenases (*28*). See Section IV for this approach to highly enantioselective and biocatalytic sulfoxidation.

C. Baeyer-Villiger Oxidation of Ketones

The Baeyer-Villiger oxidation of ketones affords esters (from open-chain ketones) and lactones (from cyclic ketones), respectively. It is typically carried out using nucleophilic oxidants, such as peracids or dihydrogen peroxide in the presence of bases (*29*). Nucleophilic oxidants are required because the first step of the reaction mechanism is the addition of the peroxidic oxidant to the ketone's carbonyl function, affording the so-called *Criegee*-intermediate (Scheme 10). The subsequent cleavage of the O–O bond in the latter, and the concerted shift of R^2 yield the ester (lactone) product. In enzymatic Baeyer-Villiger oxidations, flavin hydroperoxides such as **26** act as the nucleophilic oxidant (*30*).

Many cases of highly enantioselective enzymatic Baeyer-Villiger oxidations of prochiral ketones are known, both isolated enzymes and whole-cell systems being the most widely used bio-catalysts for this purpose (*31,32*). Flavin hydroperoxides and synthetic analogues – in the absence of an apo-enzyme – have been used in biomimetic approaches to asymmetric Baeyer-Villiger oxidation (*33*). However, no diversity-based approach (i.e., generation and screening of flavin-type libraries) appears to be reported. Section III.A describes a related organocatalytic approach to asymmetric olefin epoxidation using a library of iminium cations as catalysts (*34*). In principle, this

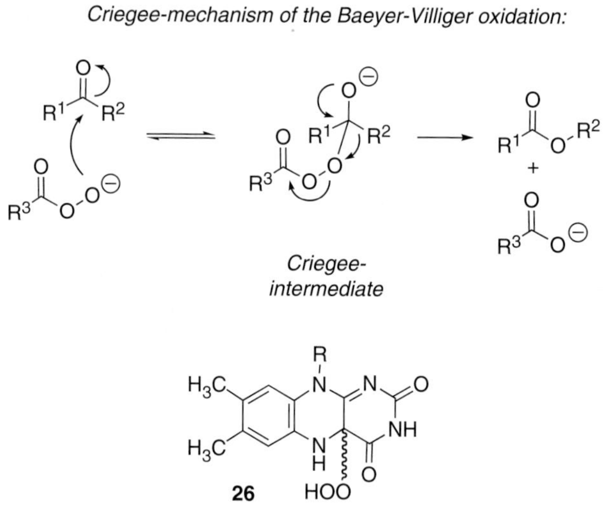

SCHEME 10. The Baeyer-Villiger oxidation of ketones.

approach might as well be used for the generation of flavin-like hydroperoxides, and might thus be useful for Baeyer-Villiger oxidations.

The mechanism shown in Scheme 10 implies further opportunities for catalyzing the Baeyer-Villiger oxidation, other than the generation of catalyst-oxidant libraries: coordination of a Lewis acidic metal ion to the ketone carbonyl oxygen atom may as well facilitate the nucleophilic addition of the oxidant (*35*). In recent years, rare-earth metal ions, and in particular trivalent scandium, have been shown to be quite effective Lewis acids for a number of preparative applications (*36*). However, no application to Baeyer-Villiger oxidations using dihydrogen peroxide appeared to be reported when we decided in 2002 to broadly screen rare earth triflates as catalysts for this purpose (Scheme 11) (*37*). The conversion of the cyclobutanones **27** to the γ-lactones **28** is a highly exothermic process, thus IR-thermography is an ideal method for parallel activity screening. As shown in Fig. 1, the lanthanide triflates employed revealed quite different catalytic activities that could hardly (or not at all) be predicted. In summary, Sc(III) and Nd(III) proved to be the most effective catalysts. There appears to be no straightforward interpretation for this result, e.g., correlation between ionic radii and catalytic activity. This case thus emphasizes that the parallel/combinatorial approach may be quite effective and may lead to successful catalyst discovery even in the absence of a solid mechanistic basis. Scheme 11 also summarizes the yields of γ-butyrolactone **28a** and the increases in temperature observed for the individual rare earth triflates after a fixed reaction time (*37*).

D. AIR OXIDATION OF ALCOHOLS

Galactose oxidase (GOase, EC 1.1.3.9.) is a mononuclear fungal copper enzyme that catalyzes the air oxidation of *D*-galactose and a number of

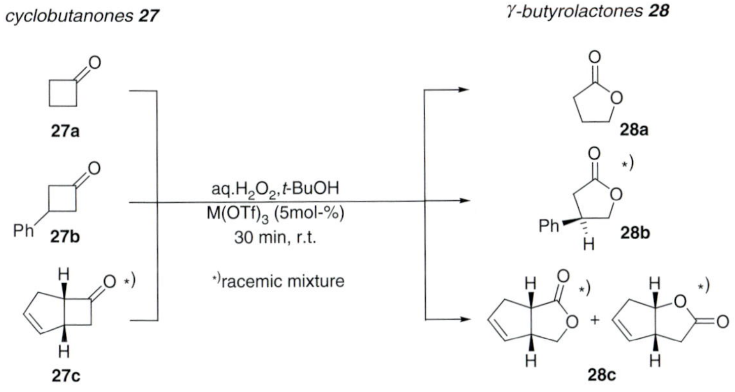

catalytic metal ion:	Eu^{3+}	La^{3+}	Lu^{3+}	Nd^{3+}	Sm^{3+}	Sc^{3+}	Y^{3+}	Ce^{4+}	Cu^{2+}
ΔT [°C] after 40 s:	1.4	1.1	1.1	2.2	1.5	2.5	1.2	0	0
yield of lactone **28a** after 30 min:	quant. in all cases							no reaction	

SCHEME 11. Rare-earth catalyzed Baeyer-Villiger oxidation of cyclobutanone (**27a**) with dihydrogen peroxide.

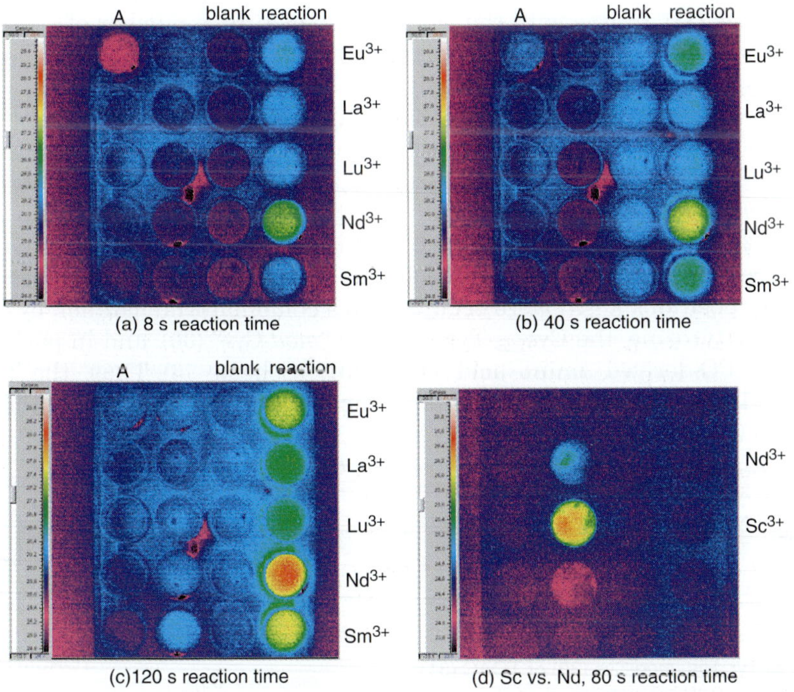

FIG. 1. IR-thermographic screening of the rare-earth catalyzed Baeyer-Villiger oxidation of cyclobutanone (**27a**). Well "A" contains cyclobutanone (**27a**) and dihydrogen peroxide, but no catalyst. The row "blank" contains dihydrogen peroxide and catalyst, but no cyclobutanone (**27a**).

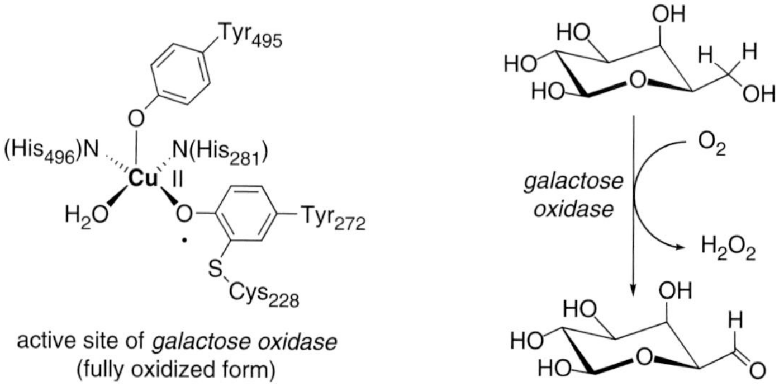

SCHEME 12. Air oxidation of alcohols effected by the copper enzyme *Galactose oxidase* (GOase).

other primary alcohols to the corresponding aldehydes (Scheme 12) (*38*). It has been a long-standing problem how one single Cu-ion can effect the two-electron oxidation of an alcohol with molecular oxygen. It is by now clear that the catalytic cycle of GOase involves the Cu(I)-Cu(II) couple *plus* the conversion of a coordinating phenoxide to the corresponding phenoxyl radical (Scheme 12). The "coordinating phenoxide" is provided by the Cys_{228}-Tyr_{272} conjugate, which stems from post-translational modification of the apoenzyme (*38*). In line with the mechanistic picture, a number of Cu-phenolate/naphtholate or TEMPO-complexes were reported to show GOase-like catalytic activity as well (*39–41*).

Berkessel *et al.* took a diversity-based approach to the discovery of functional models for GOase (*42*). Based on the known coordination pattern of the catalytically active copper ion (X-ray crystal structure of GOase from *Dactylium dendroides*, schematically in Scheme 12), the peptide/peptoid library **29** (Scheme 13) was designed. In this library, the Gly-*D*-Pro-turn positions the four amino acids X^1–X^4 such that they can coordinate a copper ion. The four positions X^1–X^4 were occupied, in a combinatorial manner, by either histidine, tyrosine, the Cys_{228}-Tyr_{272}-model "*mod*-Cys" (**30**), and in particular the TEMPO-derived amino acid TOAC (**31**) (Scheme 13). Thus, the TOAC-derived library comprised a total of 81 decapeptides, and it was synthesized

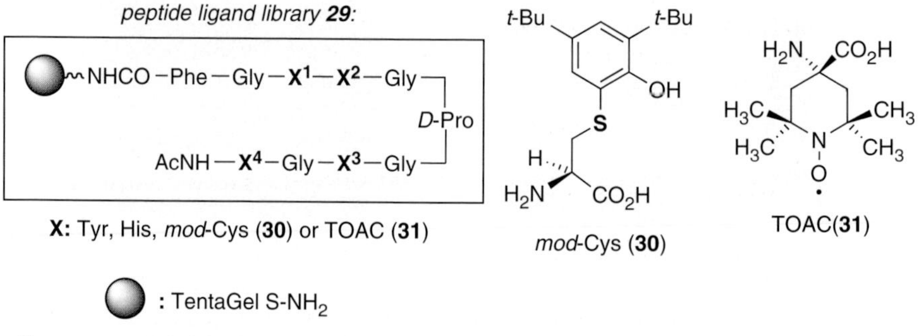

SCHEME 13. Models for *Galactose oxidase* (GOase) based on peptide libraries.

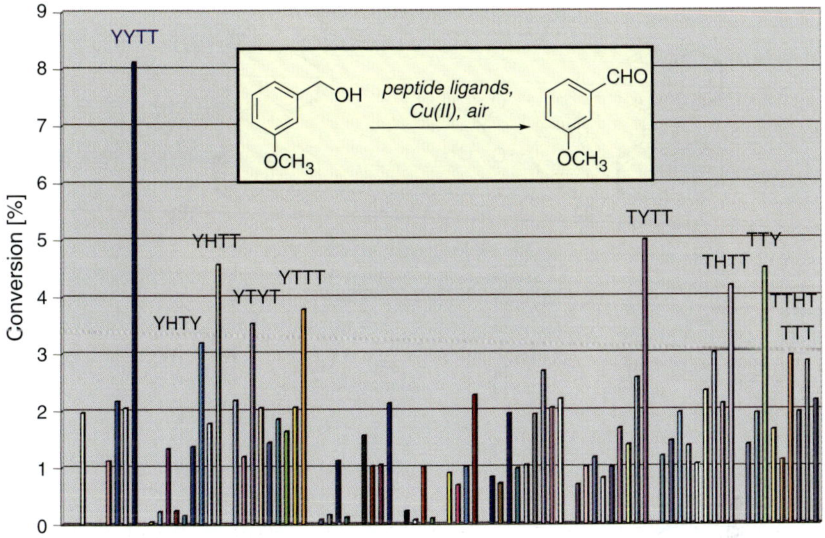

FIG. 2. Relative catalytic activities of the library members in the copper catalyzed air oxidation of 3-methoxybenzyl alcohol. T stands for TOAC (**31**), and the four-letter code denotes the amino acids in positions X^1–X^4.

on solid support using the IRORI radiofrequency encoding method (*42,43*). After incubation with copper ions, the library was screened for catalytic activity. For this purpose, 3-methoxybenzyl alcohol was used as the test substrate. The oxidation product, i.e., 3-methoxybenzaldehyde, shows a UV-maximum at $\lambda = 310$ nm, which is absent in the UV/VIS-spectrum of the starting benzyl alcohol. Thus, the progress of the substrate oxidation can be monitored by an automated titer plate reader. The result of the library screening is summarized in Fig. 2. Obviously, there are significant differences in the catalytic performance of the individual peptides, and the presence of redox-active TOAC (**31**) is necessary for efficient oxidation catalysis. Control experiments established that it is indeed the intact peptide ligand that renders the Cu-ion catalytically active. For example, simple mixtures of Cu(II), TOAC, His, or Tyr have almost no activity. Thus, the initial goal, the discovery of GOase activity in libraries of Cu-peptide complexes has been achieved. However, the rates of substrate conversion are still low. In the absence of structural information on the catalytically active Cu-peptide complexes, it is not yet possible to reasonably delineate a mechanistic proposal for the catalytic air oxidation found.

III. Biomimetic Oxidation Catalysis Based on Purely Organic Catalysts

A. Iminium-ion Catalyzed Asymmetric Epoxidation of Olefins

Chiral iminium ions can serve as mimics of flavo-enzymes in that they catalyze asymmetric oxygen transfer from dihydrogen peroxide or persulfate

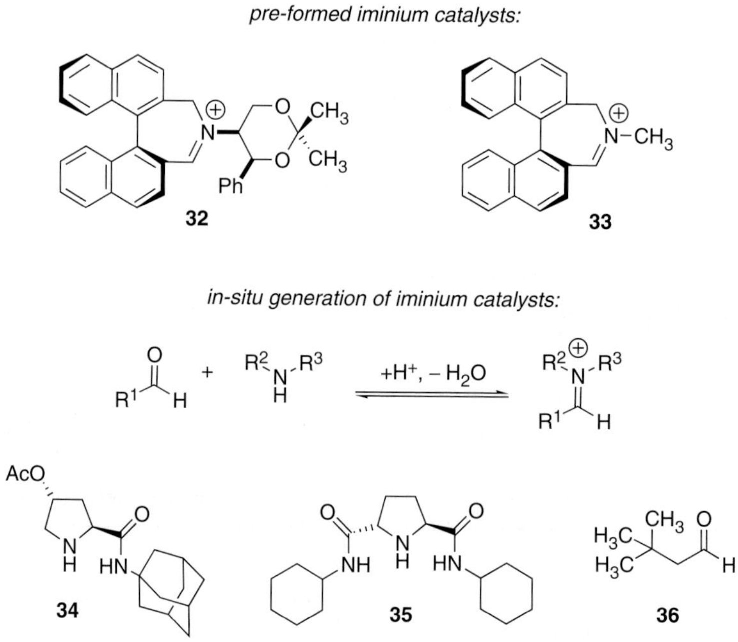

50–100mol % of *in situ*-generated iminium catalyst
persulfate as oxidant

amine	aldehyde	product epoxide	yield (%)	ee (%)
34	**36**	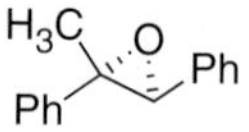	99	59
35	**36**		93	65

SCHEME 14. Iminium-ion epoxidation catalysts.

to acceptors such as olefins, thioethers, or ketones (*44*). High enantioselectivites have been achieved e.g., by the iminium system **32** (Page *et al.* (45)) or **33** (Aggarwal *et al.* (46)). Iminium ions can be formed by the reaction of an aldehyde (or ketone) with a secondary amine (Scheme 14). Thus, iminium ion catalysts are ideally suited for the combinatorial *in situ* preparation from a set of (structurally diverse) aldehydes and a set of secondary amines. In 2001, this idea was put into practice by Yang *et al.* (Scheme 14) (*34*). The *in situ* combination of 6 aldehydes with 5 secondary amines afforded a total of 30 catalyst combinations. Conventional GC-screening revealed that of all possible combinations, the two shown in Scheme 14 (**34**-**36** and **35**-**36**) afforded the highest enantioselectivities in the epoxidation of *trans*-stilbene and methyl-*trans*-stilbene. This result, i.e., the most effective combination

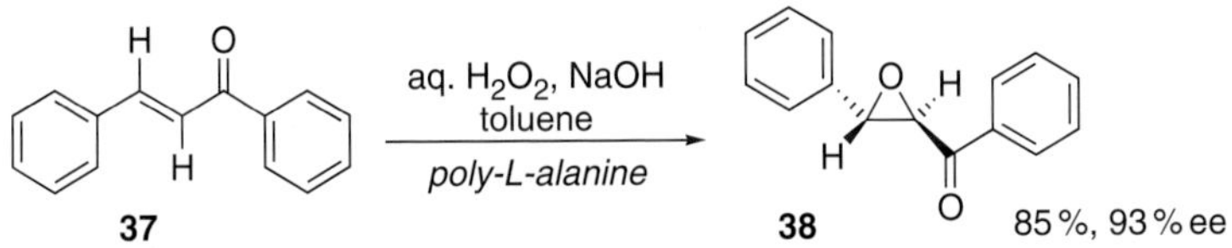

SCHEME 15. The Juliá-Colonna epoxidation of chalcones.

of aldehyde and secondary amine, could not have been predicted, and thus demonstrates the strength of the parallel/combinatorial approach to catalyst discovery.

B. PEPTIDE AND PEPTOID CATALYSTS FOR THE ASYMMETRIC EPOXIDATION OF ENONES

About one half of the enzymes known do not contain catalytically active metal ions. Of the metal-free enzymes, a significant portion does not harbor a co-factor. In other words, the latter group of enzymes are purely peptidic catalysts (*47*). Catalysis is effected by the functional amino acid side chains present in the active sites (e.g., the "catalytic triad" of the serine hydrolases), and by the overall effects of the surrounding protein (local dipole moments, overall motions, etc.). Given the high level of contemporary peptide synthesis on solid phase, and the ease of its adaptation to split-and-combine library generation, it is not surprising that combinatorial peptide synthesis has already been used in the search for biomimetic catalysts in a number of instances. However, most efforts were directed towards the generation of artificial hydrolase, and not oxidase or peroxidase activity (*48*).

A contrasting example is the peptide-catalyzed asymmetric epoxidation of chalcones, the so-called Juliá-Colonna reaction. The latter authors discovered in the early 1980s that poly-amino acids such as poly-*L*-alanine or poly-*L*-leucine catalyze the asymmetric epoxidation of chalcone (**37**) and derivatives using alkaline dihydrogen peroxide as the terminal oxidant (Scheme 15) (*49,50*).

The poly-amino acids employed for the Juliá-Colonna epoxidation are statistical mixtures, the maximum of the length distribution being around the 20–25mers (*51*). Well-defined peptides of known sequence have been used to shed light on the mechanism of catalysis in the epoxidation of enones with dihydrogen peroxide (*52*). The most fundamental question to be addressed refers to the minimal structural element (i.e., the minimal peptide length) required for catalytic activity and enantioselectivity. Berkessel *et al.* have addressed this question of the minimal structural unit, i.e., how many amino acid residues are actually required for efficient catalysis. To tackle this question, they have synthesized the whole series of *L*-leucine 1- to 20-mers on solid support (Scheme 16) (*53*). The subsequent screening of the monodisperse and solid-phase-bound peptides for activity and selectivity in the epoxidation of chalcone (**37**) was performed by regular HPLC analysis on chiral stationary phase, and the result is shown in Scheme 16 (bottom)

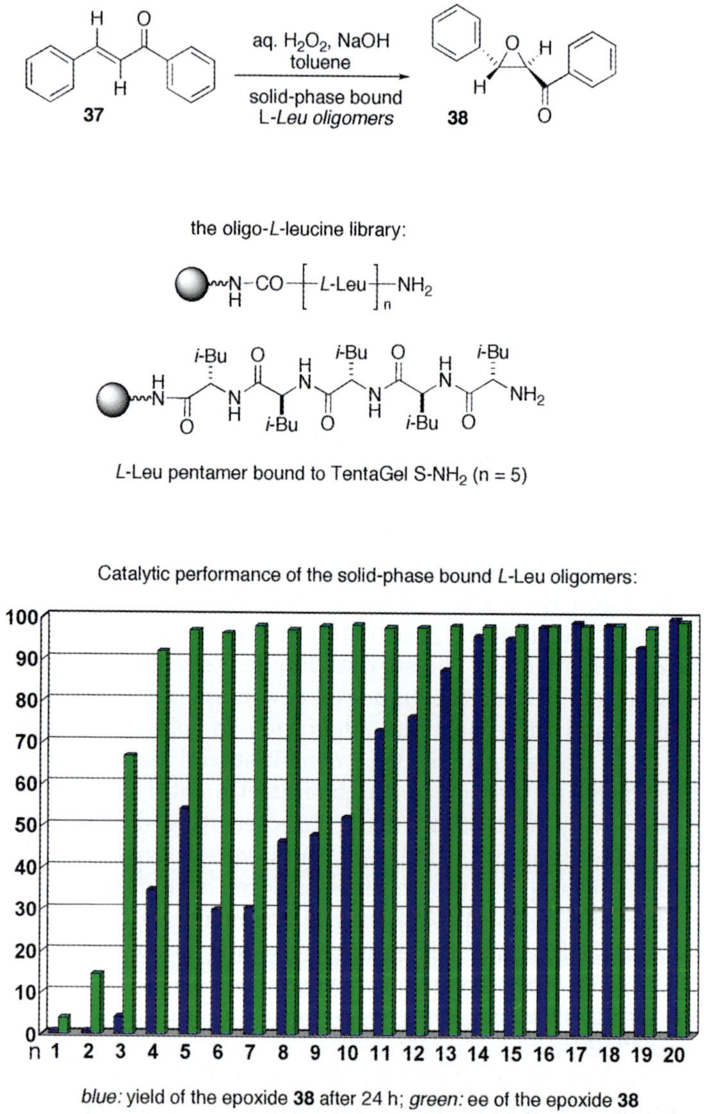

SCHEME 16. Monodisperse and solid-phase-bound peptide catalysts for the Juliá-Colonna epoxidation.

Full enantioselection (≥90 % ee) is achieved already with the 4- to 5-mers, and no further improvement results from using longer peptide chains (green columns). However, the catalytic efficiency (measured as chemical yield after 24 h reaction time) increases with increasing chain length of the peptide catalyst (blue columns). From the sum of the experiments it was concluded that the three non-intrahelical NH-bonds present at the N-terminus play a crucial role in the catalytic mechanism. Based on the results of molecular modeling studies, Berkessel *et al.* (*53*) suggested that the carbonyl oxygen atom of the enone substrate is H-bound to the NHs of the amino acids

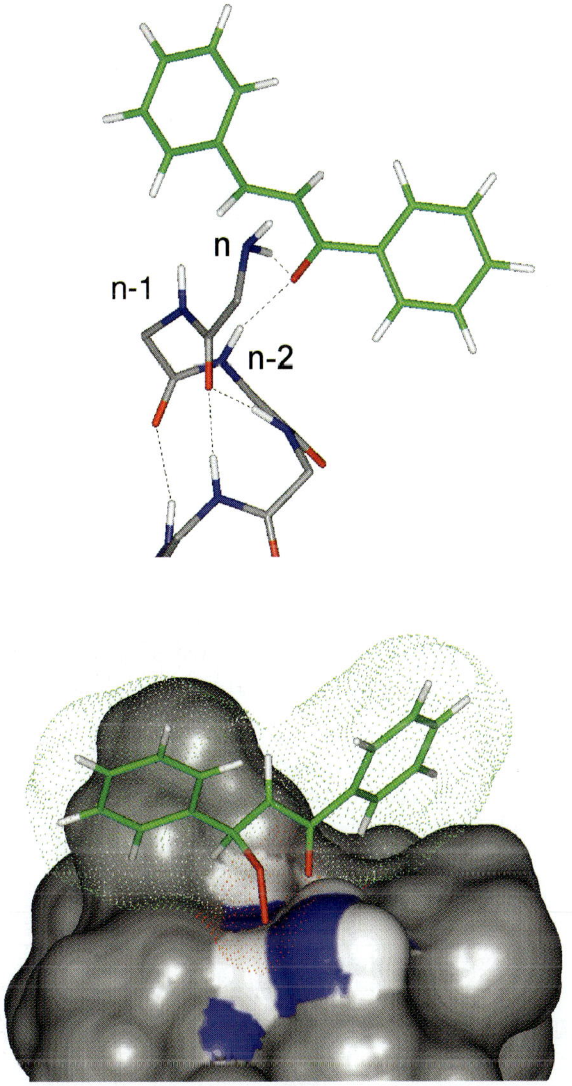

FIG. 3. Peptide-catalyzed Juliá-Colonna epoxidation of chalcone (**37**); top: binding of the substrate enone to the N-terminus of the helical peptide; bottom: β-hydroperoxy-enolate bound to the N-terminus of the peptide catalyst.

at the N-terminus and at position n-2. Furthermore, the hydroperoxide nucleophile is delivered by the NH-group of amino acid n-1. Consequently, the sense of helicity of the peptide catalyst determines the sense of induction in the epoxidation reaction. Figure 3 (top) illustrates the binding of chalcone to the N-terminus of the peptide. In Fig. 3 (bottom), it is demonstrated how a hydroperoxide anion, bound to NH of the amino acid n-1, is delivered face-selectively to the β-carbon atom of the bound substrate (chalcone **37**) (*53*). According to this model, the action of the peptide catalysts used in the

Juliá-Colonna epoxidation bears a lot of similarity to that of enzymes, in particular the binding/activation and proper orientation of the substrates which ultimately effects the excellent enantioselectivities in the overall process. In fact, the "3 NH-motif" discovered as the catalytically active site also acts as the oxy-anion hole in serine esterases and is known to bind/stabilize a variety of full or partially negatively charged entities (the β-hydroperoxyenolate, in the present case) (*54,55*).

This study illustrates how the library approach, and in particular the parallel evaluation of structurally closely related catalysts, can aid in the development of mechanistic understanding.

IV. Directed Evolution of Enzymes for Oxidation Catalysis

In many instances, naturally occurring (so-called "wild-type") enzymes show very high efficiency and selectivity for certain substrates, but not necessarily for the ones of particular interest from a synthetic point of view. In the directed evolution approach, enzymes are adapted to a given task by mutation on the DNA level (mostly by error prone PCR, epPCR, or by gene

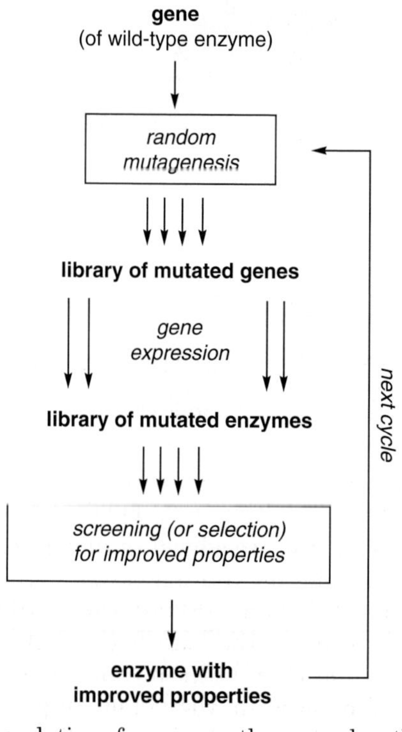

SCHEME 17. Directed evolution of enzymes – the general method.

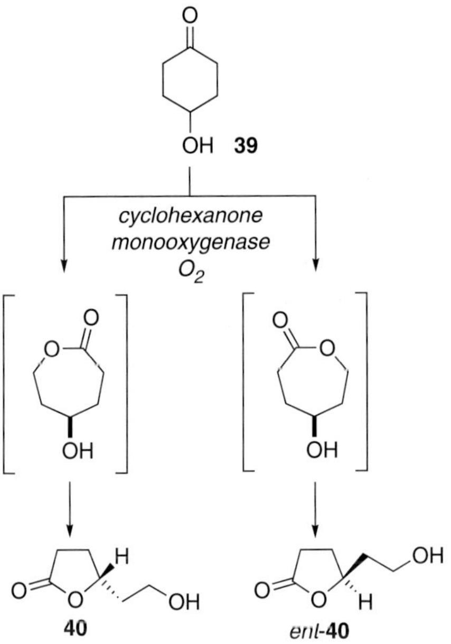

entry	amino acid exchanges in mutants	favored product enantiomer	ee [%]
1	wild-type	**40**	9
2	F432Y, K500R	**40**	34
3	L143F	**40**	40
4	F432I	**40**	49
5	L426P, A541V	**40**	54
6	L220Q, P428S, T433A	*ent*-**40**	18
7	D41N, F505Y	*ent*-**40**	46
8	K78E, F432S	*ent*-**40**	78
9	F432S	*ent*-**40**	79

SCHEME 18. Directed evolution of a cyclohexanone monooxygenase for the oxidative desymmetrization of 4-hydroxycyclohexanone.

shuffling) and subsequent screening for mutants showing improved properties (Scheme 17) (*4,9,10*).

Reetz *et al.* used this approach to adapt the cyclohexanone monooxygenase from *Acinetobacter sp.* NCIMB 9871 (EC 1.14.13.22) to the desymmetrization of 4-hydroxycyclohexanone **39** (Scheme 18) (*56*). Upon Baeyer-Villiger oxidation, the latter substrate affords (after *trans*-lactonization) the enantiomeric γ-butyrolactones **40** and *ent*-**40** (Scheme 18). The enzyme was heterologously expressed in *E. coli* and used in whole-cell preparations. The wild-type enzyme provided 9 % ee in favor of the *R*-product **40**.

After one round of mutagenesis (by epPCR), 10,000 mutants were screened, with *ca.* two dozen of them showing enhanced *R*-selectivity, while *ca.* one dozen mutants displayed reversal of enantioselection in favor of the *S*-product *ent*-**40**. As shown in Scheme 18, eight mutants, showing up to 54 % *R*-selectivity and up to 79 % *S*-selectivity, were sequenced. One further round of evolution, starting from the mutant shown in entry 3, led to an enzyme that had three additional amino acid exchanges and afforded an ee of the product **40** of 90 %. 4-Methoxycyclohexanone was studied as substrate as well. Whereas the wild-type cyclohexanone monooxygenase affords 4-methoxycaprolactone with *S*-selectivity (78 %), the mutant shown in entry 8 of the Table in Scheme 18 gave 99 % ee (and the same sense of induction). This enzyme provides 95–99 % ee also in the Baeyer-Villiger oxidation of 4-methyl-, ethyl-, chloro-, bromo-, and iodo-cyclohexanone (as does the wild-type enzyme). Please note that the mutant shown in entry 8 differs from the wild-type enzyme by one amino acid exchange only (F432S). As amino acid 432 seems to be particularly important for selectivity, saturation mutagenesis of this position was performed, i.e., all 20 natural amino acids were placed into this position. In fact, two moderately *R*-selective mutants were identified, but the enantioselectivity of the F432S mutation (entry 8) was not surpassed. In summary, the directed evolution approach provided an enzyme mutant that provided a remarkable 90 % ee in the Baeyer-Villiger desymmetrization of 4-hydroxycyclohexanone (**30**), as compared to 9 % ee of the wild-type enzyme. With 4-methoxycylohexanone as the substrate, the product ee could be raised from 78 % (wild-type) to 99 %.

The same whole cell system/monooxygenase was used for a directed evolution approach to the asymmetric sulfoxidation of prochiral thioethers, such as **41** shown in Scheme 19 (*28*). The wild-type cyclohexanone monooxygenase from *Acinetobacter sp.* NCIMB 9871 affords an ee of only 14 % in favor of the *R* product **42**. In this approach, the whole 10,000 mutant library generated for the Baeyer-Villiger study above was screened for the enantioselective oxidation of the thioether **41** (Scheme 19), using automated HPLC. More than 20 mutants affording 85–99 % ee were identified (some *R*-, some *S*-selective). Five mutants giving >95 % ee were sequenced (Table in Scheme 19). It is interesting to note that very high enantioselection was achieved, both in favor of the *R*-product **42** (entries 2,3) and the *S*-product *ent*-**42** (entries 4-6). Four of the "best" mutants had amino acid exchanges different from the monooxygenase mutants optimized for the Baeyer-Villiger reaction shown in Scheme 18. Please note, however, that the mutant in entry 3 (F432S) is *exactly* the one that gave the best ee in the first mutation round for the Baeyer-Villiger reaction (Scheme 18, Table, entry 9)! In other words, the single mutation at position 432 gave a surprisingly versatile biocatalyst, and it was identified twice by screening the same 10,000-membered mutant library in two different reactions!

The enzyme mutants used for the enantioselective sulfoxidation of the thioether **41** were also tested in the oxidative kinetic resolution of the racemic sulfoxide *rac*-**42** (*28*). For example, the mutant shown in entry 2 of the Table in Scheme 19 gave 98.7 % ee of the *R*-sulfoxide **42** after 43 % conversion

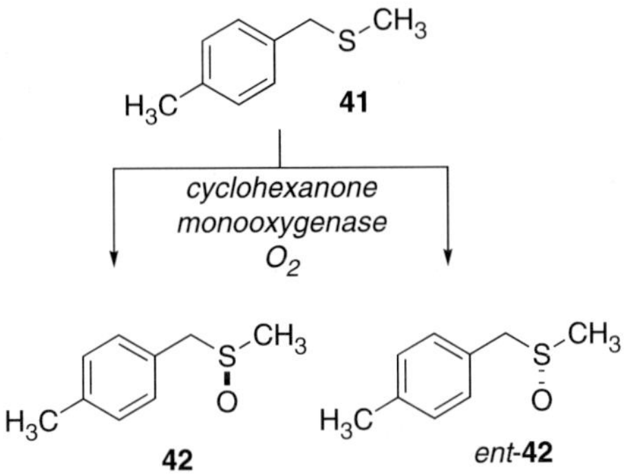

entry	amino acid exchanges in mutants	favored product enantiomer (yield [%])	ee [%]	sulfone [%]
1	wild type	**42** (75)	14.0	< 1
2	D384H	**42** (75)	98.9	7.9
3	F432S	**42** (55)	98.7	20.0
4	K229I, L248P	*ent*-**42** (77)	98.1	5.6
5	Y132C, F246I, V361A, T415A	*ent*-**42** (52)	99.7	26.6
6	F16L, F277S	*ent*-**42** (84)	95.2	5.6

SCHEME 19. Directed evolution of a cyclohexanone monooxygenase for the asymmetric sulfoxidation of a prochiral thioether.

to the sulfone (*57*). This means that asymmetric sulfoxidation and kinetic resolution are effected by this enzyme in a cooperative manner. The same was found for the mutant shown in entry 5. In contrast, the wild-type enzyme, which is only a poorly enantioselective catalyst for the sulfoxidation, also fails in the kinetic resolution of the racemic sulfoxide *rac*-**42**.

In a next round of random mutagenesis, starting from the mutant shown in entry 5 of the Table in Scheme 19, the resulting library of 1600 clones was screened for both high *S*-selectivity in the sulfoxidation *and* low sulfone production. A mutant was identified which provided the sulfoxide product **42** with 99.8 % ee and at negligible (< 5 %) sulfone formation.

With regard to other thioether substrates, it was reported that the mutant shown in entry 6 effects the oxidation of ethylphenyl thioether with a preference for the *R*-sulfoxide (88 % ee), whereas the mutant shown in entry 4 gives predominantly the *S*-sulfoxide with excellent enantiomeric purity (98.9 % ee) (*28*).

The examples presented here clearly demonstrate that the directed evolution of enzymes is a combinatorial approach to novel catalysts *par excellence*: Large libraries can be generated, and the methodology for the rapid assessment of the libraries with respect to activity and selectivity has been established during the recent years. Limitations may still be seen in the need to either use whole cell catalyst systems, or to rely on enzyme excretion to the medium. Nevertheless, the directed evolution of enzymes will definitely keep on revolutionizing the use of enzymes in selective oxidation processes, and in many other fields.

Epilogue

Chemists have long been intrigued by the fascinating efficiency and selectivity of enzymes. In its original sense, the term biomimetic catalysis refers to the attempt to generate enzyme-like efficiency and selectivity by mimicking the enzymes' active sites by low-molecular weight models. Usually, the design approach has been taken, i.e., the model systems were designed according to structural information available e.g., from X-ray crystallographic studies of the corresponding enzyme(s). In recent years, two very important developments have contributed to the rapid further development of the field: Combinatorial chemistry and the directed evolution of enzymes. The former approach significantly enhances the chemist's ability to generate and screen novel catalyst candidates. The latter method allows the adaptation of existing enzymes to novel tasks even in the absence of any structural and/or mechanistic information. Altogether, biomimetic oxidation catalysis has significantly profited from both of these novel and diversity-based approaches. The author sincerely hopes that, one day, it may also be possible to evolve "chemical" catalysts, without the need for the biological reproduction and amplification mechanisms of the cell, and thus without the limitation to peptides (enzymes) or nucleic acids (ribozymes).

Acknowledgments

Our own work in the area of (diversity based) biomimetic oxidation catalysis was financially supported by the Fonds der Chemischen Industrie, by the BASF AG, Ludwigshafen, and by the Deutsche Forschungsgemeinschaft (priority program "Radicals in Enzymatic Catalysis").

References

1. Sheldon, R. A. "*Oxidations in Aqueous Phase Organometallic Catalysis*"; 2nd edn.; Eds. Cornils, B.; Herrmann, W. A., Wiley-VCH. Weinheim, **2004**, pp. 473–480.
2. Blaser, H. U.; Schmidt, E. (Eds.) "*Asymmetric Catalysis on Industrial Scale*"; Wiley-VCH: Weinheim, **2004**.
3. Meunier, B. (Ed.) "*Biomimetic Oxidations Catalyzed by Transition Metal Complexes*"; Imperial College Press: London, **2000**.
4. (a) Gennari, C.; Piarulli, U. *Chem. Rev.* **2003**, *103*, 3071; (b) Reetz, M. T. *Angew. Chem.* **2001**, *113*, 292; *Angew. Chem. Int. Ed.* **2001**, *40*, 284; (c) Jandeleit, B.; Schaefer, D. J.; Powers, T. S.; Turner, H. W.; Weinberg, W. H. *Angew. Chem.* **1999**, *111*, 2648; *Angew. Chem. Int. Ed. Engl.* **1999**, *38*, 2494.

5. It is not intended to discredit "classical" chemical synthesis, product purification and characterization, as this approach still represents our major source of new (and often unexpected) chemical knowledge.
6. Balkenhohl, F.; von dem Bussche-Hünnefeld, C.; Lansky, A.; Zechel, C. *Angew. Chem.* **1996**, *108*, 2436; *Angew. Chem. Int. Ed. Engl.* **1996**, *35*, 2288.
7. Nicolaou, K. C.; Hanko, R.; Hartwig, W. (Eds.) "*Handbook of Combinatorial Chemistry*"; Wiley-VCH: Weinheim, **2002**.
8. Still, W. C. *Acc. Chem. Res.* **1996**, *29*, 155.
9. (a) Arnold, F. H. *Acc. Chem. Res.* **1998**, *31*, 125; (b) Reetz, M. T.; Jaeger, K.-E. *Chem. Eur. J.* **2000**, *6*, 407.
10. Reetz, M. T.; Zonta, A.; Schimossek, K.; Liebeton, K.; Jaeger, K.-E. *Angew. Chem.* **1997**, *109*, 2961; *Angew. Chem. Int. Ed.* **1997**, *36*, 2830.
11. See ref. 4b for an overview of current high-throughput screening methodology.
12. See ref. 48c for an overview of screening methods for bead-bound libraries.
13. See refs. 48a-c for the use of insoluble dyes in the bulk screening of bead-bound libraries for catalytic activity.
14. (a) Müller, M.; Mathers, T. W.; Davis, A. P. *Angew. Chem.* **2001**, *113*, 3929; *Angew. Chem. Int. Ed. Engl.* **2001**, *40*, 3813; (b) Harris, R. F.; Nation, A. J.; Copeland, G. T.; Miller, S. J. *J. Am. Chem. Soc.* **2000**, *122*, 11270.
15. For an example of the identification of catalytic activity in a bead-bound library by IR-thermography, see: Taylor, S. J.; Morken, J. P. *Science* **1998**, *280*, 267.
16. Reetz, M. T.; Becker, M. H.; Liebl, M.; Fürstner, A. *Angew. Chem.* **2000**, *112*, 1294; *Angew. Chem. Int. Ed. Engl.* **2000**, *39*, 1236.
17. Holzwarth, A.; Schmidt, H.-W.; Maier, F. W. *Angew. Chem.* **1998**, *110*, 2788; *Angew. Chem. Int. Ed. Engl.* **1998**, *37*, 2644.
18. de Montellano, O. (Ed.) "*Cytochrome P-450*"; Plenum Press: New York, **1986**.
19. McMurry, T. J.; Groves, J. T. in ref. 18, p. 1.
20. Francis, M. B.; Jacobsen, E. N. *Angew. Chem.* **1999**, *111*, 987; *Angew. Chem. Int. Ed.* **1999**, *38*, 937–941.
21. (a) Fernández, I.; Khiar, N. *Chem. Rev.* **2003**, *103*, 3651; (b) Carreno, M. C. *Chem. Rev.* **1995**, *95*, 1717.
22. Legros, J.; Dehli, J. R.; Bolm, C. *Adv. Synth. Catal.* **2005**, *347*, 19.
23. Holland, H. L. *Chem. Rev.* **1998**, *108*, 472.
24. (a) Bolm, C.; Bienewald, F. *Angew. Chem.* **1995**, *107*, 2883; *Angew. Chem. Int. Ed. Engl.* **1995**, *34*, 2640; (b) Bolm, C.; Bienewald, F. *Synlett* **1998**, 1327.
25. Improved enantioselectivity (up to 88% ee) for simple thioethers was achieved with ligands harboring two elements of chirality, see: (a) Vetter, A. H.; Berkessel, A. *Tetrahedron Lett.* **1998**, *39*, 1741; Ohta, C.; Shimizu, H.; Kondo, A.; Katsuki, T. *Synlett* **2002**, 161.
26. Pelotier, B.; Anson, M. S.; Campbell, I. B.; Macdonald, S. J. F.; Priem, G.; Jackson, R. F. W. *Synlett.* **2002**, 1055.
27. Green, S. D.; Monti, C.; Jackson, R. F. W.; Anson, M. S.; Macdonald, S. J. F. *Chem. Commun.* **2001**, 2594.
28. Reetz, M. T.; Daligault, F.; Brunner, B.; Hinrichs, H.; Deege, A. *Angew. Chem.* **2004**, *116*, 4170; *Angew. Chem. Int. Ed.* **2004**, *43*, 4078.
29. Krow, G. R. *Org. React.* **1993**, *43*, 251.
30. Renz, M.; Meunier, B. *Eur. J. Org. Chem.* **1999**, 737.
31. Drauz, K.; Waldmann, H. (Eds.) "*Enzyme Catalysis in Organic Synthesis*"; Wiley-VCH: Weinheim, **2002**.
32. Faber, K. "*Biotransformations in Organic Chemistry*"; 4th edn; Springer: Berlin, **2000**.
33. Mihovilovic, M.D.; Müller, B.; Stanetty, P. *Eur. J. Org. Chem.* **2002**, 3711.
34. Wong, M.-K.; Ho, L.-M.; Zheng, Y.-S.; Ho, C.-Y.; Yang, D. *Org. Lett.* **2001**, *3*, 2587.
35. Strukul, G. *Angew. Chem.* **1998**, *110*, 1256; *Angew. Chem. Int. Ed.* **1998**, *37*, 1198.
36. Kobayashi, S.; Sugiura, M.; Kitagawa, H.; Lam, W. W.-L. *Chem. Rev.* **2002**, *102*, 2227.
37. Berkessel, A.; Ashkenazi, E.; Andreae, M. R. M. *Appl. Catal. A: General* **2003**, *254*, 27.
38. Gamez, P.; Koval, I. A.; Reedijk, J. *Dalton Trans.* **2004**, 4079.

39. Wang, Y.; DuBois, J. L.; Hedman, B.; Hodgson, K. O.; Stack, T. D. P. *Science* **1998**, *279*, 537.
40. Chaudhuri, P.; Hess, M.; Müller, J.; Hildenbrand, K.; Bill, E.; Weyhermüller, T.; Wieghardt, K. *J. Am. Chem. Soc.* **1999**, *121*, 9599.
41. Gamez, P.; Arends, I. W. C. E.; Reedijk, J.; Sheldon, R. A. *Chem. Commun.* **2003**, 2414.
42. Berkessel, A.; Dousset, M.; Bulat, S.; Glaubitz, K. *Biol. Chem.* **2005**, *386*, 1035–1041.
43. Nicolaou, K. C.; Xiao, X.-Y.; Parandoosh, Z.; Senyei, A.; Nova, M. P. *Angew. Chem.* **1995**, *107*, 2476; *Angew. Chem. Int. Ed. Engl.* **1995**, *34*, 2289.
44. For a general review of metal-free oxidation methods, including achiral and chiral catalysts, see: Adam, W.; Saha-Möller, C. R.; Ganeshpure, P. A. *Chem. Rev.*, **2001**, *101*, 3499.
45. Page, P. C. B.; Buckley, B. R.; Blacker, A. J. *Org. Lett.* **2004**, *6*, 1543.
46. Aggarwal, V. K.; Wang, M. F. *Chem. Commun.* **1996**, 191.
47. White, J. S.; White, D. C. "*Source Book of Enzymes*"; CRC Press: New York, **1997**.
48. (a) Berkessel, A.; Hérault, D. A. *Angew. Chem.* **1999**, *111*, 99; *Angew. Chem. Int. Ed. Engl.*, **1999**, *38*, 102; (b) Berkessel, A.; Riedl, R. *J. Comb. Chem.* **2000**, *2*, 215; (c) Berkessel, A. *Curr. Opin. Chem. Biol.* **2003**, *7*, 409; (d) De Muynck, H.; Madder, A.; Farcy, N.; De Clercq, P. J.; Pérez-Payan, M. N.; Öhberg, L. M.; Davis, A. P. *Angew. Chem.* **2000**, *112*, 149; *Angew. Chem. Int. Ed. Engl.* **2000**, *39*, 145; (e) Madder, A.; Li, L.; De Muynck, H.; Farcy, N.; Van Haver, D.; Fant, F.; Vanhoenacker, G.; Sandra, P.; Davis, A. P.; De Clercq, P. J. *J. Comb. Chem.* **2002**, *4*, 552.
49. For an overview of the Juliá-Colonna epoxidation of enones, see: Berkessel, A.; Gröger, H. "*Asymmetric Organocatalysis*"; Wiley-VCH: Weinheim, **2005**, Chapter 10.2.
50. (a) Juliá, S.; Masana, J.; Vega, J. C. *Angew. Chem.* **1980**, *92*, 968; *Angew. Chem. Int. Ed. Engl.* **1980**, *19*, 929; (b) Juliá, S.; Guixer, J.; Masana, J.; Rocas, J.; Colonna, S.; Annuziata, R.; Molinari, H. *J. Chem. Soc. Perkin Trans.* **1982**, 1317.
51. Bentley, P. A.; Kroutil, W.; Littlechild, J. A.; Roberts, S. M. *Chirality* **1997**, *9*, 198.
52. (a) Flood, R. W.; Geller, T. P.; Petty, S. A.; Roberts, S. M.; Skidmore, J.; Volk, M. *Org. Lett.* **2001**, *3*, 683; (b) Bentley, P. A.; Cappi, M. W.; Flood, R. W.; Roberts, S. M.; Smith, J. A. *Tetrahedron Lett.* **1998**, *39*, 9297; (c) Takagi, R.; Shiraki, A.; Manabe, T.; Kojima, S.; Ohkata, K. *Chem. Lett.* **2000**, 366.
53. Berkessel, A.; Gasch, N.; Glaubitz, K.; Koch, C. *Org. Lett.* **2001**, *3*, 3839.
54. For recent calculations and kinetic studies, leading to a mechanistic model related to the one described by Berkessel *et al.* in ref. 53, see: (a) Kelly, D. R.; Roberts, S.M. *Chem. Commun.* **2004**, 2018; (b) Kelly, D. R.; Caroff, E.; Flood, R. W.; Heal, W.; Roberts, S. M. *Chem. Commun.* **2004**, 2016; (c) Carrea, G.; Colonna, S.; Meek, A. D.; Ottolina, G.; Roberts, S. M. *Chem. Commun.* **2004**, 1412.
55. For the general concept of "the nest", the anion-binding motif generated by three peptide main-chain NH groups, see: (a) Watson, J. D.; Milner-White, E. J. *J. Mol. Biol.*, **2002**, *315*, 171; (b) Watson, J. D.; Milner-White, E. J. *J. Mol. Biol.* **2002**, *315*, 183.
56. Reetz, M. T.; Brunner, B.; Schneider, T.; Schulz, F.; Clouthier, C. M.; Kayser, M. M. *Angew. Chem.* **2004**, *116*, 4167; *Angew. Chem. Int. Ed.* **2004**, *43*, 4075.
57. The apparent inconsistency of the values given for the ee of the remaining sulfoxide and the conversion of the racemate is most likely due to the use of a whole-cell system (Reetz, M. T.; private communication).

SELECTIVE CONVERSION OF HYDROCARBONS WITH H_2O_2 USING BIOMIMETIC NON-HEME IRON AND MANGANESE OXIDATION CATALYSTS

STEFANIA TANASE and ELISABETH BOUWMAN*

Leiden Institute of Chemistry, Gorlaeus Laboratories, Leiden University, PO Box 9502, 2300 RA Leiden, The Netherlands

I. General Introduction

Oxidation is one of the most important industrial reactions. In the past decades, impressive progress has been made in the development of oxidation catalysis, resulting in many novel industrial technologies. Oxidation catalysis is particularly an important technological area that lies at the heart of a variety of processes for producing bulk and fine chemicals and for eliminating pollution. In technical chemistry, oxidation processes play a central role and serve as a basis for the generation of a number of major intermediates and final products. More than 20 % of all industrial organic materials are obtained by catalytic oxidation. Major examples are the large-scale production of ethylene oxide and propylene oxide from ethylene or propylene, of maleic anhydride from butane, of acrolein and acrylic acid from propylene, xylene oxidation to phthalic anhydride or terephthalic acid, or cyclohexane oxidation to

ADVANCES IN INORGANIC CHEMISTRY
VOLUME 58 ISSN 0898-8838 / DOI: 10.1016/S0898-8838(05)58002-4

cyclohexanol, cyclohexanone, and adipic acid. Use of suitable catalysts is indispensable for the economically efficient and sustainable implementation of these processes. For example, the predominant method of cyclohexane oxidation is a metal-catalyzed oxidation, which employs small amounts of Co, Cr, and/or Cu as catalyst (*1*). However, the suspected toxicity and hepatocarginogenicy of cobalt makes it unacceptable catalyst for further industrial applications. An increase in the application of oxidation catalysis in fine chemistry is expected to make the production in this field of chemistry more efficient, resulting in less expensive and more environmentally acceptable processes (*2*). Interesting oxidation reactions in fine chemistry are epoxidation reactions, because the resulting materials can be used for the synthesis of a wide variety of products. Asymmetric catalysis is a very attractive route to chiral, non-racemic molecules, because catalyst tuning ideally allows for efficient and highly selective reactions to occur (*3*). Apart from the use in the synthesis of chemicals, oxidation catalysis can also become important for other applications, such as bleaching processes. In these processes, the use of iron and manganese catalysts with dioxygen or dihydrogen peroxide as the primary oxidant is required because of environmental reasons.

In nature, catalysis is of vital importance in a wide variety of biochemical processes catalyzed by metalloenzymes. The functions performed by the metalloenzymes and their significance have inspired a range of biomimetic studies, and intense efforts have been concentrated on the synthesis of low molecular weight complexes to model the spectroscopic features of such enzymes (*4–6*). Concomitantly, a challenging frontier in the field of biomimetic chemistry is the development of highly reactive model compounds that are able to effect rapid and selective substrate conversion. Some important catalytic transformations that have gained attention in recent years involve the selective oxidation of non-activated C–H bonds (*7,8*). Iron- and manganese-catalyzed alkane and alkene oxidations have attracted a great deal of interest in organic synthesis because the development of fundamentally new selective routes from hydrocarbons to more valuable products (such as alcohols, ketones, acids, and peroxides) is necessary (*9–11*). Furthermore, using inexpensive, environmentally friendly metals in coordination environments that can be tuned with sterically and electronically tunable ligands, many catalytic reactions have been elaborated to a high level of enantioselectivity (*12–14*).

The oxidation of saturated hydrocarbons in the presence of iron- or manganese-containing catalysts can be achieved by using a variety of oxidants including alkyl hydroperoxides, peroxycarboxylic acids, iodosylbenzene, dihydrogen peroxide, and dioxygen (*9–11*). It has been shown that chiral iron- and manganese-porphyrin complexes catalyze the asymmetric epoxidation of unfunctionalized alkenes (*15*). Except for a number of experiments in which up to 96 % enantiomeric excess (*ee*) has been reported (*16,17*), in most epoxidation reactions with chiral porphyrins only a low to moderate enantiomeric excess of the product is obtained (*18,19*). In association with these catalysts, alkyl hydroperoxides and iodosylbenzene are often used as primary oxidants (*18,19*).

Non-heme iron and manganese complexes are preferred for catalytic oxidation reactions because they are more easily accessible and their use as catalysts results generally in significantly higher enantiomeric excess of epoxides (*9*). For possible commercial applications, the use of dihydrogen peroxide as an oxidant is the most attractive primary oxidant in terms of costs, ease, and safety of handling. Dihydrogen peroxide has relatively high oxygen content, can be safely used in concentrations up to 60 % and leaves only water as a waste product (*2,20*). As a consequence, the functionalization of hydrocarbons with dihydrogen peroxide in the presence of non-heme iron or manganese-based catalysts has received considerable attention in the last years (*9,10*). Herein, an overview is given on the successful modeling of the functional aspects of the non-heme iron and manganese redox enzymes. The focus of this review is on complexes that are able to catalyze hydrocarbon oxidation reactions using dihydrogen peroxide as oxidant.

II. Biological Enzymes and Catalytic Oxidation

A. Introduction

In nature, oxidation reactions are essential for aerobic life. Energy for cells is provided by the combustion of carbohydrates and fatty acids with dioxygen. Oxidation reactions are also involved in biosynthesis, metabolism reactions, and the detoxification of harmful compounds. In several of these reactions, iron or manganese enzymes are involved. These manganese and iron enzymes have frequently been used as a source of inspiration for the development of manganese- and iron-based oxidation catalysts.

Various iron-containing biomolecules are involved in the transport, storage, and activation of dioxygen or in the activation of reduced derivatives of dioxygen, such as the superoxide radical anion ($O_2^{\bullet -}$) and dihydrogen peroxide (H_2O_2) (*21,22*). These iron proteins can be divided into heme (containing a porphyrin group with an iron(III) ion) (*23*) and non-heme proteins (*4*). For some identical biological functions, both heme and non-heme proteins have been characterized. Apart from the iron enzymes, various manganese-containing enzymes are also known to use dioxygen and its reduced derivatives for redox reactions (*24*).

B. Non-Heme Iron Enzymes

Metalloenzymes containing non-heme iron centers are widespread in nature. Several members of this family isolated from mammals, plants, or bacteria, have now been structurally characterized (*4*). In addition to the crystal structures for the isolated resting states, an increasing amount of spectroscopic information has become available concerning the active sites of these iron enzymes (*25–27*). The non-heme iron enzymes perform a broad range of functions, but most important is their role in the activation of dioxygen for

TABLE I

SELECTION OF IMPORTANT MONO- AND DIIRON METALLOENZYMES WITH CATALYTIC FUNCTION IN OXIDATION REACTIONS

Enzyme class	Function	Ref.
α-keto acid dependent oxygenases	Unactivated C–H bond oxidation	(*4*)
Aromatic amino acid hydroxylases	Aromatic amino acids hydroxylation	(*174*)
Catechol dioxygenase – intradiol cleavage	Arene dihydroxylation	(*175*)
Fe-superoxide dismutase	Superoxide radical redox dismutation	(*176*)
Isopenicillin N synthase	Isopenicillin N formation	(*177*)
Lipoxygenases	Unsaturated fatty acid oxidation	(*178*)
Nitrile hydratase	Nitrile to amide hydration	(*179*)
Rieske oxygenase	Arene *cis*-dihydroxylation	(*29*)
Bleomycin	DNA cleavage	(*6*)
Ribonucleotide reductase	Tyrosyl radical generation	(*180*)
Methane monooxygenase	Methane to methanol oxidation	(*181*)
Purple acid phosphatase	Phosphate ester hydrolysis	(*182*)
Δ^9 desaturase	Alkane to alkene conversion	(*183*)
Toluene monooxygenase	Toluene to cresol oxidation	(*184*)
Phenol hydroxylase	Phenol to catechol oxidation	(*185*)
Alkene monooxygenase	Alkene epoxidation	(*186*)
Butane monooxygenase	Butane to butanol oxidation	(*187*)
ω-alkane hydroxylase	Alkane to alcohol oxidation	(*188*)
DMQ monooxygenase	Quinone generation	(*189*)

the oxidation of different substrates. Some of these oxidation reactions have little or no precedent in organic chemistry; hence there is a considerable interest in unraveling the catalytic mechanism of action of these enzymes.

Enzymes performing oxidation catalysis can generally be divided into two classes: the oxidases, which act through electron transfer with concomitant reduction of dioxygen to water, and oxygenases, which catalyze the reaction of dioxygen with organic or inorganic substrates. Mono-oxygenases catalyze the incorporation of one atom of dioxygen into the product, while dioxygenases incorporate both atoms of dioxygen into the product. In Table I, a list is given of the non-heme metalloenzymes that have been identified to contain a mono- or diiron active site motif and perform a catalytic function in oxidation reactions.

The mononuclear non-heme site is typically penta- or hexacoordinate with one mono- or didentate carboxylate group derived from asparagines, glutamate, cysteine, or tyrosinase and at least two histidines around the iron center; there are one or two reactive coordination sites that are occupied by water molecules. The field of mononuclear non-heme iron active sites including enzymes, models, and intermediates has recently been reviewed (*4*). Intradiol-cleaving enzymes represent the first and the most thoroughly studied subclass of the mononuclear non-heme iron enzymes. The catechol dioxygenases catalyze the oxidative cleavage of catechols and are used by nature to degrade aromatic molecules in biosphere. The intradiol catechol dioxygenases catalyze the oxidation of catechols to *cis,cis*-muconic acids,

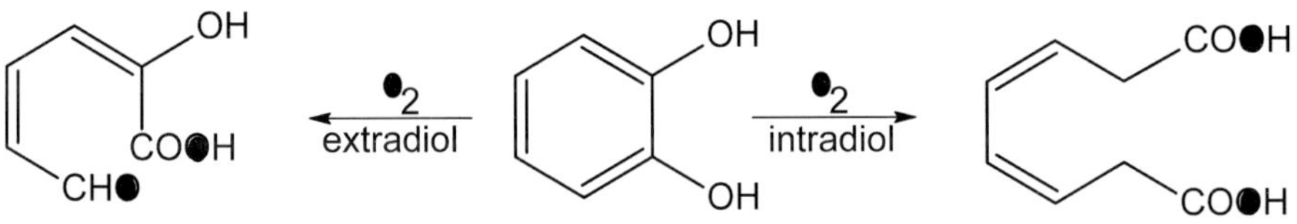

FIG. 1. Intradiol vs. extradiol oxidative cleaving reactions.

whereas the extradiol catechol dioxygenases catalytically convert catechols to muconic semialdehydes (Fig. 1). Lipoxygenases are a group of non-heme iron-containing dioxygenases catalyzing the addition of molecular dioxygen to polyunsaturated fatty acids in a stereospecific as well as regiospecific way and are involved in the biosynthesis of inflammatory mediators (*28*). The initial step in the bacterial degradation of aromatic hydrocarbons is usually a *cis*-dihydroxylation of the aromatic ring, to give a *cis*-diol, in which two atoms of dioxygen (O_2) are incorporated into the substrates. This reaction is catalyzed by a family of non-heme iron dioxygenases, of which the best studied is naphthalene 1,2-dioxygenase (*29*). Besides the insertion of O_2 into an aromatic C–C bond, non-heme iron dioxygenases can also carry out the *cis*-dihydroxylation of olefinic C–C bonds.

The iron complex of bleomycin (BLM), a glycopeptide-derived antibiotic, is an important example of the monoxygenases (*6*). *In vivo*, BLM iron complexes show anti-tumor activity that is the result of oxidative DNA cleavage. The iron complex of BLM is proposed to be square-pyramidal in which the iron(II) atom is coordinated by five nitrogen donors (Fig. 2) (*6,30*). The vacant axial position *trans* to the primary amine can be occupied by either a part of the sugar side chain of BLM or by a solvent molecule. This position is also the assumed coordination site for dioxygen. To allow catalysis, the dioxygen molecule is reduced, resulting in the formation of an Fe^{III}–OOH species. (*6*) Apart from dioxygen, dihydrogen peroxide has also been used to generate the Fe^{III}–OOH complex (*6*).

In the diiron non-heme enzymes, the iron coordination sphere typically is composed of a single histidine nitrogen, several carboxylate oxygens and oxo/hydroxo bridge(s); a relatively symmetric ligand environment is present

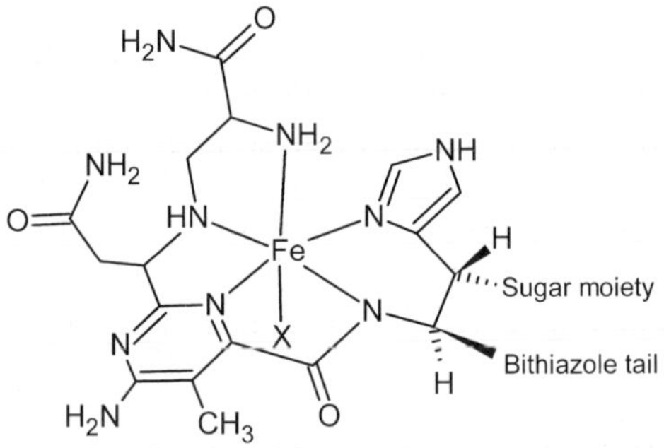

FIG. 2. Proposed structure of iron bleomycin; X = CO, NO, RNC (*6*).

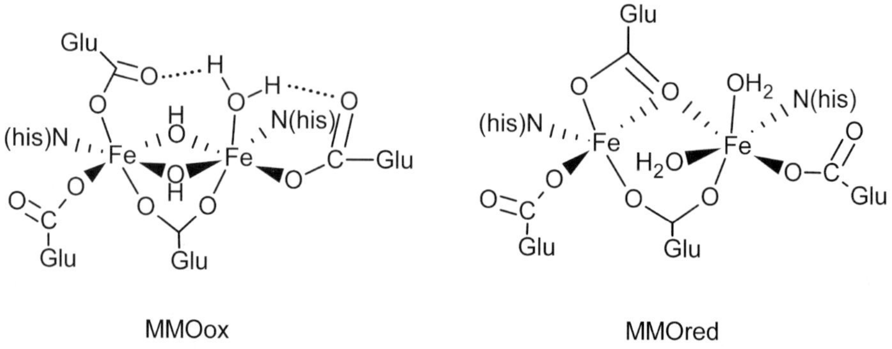

FIG. 3. Active site structures of MMO_{ox} and MMO_{red} (*32*).

for the two iron centers, with a flexible ligand coordination that allows rearrangements upon reduction. Within this class of enzymes, methane monooxygenases (MMOs) are widely investigated dinuclear non-heme iron enzymes that catalyze the oxidation of methane to methanol with use of dioxygen. This unique function achieves the first step in the metabolism of methanotrophic bacteria, which consume methane and dioxygen as their sole sources of carbon and energy. Many other hydrocarbon substrates are oxidized by MMO (*10*). The structure of both oxidized (MMO_{ox}) and reduced (MMO_{red}) enzymes have been elucidated and schematic drawings of the active sites are shown in Fig. 3. Dioxygen activation by enzymes containing a dinuclear non-heme iron site, as well as the most recent mechanistic studies on the hydroxylation of methane by methane monooxygenase have been reviewed (*31,32*). A recent review has been focused on the synthetic structural models for non-heme carboxylate-bridged diiron metalloproteins (*33*).

C. Manganese Redox Enzymes

Several enzymes are known, that naturally contain manganese in the active site and include classes as oxidoreductases, transferases, hydrolases, lyases, isomerases, ligases, lectins, and integrins (*24*). Some representative enzymes that are involved in catalytic transformations are presented in Table II. The dioxygen-evolving complex of Photosystem II (PSII) and manganese catalase are the most intensively studied manganese enzymes (*5,24,34,35*). Except for the catalase enzymes and the dioxygen-evolving complex, the manganese enzymes seem to be structurally related to the iron enzymes, but are often found in less sources than the iron analogues.

The dioxygen-evolving complex (OEC) is involved in the earliest stages of the overall photosynthetic processes, in which NADPH is produced for carbohydrate biosynthesis. The OEC catalyzes the photoinduced oxidation of water, thereby releasing molecular dioxygen. This four-electron oxidation of water (Eq. (1)) is accomplished by a cubane-like Mn_3CaO_4 cluster linked to a fourth manganese center by a single μ-oxo bridge as shown by the recent X-ray analysis of PSII (Fig. 4) (*36,37*). On the basis of EPR and ENDOR

TABLE II

SELECTION OF IMPORTANT MANGANESE REDOX ENZYMES INVOLVED IN OXIDATIVE CONVERSIONS

Enzyme class	Function	Ref.
Superoxide dismutase	Superoxide decomposition	(*190*)
Mn catechol dioxygenase	Catechol oxidation	(*191*)
Mn catalase	Dihydrogen peroxide decomposition	(*192*)
Oxygen evolving complex	Water to dioxygen conversion	(*37*)
Mn lipoxygenase	Polyunsaturated fatty acid oxidation	(*193*)

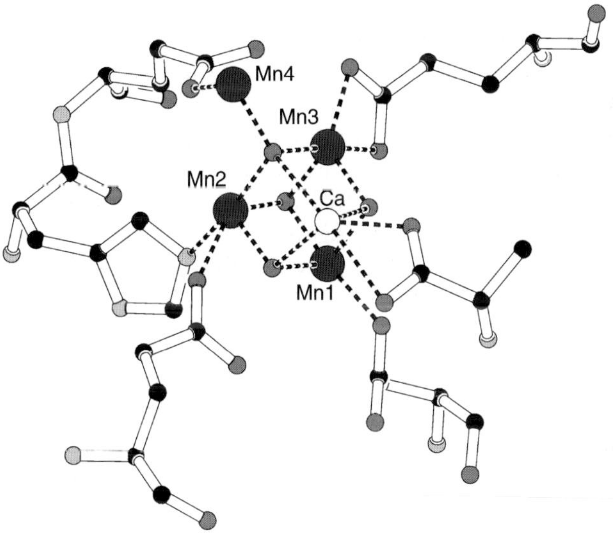

FIG. 4. The dioxygen-evolving center of Photosystem II (*36*).

studies, the most relevant manganese oxidation states have been assigned to be $Mn^{III}Mn_3^{IV}$ (*38*), although an alternate proposal involving $Mn_3^{III}Mn^{IV}$ has been offered (*38*). Concerning the mechanism of water oxidation, all studies suggest that two water molecules are bound to two neighboring manganese sites and subsequently a deprotonation step occurs that is followed by O–O bond formation (*38*). It is generally accepted that Ca^{II} is an essential cofactor for the photosynthetic water oxidation and the presence of an $Mn^{V}{=}O$ intermediate is proposed on the structural basis of the various intermediates derived from inorganic model chemistry (*38*). The OEC is also able to generate O_2 from H_2O_2 by a catalase-like two-electron oxidation (Eq. (2)) (*6*). Nevertheless, the exact mechanism of water oxidation has not yet been elucidated.

$$2H_2O \longrightarrow O_2 + 4e^- + 4H^+ \quad (1)$$

$$H_2O_2 \longrightarrow O_2 + 2e^- + 2H^+ \quad (2)$$

$$2H_2O_2 \longrightarrow 2H_2O + O_2 \quad (3)$$

Catalase enzymes are present in nearly all aerobic cells and their main function is to protect the cell from the toxic effects of dihydrogen peroxide by catalyzing its decomposition into molecular oxygen and water without the production of free radicals (Eq. (3)). Catalase structures from many different species have been studied by X-ray diffraction and they revealed the presence of a dinuclear manganese center (*5,24,39*). Although it is clear that all catalase enzymes share a general structure, some of them differ in the number and identity of domains. Spectroscopic studies on catalase mimics have shown that both Mn_2^{II} and Mn_2^{III} are involved in the catalytic cycle (Fig. 5) (*40,41*). In the proposed mechanism for catalase activity, dihydrogen peroxide decomposition is initiated by the binding of H_2O_2 to the $Mn^{III}Mn^{III}$ dinuclear center followed by reduction to the $Mn^{II}Mn^{II}$ intermediate and concomitant oxidation of the peroxide to O_2. Subsequent binding of a second molecule of H_2O_2 to the $Mn^{II}Mn^{II}$ species results in the formation of a new species containing a terminal hydroperoxo group with the release of a water molecule. The switch of the terminal hydroperoxo group into a bridging mode and release of the second manganese-bound water molecule results in the oxidation of the $Mn^{II}Mn^{II}$ species, thus closing the catalytic cycle.

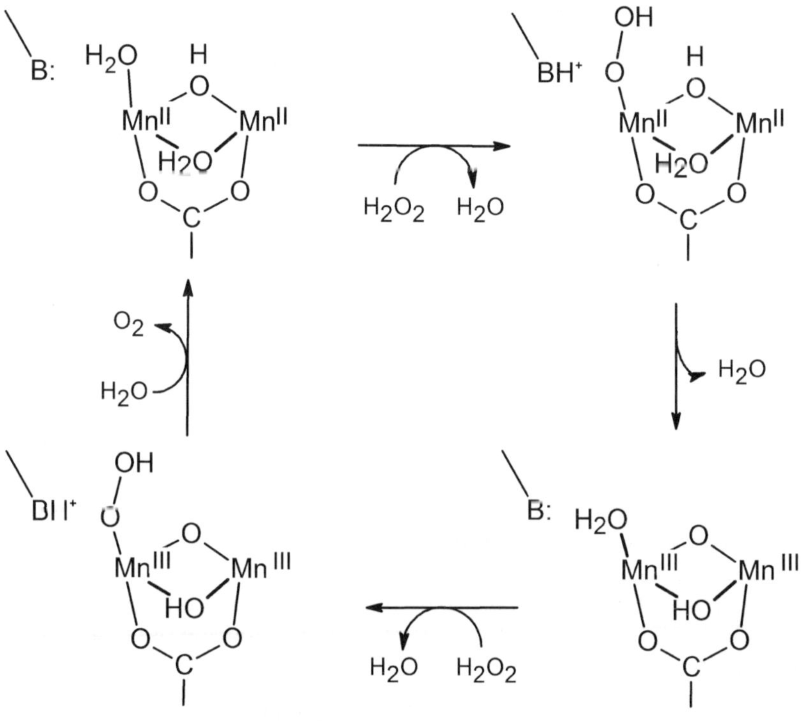

FIG. 5. Proposed mechanism for H_2O_2 decomposition by manganese catalase (*5*).

The structural, spectroscopic, and reactivity models for the manganese catalases have been recently reviewed (*5*).

D. Ligand Design for Biomimetic Non-Heme Iron and Manganese Complexes

Knowledge of homogeneous catalysis at the metal centers of model complexes is based upon the progress and understanding of their chemistry and reactivity. The stability and the selectivity of a catalyst are strongly related to its molecular structure. Consideration of steric, electronic, and conformational properties is necessary to design appropriate ligands for the synthesis of effective catalysts. As a result, an important goal in the development of practical biomimetic catalysts is to design sterically demanding polydentate ligands that can bind one or two metal centers and hold them in close proximity. These ligands must be resistant to oxidation and also be strongly electron donating, in order to achieve high oxidation states. In addition, the ligands should have the versatility to bind to different metal centers, in order to be able to control the reactivity by changing the metal ions. Another option to modulate the reactivity of the catalyst can be accomplished by modification of the ligand donor properties.

As far as hydrocarbon oxidation is concerned, a selective and rapid C–H bond activation is required. In the particular case of iron or manganese/H_2O_2-catalyzed oxidations, preferentially a reactive metal-oxo species is generated, which reacts rapidly with the C–H bond *via* "hydrogen abstraction" without the need for high-energy reagents. This can be achieved by using strongly electron-donating ligands, which allow stabilization of high-valent metal complexes. The ancillary N-donor and O-donor ligands play a critical role in the formation and stabilization of high-valent iron or manganese species. High-valent manganese or iron species have been detected by extensive EPR, EXAFS, and X-ray studies. From the results of synthetic studies, it has been realized that iron or manganese ions in the oxidation state IV are primarily stabilized with bridging oxide (O^{2-}) groups.

A wide variety of ligands from didentate to pentadentate as well as dinucleating ligands have been designed and synthesized for modeling the active site of iron and manganese enzymes exhibiting a catalytic function. Relevant iron and manganese complexes of these ligands together with their catalytic activity in the oxidation of hydrocarbons will be discussed in the next sections.

III. Biomimetic Iron Catalysts

A. Alkane Hydroxylation

A.1. Mononuclear Iron Complexes

A large number of mononuclear iron(II) complexes with polydentate ligands (Fig. 6) have been studied as catalysts in hydroxylation reactions of alkanes employing H_2O_2 as terminal oxidant; in Table III the reported catalytic activities associated with these catalysts are summarized. In addition, a few mononuclear iron(III) biomimetic catalysts have also been reported, containing deprotonated carboxamido-nitrogen based ligands that prevent the formation of hydroxo- or oxo-bridged polymeric species (Fig. 6) (*42*).

The reactivity patterns for the alkane functionalization by iron(II) mononuclear complexes using H_2O_2 as terminal oxidant suggest that two reaction pathways are mainly associated with this type of chemistry: one involving uncontrolled hydroxyl radicals, in particular produced by Haber-Weiss

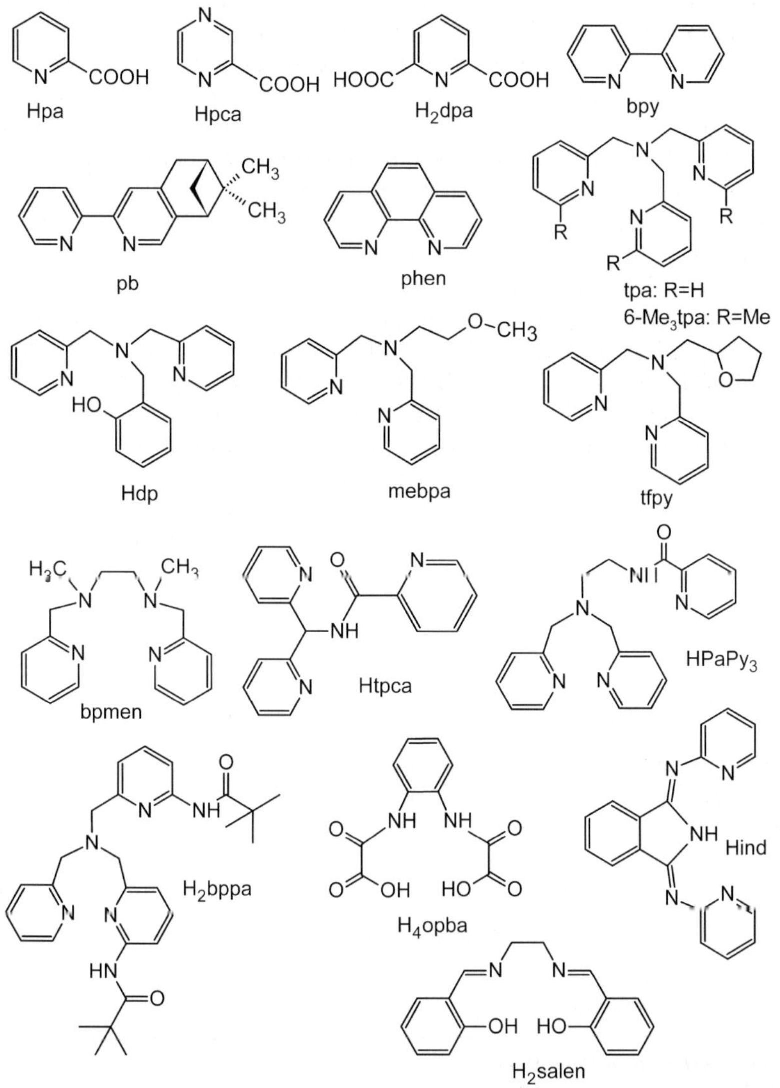

FIG. 6. Ligands used for the synthesis of mononuclear and dinuclear non-heme iron catalysts.

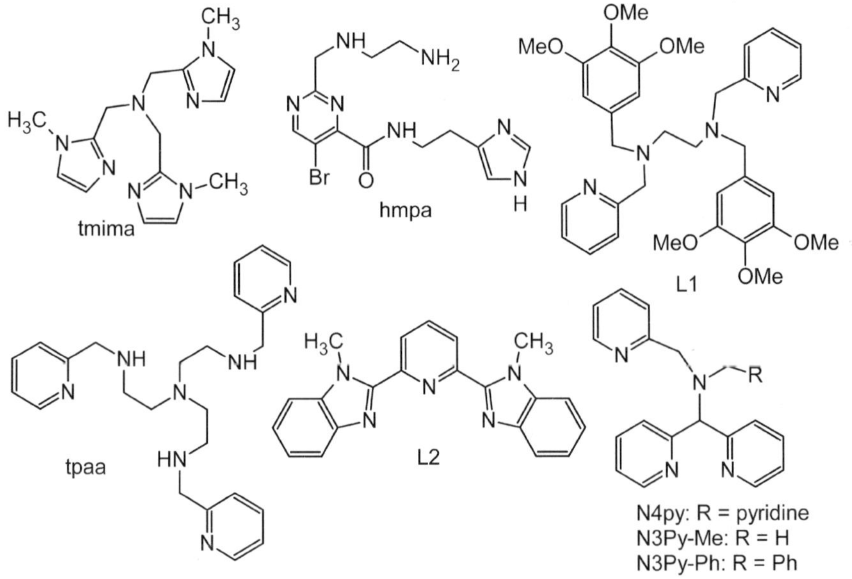

FIG. 6. Continued

TABLE III

CATALYTIC OXIDATION OF CYCLOHEXANE CATALYZED BY MONONUCLEAR IRON(II) AND IRON(III) COMPLEXES USING H_2O_2 AS THE PRIMARY OXIDANT[a]

Catalyst	A/K[b]	KIE[c]	Eff.(%)[d]	Ref.
$[Fe(pa)_2]$	1.5	–	9	*(47)*
$[Fe(pca)_2(py)_2]$	2.4	–	35	*(194)*
$[Fe(bpy)_2](ClO_4)_2$	0.8	1.4	9	*(195)*
$[Fe(dpa)_2]$	2.3	–	16	*(196)*
$[Fe(Hdp)Cl_2]$	1.2	–	3	*(76)*
$[Fe(tpa)(CH_3CN)_2](ClO_4)_2$	4.3	3.5	37	*(57)*
$[Fe(5\text{-}Metpa)(CH_3CN)_2](ClO_4)_2$	9.0	3.8	40	*(44)*
$[Fe(bpmen)(CH_3CN)_2](ClO_4)_2$	8.0	3.2	63	*(45)*
$[Fe(L_1)(CH_3CN)_2](ClO_4)_2$	8.0	–	38	*(197)*
$[Fe(N4py)(CH_3CN_2](ClO_4)_2$	1.4	1.5	31	*(60)*
$[Fe(N3py\text{-}Me)(CH_3CN_2](ClO_4)_2$	4.6	3.5	19	*(62)*
$[Fe(N3py\text{-}Ph)(CH_3CN_2](ClO_4)_2$	5.8	2.4	19	*(62)*
$[Fe(pma)(CH_3CN)](ClO_4)_2$	0.9	–	10	*(94)*
$[Fe(tpaa)](ClO_4)_2$	–	–	–	*(109)*
$[Fe(H_2bppa)(HCOO)](ClO_4)_2$	–	–	7	*(63)*
$[Fe(Htpca)(CH_3CN)_2](ClO_4)_2$	7	–	6	*(83)*
$[Fe(HPaPy_3)(CH_3CN)](ClO_4)_2$	1	–	12	*(42)*

[a]Reactions performed in acetonitrile.
[b]A/K = cyclohexanol to cyclohexanone ratio.
[c]Kinetic isotope effect of cyclohexanol formation.
[d]Efficiency based on H_2O_2.

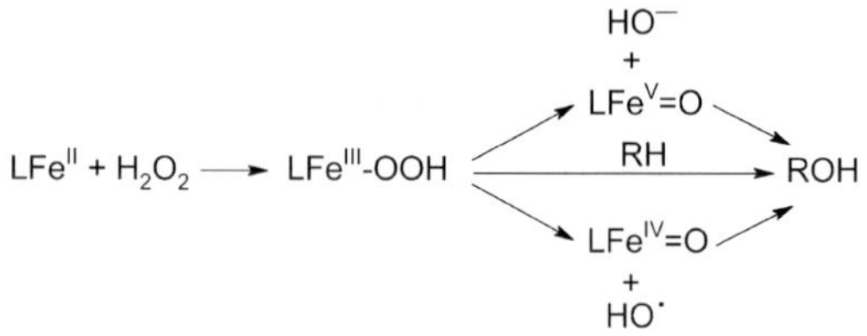

FIG. 7. Possible pathways for the decomposition of Fe^{III}–OOH transient intermediates (*10*).

decomposition of H_2O_2 (Eqs. (4) and (5)), and the other involving a metal-based oxidant formed by decomposition of a Fe^{III}–OOH transient intermediate (Fig. 7) (*43–45*). In the presence of an organic substrate, the highly reactive hydroxyl radical generated by Eq. (4) and both Fe^{II} and Fe^{III} ions (Eqs. (4) and (5)) enters into a series of consecutive reactions that lead to substrate oxidation (*46*). Usually, the oxidation reactions that are carried out using excess of H_2O_2 relative to the iron catalyst typically yield large amounts of free radicals due to the rapid reaction of the iron intermediates with H_2O_2. These free radicals tend to initiate radical chain oxidations and as a result, non-selective oxidations are observed. Furthermore, the oxidation of alkanes lacking tertiary C–H bonds is difficult because the primary or secondary radicals that are generated cannot renew the redox cycle easily (*46*). In fact, if not carried out in suitable organic solvents, the simple Fenton reaction will take place, involving Fe^{II} and dihydrogen peroxide, which is not very effective to carry out the oxidation of cyclohexane and the yield of oxygenated products is marginal with respect to the used amount of H_2O_2 (*47*). Therefore, the oxidation of alkanes is carried out preferably in acetonitrile or acetone. However, for industrial application the combination acetone – dihydrogen peroxide is not desired due to the formation of the acetone peroxide adduct, which is a primary explosive (*48*).

$$Fe^{II} + HOOH \longrightarrow Fe^{III} + HO^{\bullet} + HO^{-} \quad (4)$$

$$Fe^{III} + HOOH \longrightarrow Fe^{II} + HOO^{\bullet} + H^{+} \quad (5)$$

On the other hand, the Fe^{III}–OOH intermediate is chemically reactive, due to the weakening of the O–O bond, and can react in several different ways:

(i) as oxidant, when the cleavage of the O–O bond occurs *via* the attack of a nucleophile,
(ii) it may decompose by the homolytic cleavage of the Fe^{III}–O bond to form Fe^{II} and an $ROO^{\bullet}$ radical,
(iii) it may undergo O–O bond heterolysis to form a species that is formally Fe^{V}=O, or
(iv) it can be activated by O–O bond homolysis to form an Fe^{IV}=O species and the short-lived highly reactive $HO^{\bullet}$ radical, which is immediately transferred to the substrate (*44*). With suitable solvents or ligands, the Fe^{IV}=O species may be sufficiently stabilized to act as the key oxidizing species. Important research efforts have been directed towards the

isolation and characterization of Fe^{III}–OOH intermediates, mainly containing pyridine-based ligands and they have been recently reviewed by Girerd and co-workers (*49*). Electrochemistry and EPR studies suggest that the formation of the Fe^{III}–OOH intermediate involves the rapid oxidation of the Fe^{II} complex to the corresponding Fe^{III} species, followed by a rearrangement to give a new Fe^{III}–OH species. Subsequently, the Fe^{III}–OH species react with H_2O_2 to form the Fe^{III}–OOH intermediate with the release of a water molecule (*50,51*).

To confirm the electronic character and direct reactivity of the iron-oxo intermediates generated from the O–O bond cleavage of an Fe^{III}–OOH intermediate, mechanistic investigation of alkane oxidation requires certain strict conditions. Thus, the presence of a large excess of substrate (RH) is needed with respect to the employed oxidant (H_2O_2) to make sure that the concentration of the oxidized substrate (ROH) is substantially lower than that of the substrate itself, in order to minimize oxidation of (the more reactive) ROH (*52*). When increasing the concentration of the oxidant, the oxidant competes with the substrate for the metal-based oxidant and thereby initiates a radical chain process generating $HO^•$ and O_2 (*53*). Some criteria utilized to address the participation of $HO^•$ radicals vs. $Fe^{IV}{=}O$ species in the oxidation of hydrocarbons include:

(i) *alcohol to ketone ratio* (A/K) in the oxidation of cyclohexane, which reflects the life time of the alkyl radicals. An A/K ratio of 1 suggests that long-lived alkyl radicals, such as the cyclohexyl radical, are trapped by dioxygen at a diffusion-controlled rate to form alkylperoxyl radicals (Eqs. (6) and (7)). Following a Russell-type termination step, recombination of these radicals will result in the formation of equimolar amounts of cyclohexanol (CyOH) and cyclohexanone (CyO; Eq. 8) (*46*). When A/K > 1, the $HO^•$ radicals formed by a metal-based oxidant (Fig. 7) react quickly with the metal center to form the alcohol as main product (*54*).

$$c-C_6H_{12} + HO^\bullet \longrightarrow c-C_6H_{11}^\bullet + H_2O \tag{6}$$

$$c-C_6H_{11}^\bullet + O_2 \longrightarrow c-C_6H_{11}OO^\bullet \tag{7}$$

$$c-C_6H_{11}OO^\bullet + c-C_6H_{11}OO^\bullet \longrightarrow C_6H_{11}OH + C_6H_{11}O + O_2 \tag{8}$$

(ii) *the kinetic isotope effect* (KIE), which is a competition reaction between protio- and deuteroalkanes based on the difference of C–H and C–D bond strength; KIE values between 1 and 2 have been reported for oxidation reactions that involved the participation of $HO^•$ radicals, while higher KIE values are typical for oxidations that involved metal-based oxidants (*10,55*).

(iii) *regioselectivity* (3°/2°), the intramolecular competition reaction in the oxidation of adamantane, which has both secondary and tertiary C–H bonds; reactions in which HO• radicals are involved lead to 3°/2° values close to 2, while the participation of metal-based oxidants affords 3°/2° ratios higher than 15 (*56*).

(iv) *retention of configuration* (RC), in the oxidation of the tertiary C–H bonds of *cis*-1,2-dimethylcyclohexane; short-lived HO• radicals – for which the "oxygen-rebound" mechanism or C–O formation step is extremely fast – afford a tertiary alcohol in which the original configuration is retained. In this case, Fe^{IV}=O species are the oxidizing species and stereospecific hydroxylation is achieved (*10*).

The first stereospecific alkane hydroxylation reaction catalyzed by a mononuclear iron model compound, $[Fe(tpa)(CH_3CN)_2](ClO_4)_2$ (Fig. 8a), was reported by Que and co-workers (*57*). In combination with H_2O_2, the compound catalyzes the oxidation of *cis*- and *trans*-1,2-dimethylcyclohexane, to afford the product tertiary alcohol with 99% retention of stereochemistry. An enhanced oxidation yield (40% based on H_2O_2 in the course of 15 min) was observed in the oxidation of cyclohexane with an A/K ratio of 4.3 and a KIE value of 3.5 (*57*). By reacting $[Fe(tpa)(CH_3CN)_2](ClO_4)_2$ with 10 equiv. of H_2O_2 and using acetonitrile as the solvent, Que and co-workers were able to trap a possible reaction intermediate at −40°C (*57*). The spectroscopic properties of this intermediate (λ_{max} = 538 nm in UV-VIS and g = 2.19, 2.15, 1.97 characteristic for S = 1/2 in EPR) indicate the presence of a low-spin Fe^{III}–OOH species (*57*).

To gain more insight into the factors controlling the formation and stability of non-heme Fe^{III}–OOH intermediates, Feringa and co-workers have

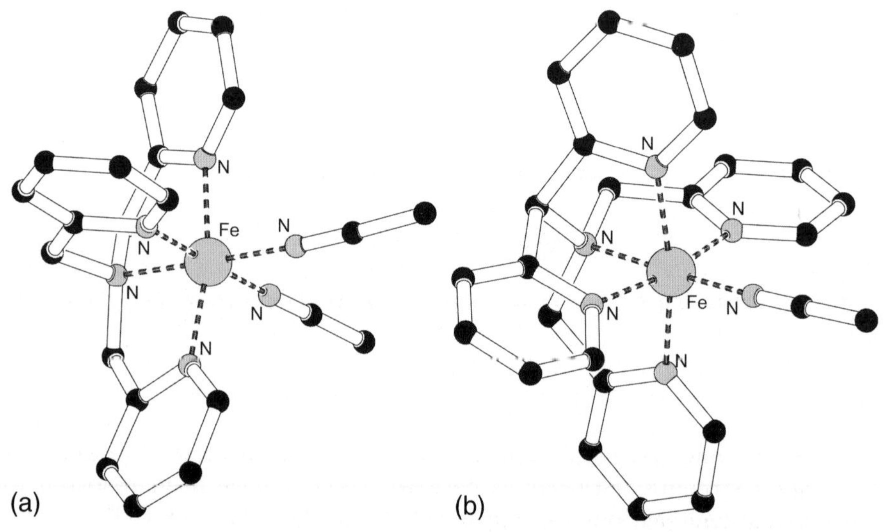

FIG. 8. Molecular structures of the cations $[Fe(tpa)(CH_3CN)_2]^{2+}$ (a) and $[Fe(N4py)(CH_3CN)]^{2+}$ (b).

designed a pyridine-based pentadentate ligand (N4py) (*58*). The iron(II) complex $[Fe(N4py)(CH_3CN)](ClO_4)_2$ (Fig. 8b) in reaction with H_2O_2 generates a low-spin Fe^{III}–OOH transient purple species, which has been spectroscopically (UV/VIS, EPR, resonance, Raman, and ESI-MS) characterized to be $[Fe^{III}(N4Py)OOH]^{2+}$ (*59*). Using acetonitrile as solvent, the complex is capable of the oxidation of cyclohexane, cyclooctane, and adamantane leading to the corresponding alcohols and ketones in up to 30 % yield (*60*). Catalytic experimental results as well as theoretical studies suggest that the oxidation products are formed *via* reactivity of the putative Fe^{IV}=O species and the action of $HO^{\bullet}$ radicals formed during the homolytic cleavage of the Fe^{III}–OOH intermediate (*60,61*). Substitution of one picolyl group in the N4py ligand with a non-coordinating moiety (CH_3, Ph) has led to the isolation of iron complexes that are able of stereoselective hydroxylation of alkanes, but that are less reactive than the parent complex $[Fe(N4py)(CH_3CN)](ClO_4)_2$ (yields up to 19 % based on H_2O_2) (*62*). A similar reactivity was observed in acetone, but a striking difference was observed for the KIE values in cyclohexanol formation (KIE = 2.4–3.5 in acetonitrile, KIE = 4.4–5 in acetone) and the lower stereoselectivity in acetone (*62*). The oxidation of *cis*- and *trans*-1,2-dimethylcyclohexane to the corresponding tertiary alcohols was studied to asses the stereoselectivity of the oxidation reaction. In acetonitrile, the oxidation of *cis*-1,2-dimethylcyclohexane afforded rather low yields of tertiary alcohols with up to 85 % retention of configuration at the tertiary carbon (*62*). Using *trans*-1,2-dimethylcyclohexane low yields of the tertiary alcohol and up to 73 % retention of configuration were observed (*62*).

Isolation and spectroscopic characterization of a mononuclear hydroperoxo-iron(III) complex of a ligand containing amido groups, $[Fe(H_2bppa)(OOH)]^{2+}$, was reported for the first time by Masuda and co-workers (*63*). The isolated complex is relatively stable at −70°C and the slow warming of its acetone solution to 20°C results in its gradual decomposition and the acceleration of the oxidation of cyclohexane, affording cyclohexanol as the main product (*63*). The observed high alcohol over ketone ratio reveals that free radicals and molecular oxygen, which can initiate non-selective radical chain oxidations, do not participate in the oxidation reaction.

Excellent oxidant-to-product conversion (70 % based on H_2O_2 in 30 min) and highly stereospecific alkane hydroxylation was observed in the oxidation reaction catalyzed by $[Fe(bpmen)(CH_3CN)_2](ClO_4)_2$ (*45*). The oxidation of *cis*-1,2-dimethylcyclohexane with this catalyst and H_2O_2 affords 4.6 TON of *cis*-1,2-dimethylcyclohexanol and no isomeric *trans*-alcohol product is observed. In addition, 26 % of the *cis*-alcohol product is ^{18}O-labeled when the reaction is carried out in the presence of 10 equiv. H_2O_2 and 1000 equiv. $H_2^{18}O$ indicating that the oxidant responsible for stereospecific alkane hydroxylation can undergo oxygen atom exchange with water. Thus, the incorporation of ^{18}O from $H_2^{18}O$ into the oxidation product provided the first evidence that a non-heme iron catalyst can hydroxylate alkane stereospecifically *via* a high-valent iron-oxo species (Fig. 9) (*45*).

Mechanistic studies based on ^{18}O labeling experiments with $[Fe(tpa)(CH_3CN)_2](ClO_4)_2$ and its α- and β-methylated analogues have shown

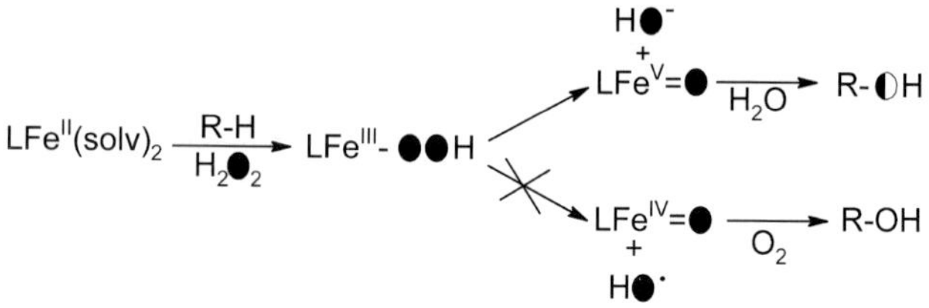

FIG. 9. Pathways for oxygen exchange with the high-valent iron-oxo species (*45*).

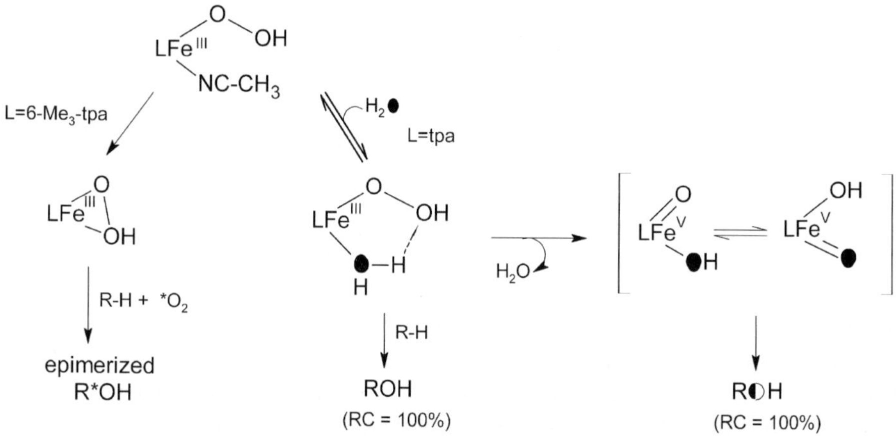

FIG. 10. General mechanism proposed for the oxidation of alkanes catalyzed by Fe^{II}(tpa) family of catalysts (*44*).

that both the Fe^{III}–OOH species and the Fe^{V}=O oxidant derived from O–O bond heterolysis play a role in alkane hydroxylation (*44*). It has been proposed that the heterolysis pathway is promoted by two factors:

(i) the low-spin iron(III) center that weakens the O–O bond and
(ii) the binding of a water ligand adjacent to the iron center, which can form a hydrogen bond to the terminal oxygen of the hydroperoxo group and thereby facilitate the departure of the hydroxide ion (Fig. 10). The Fe^{III}–OOH key intermediate may be a high-spin or low-spin species depending on the particular tpa ligand. The O–O bond of the high-spin intermediates is stronger than that of the low-spin counterparts, giving rise to different chemistry as reflected by the intermolecular kinetic isotope effect in the formation of cyclohexanol (KIE = 5 for $[Fe(tpa)(CH_3CN)_2](ClO_4)_2$ vs. KIE = 9 for $[Fe(5\text{-}Me_3\text{-}tpa)(CH_3CN)_2](ClO_4)_2$) and the product distribution in the oxidation of adamantane ($3°/2° = 17$ for $[Fe(tpa)(CH_3CN)_2](ClO_4)_2$ vs. $3°/2° = 21$ for $[Fe(5\text{-}Me_3\text{-}tpa)(CH_3CN)_2](ClO_4)_2$) (*44*). The highly stereospecific alkane hydroxylation observed with $[Fe(tpa)(CH_3CN)_2](ClO_4)_2$ and $[Fe(bpmen)(CH_3CN)_2](ClO_4)_2$ as compared with $[Fe(N4py)(CH_3CN)](ClO_4)_2$ clearly shows the importance of the accessibility to two labile sites on the

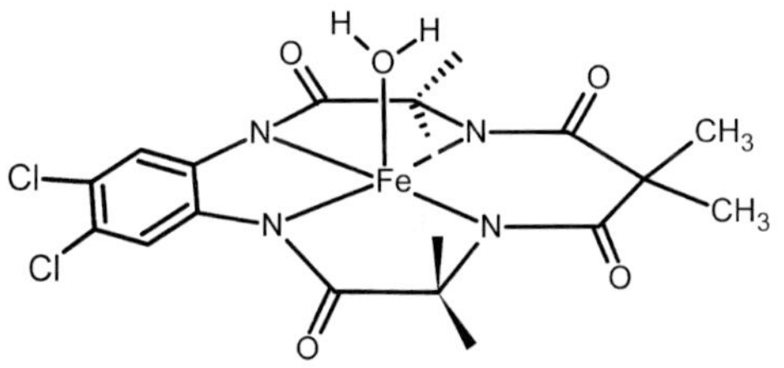

FIG. 11. Schematic structure of the $[Fe(taml)(H_2O)]^-$ anion.

metal center, that are required for the formation of an intramolecular hydrogen bond in order to promote the O–O bond heterolysis (*44*).

Several other mononuclear iron(II) complexes bearing picolinate, 2-pyrazine carboxylate or polydentate ligands have been reported to catalyze the hydroxylation of alkanes by H_2O_2 (Table III). These complexes show a fair activity in the oxidation of alkane with product yields up to 40% based on H_2O_2.

Collins and co-workers have reported an iron(III) complex of a macrocyclic tetraamido ligand (taml, Fig. 11) which is very resistant to oxidative degradation and also to hydrolysis (*64,65*). In combination with dihydrogen peroxide, this complex was found to be a powerful catalyst for dye bleaching reactions (*64*). Several azo- and quinone-based dyes at both pH 7.4 and 10 could be bleached in less than 15 min, with up to 195 turnovers based on catalyst. By changing the substituents on the six-membered chelate ring, the effective catalyst lifetime can be changed (*64*). Thus, the replacement of methyl groups with ethyl groups was found to result in a shorter-lived H_2O_2 activator due to the ligand degradation initiated by a hydrogen abstraction from the methylene C–H bond of one of the two ethyl groups (*64*). High-valent iron-oxo species were proposed to be involved in the oxidations catalyzed by the Fe-taml complex in combination with dihydrogen peroxide, but this has not yet been corroborated by spectroscopic studies (*66*). A stable catalytically active diiron(IV)-μ-oxo compound, obtained by reaction of Fe-taml complex with dioxygen, was recently isolated and spectroscopically characterized (*67*).

Mascharak and co-workers have designed and synthesized the pentadentate ligand $HPaPy_3$, which stabilizes the mononuclear iron(III) species in $[(PaPy_3)Fe(CH_3CN)](ClO_4)_2$ (Fig. 12) and which affords the intermediate species $[(PaPy_3)Fe(OOH)]^+$ *via* substitution of the solvent molecule by HOO^- at $-60\,^\circ C$ (*42*). The transient intermediate is quite reactive and it is decomposed within minutes when the reaction mixture is warmed to room temperature. The complex $[(PaPy_3)Fe(CH_3CN)](ClO_4)_2$ is able to catalyze the oxidation of cyclohexane in the presence of 150 equiv. of H_2O_2 with a resulting A/K ratio of 1 which suggests the involvement of hydroxyl radicals or species having properties like $HO^\bullet$ radicals. This conclusion is supported by the mechanistic studies using 2-methyl-1-phenylprop-2-yl hydroperoxide as oxidant, for which no oxidation of cyclohexane was observed (*42*). Que and co-workers have studied the reaction of $[(PaPy_3)Fe(OCH_3)](ClO_4)$ with H_2O_2 and were also able to trap a red species (λ_{max} = 480 nm with an S = 1/2 EPR

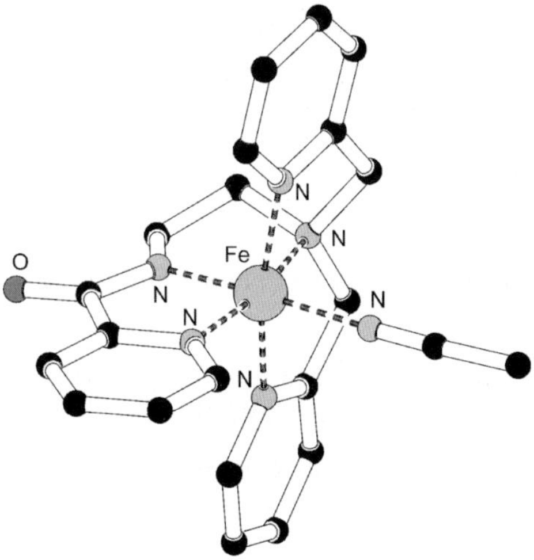

FIG. 12. Molecular structure of $[Fe(PaPy_3)(CH_3CN)](ClO_4)_2$.

signal at g = 2.25, 2.17, and 1.95) that resembled an $[(PaPy_3)Fe(OOH)](ClO_4)$ intermediate (*68*). This species has further been identified by electronspray ionization mass spectrometry and was further characterized by resonance Raman and EXAFS analysis (*68*).

A.2. Dinuclear Iron Complexes

A large number of dinuclear iron complexes that display catalytic activity in the hydroxylation of alkanes has been reported. These complexes actually mimic MMO in a functional manner. However, there are only a few iron complexes that display both the catalytic and the structural features of MMO and in this section we will include only these model compounds. Representative results of alkane functionalization catalyzed by dinuclear iron complexes using H_2O_2 as oxidant are shown in Table IV.

The first reports of oxidation reactions of a variety of hydrocarbons with the biomimetic dinuclear iron complexes $[Fe_2O(OAc)_2(bpy)_2Cl_2]$ and $[Fe_2O(OAc)(tmima)_2](ClO_4)_3$ using H_2O_2 as oxidant appeared in 1991 by Fish and co-workers (*69*). The reactivity of these complexes was rather low, with maximum turnover numbers in the oxidation of cyclohexane of 15 after 6–18 h reaction time. Enhanced catalytic activity in the oxidation of cyclohexane (46 mmol of cyclohexanol + cyclohexanone per mmol catalyst in 10 min) was observed for $\{[Fe(bpy)_2(H_2O)]_2O\}(ClO_4)_4$, in which the presence of labile water molecules on the iron centers play an essential role for acceptable catalytic activity (*53*). The system is also able to oxidize adamantane with high selectivity for the tertiary carbon, as indicated by a high 3°/2° ratio of 9. Low catalytic activity for cyclohexane hydroxylation was also found for the phenanthroline analogue $\{[Fe(phen)_2(H_2O)]_2O\}(ClO_4)_4$, with very low yields based on the oxidant (about 10 %) (*70*). However, the major drawback of these systems was the fast accumulation of degraded and inactive catalyst in the form of $[Fe(bpy)_3]^{2+}$ and $[Fe(phen)_3]^{2+}$ species, and the fast

TABLE IV

CATALYTIC OXIDATION OF CYCLOHEXANE AND ADAMANTANE BY (μ-OXO)DIIRON COMPLEXES USING H_2O_2 AS PRIMARY OXIDANT[a]

Catalyst	A/K[b]	Eff.(%)[c]	KIE[d]	3°/2°[e]	Ref.
$\{[Fe(bpy)_2(H_2O)\}_2O\}(ClO_4)_4$	1.3	9	–	–	(*53*)
$\{[Fe(bpy)_2(H_2O)]_2O\}(ClO_4)_4$	–	–	2.4	4.5	(*70*)
$\{[Fe(4,4'\text{-}Me_2bpy)_2(H_2O)]_2O\}(ClO_4)_4$	–	–	3.1	6.2	(*70*)
$\{[Fe(5\text{-}NO_2phen)_2(H_2O)]_2O\}(ClO_4)_4$	–	–	2	3.5	(*70*)
$\{[Fe(pb)_2(H_2O)]_2O\}(ClO_4)_4$	2	30	3.2	3.5	(*71,72*)
$\{[Fe(L_2)(NO_3)(CH_3OH)]_2O\}(NO_3)_2$	1.1	36	–	–	(*98*)
$\{[Fe(mebpa)Cl]_2O\}(ClO_4)_2$	1	–	–	6.7	(*77*)
$\{[Fe(mebpa)Cl]_2O\}(ClO4)_2$	3.5	53	–	–	(*79*)
$\{[Fe(tfpy)Cl]_2O\}(ClO4)_2$	2	3	–	–	(*78*)
$\{[Fe(ind)Cl]_2O\}$	~1	29	–	4	(*198*)
$\{[Fe(opba)]_2O\}$	0.9	5	2.4	3	(*84*)
$[Fe_2O(OAc)(bpmen)_2](ClO_4)_3$	0.8	2.5	1.4	3.6	(*199*)
$[Fe_2O(OAc)Cl_2(bpy)_2]$	~1	~8	1.6	3.4	(*69*)
$[Fe_2O(OAc)(tmima)_2](ClO_4)_3$	<1	~9	–	–	(*69*)
$\{[Fe(tpca)(H2O]_2O\}(ClO4)_2$	7	10	97	6	(*83*)

[a]Reactions performed in acetonitrile.

[b]A/K = cyclohexanol to cyclohexanone ratio in the oxidation of cyclohexane

[c]Efficiency based on H_2O_2.

[d]Kinetic isotope effect of cyclohexanol formation.

[e]1-adamantanol/(2-adamantanol + 2-adamantanone) in the oxidation of adamantane corrected for the number of C–H bonds in a group.

dismutation of H_2O_2 (*53*). To prevent the formation of inactive iron(II) species during the catalytic cycle, Fontecave and co-workers have isolated a μ-oxo diiron(III) complex containing a hindered chiral bipyridine derivative (*71*). The compound $\{[Fe(pb)_2(H_2O)]_2O\}(ClO_4)_4$ is able to effect selectively the oxidation of cyclohexane (A/K = 2) with a total yield of 30 % based on H_2O_2 (*71*). The most impressive achievement is the oxidation of 1,2-dimethylcyclohexane, which occurs stereospecifically and for which the *cis*- and *trans*- configuration was retained in the tertiary alcohol product (*72*). In all cases, the presence of a metal-based oxidant has been proposed (*73*). On the basis of Raman and EPR spectroscopic studies and ESI mass spectroscopic analysis, it has been proposed that a (μ-peroxo)diiron(III) intermediate is the active species in the catalytic cycle. (*73*) The involvement of a metal-based oxidant was further confirmed by the iron complex containing the same chiral ligand $\{[Fe(pb)_2(H_2O)]_2O\}(ClO_4)_4$, which is able to catalyze the asymmetric oxidation of sulfide to sulfoxides by dihydrogen peroxide, with enantioselective excess up to 38 % (*71*).

Structural mimics for MMO and alkane functionalization studies with μ-oxo diiron(III) complexes of a series of tripodal-based ligands (mebpa, tfpy) have been reported by the group of Nishida (74–78). In most of the cases, no selectivity in the oxidation products (A/K = 1) was observed as a result of the conditions used for this reaction, which employed dihydrogen

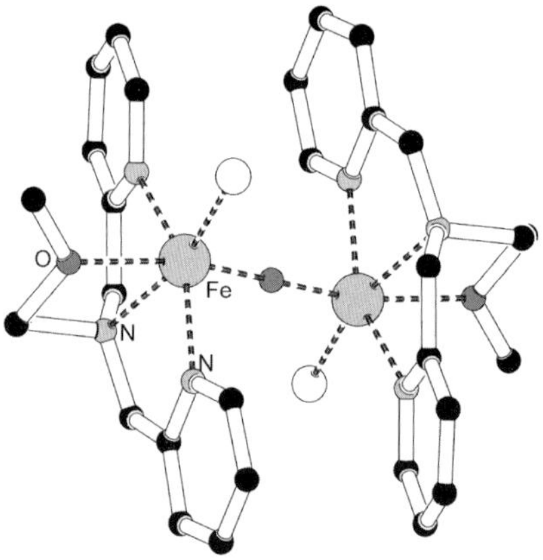

FIG. 13. Molecular structure of $\{[Fe(mebpa)Cl]_2O\}(ClO_4)_2$.

peroxide in high concentration and because of which it was catalytically decomposed. In contrast, Reedijk and co-workers have recently found that $\{[Fe(mebpa)Cl]_2O\}(ClO_4)_2$ (Fig. 13) efficiently catalyzes the oxidation of cyclohexane with an A/K ratio of 3.5 and the catalytic efficiency is dependent not only on the catalyst and additive, but also on the reaction conditions (*79*). In addition, the compound catalytically oxidizes adamantane to the corresponding secondary and tertiary products with a 3°/2° ratio of 6. The catalytic behavior of the compound in acetone as solvent as well as the catalytic results obtained in olefin epoxidation studies suggest that two competitive mechanistic pathways might be involved: one leading to oxidation products formed via radicals/$H^{\bullet}$ abstraction mechanisms, and the second one leading to oxygen-transfer processes (*79*). Also, the participation of putative $Fe^{IV}{=}O$ species have been suggested, which are formed during the homolytic cleavage of the Fe^{III}–OOH bond, and that are also able of hydrogen abstraction and can hydroxylate further the C–H bond of alkanes.

Que and co-workers have observed the formation of green, transient high-valent species in the reaction of a μ-oxo diiron(III)-containing tpa complex (Fig. 14) with H_2O_2 in CH_3CN at −40°C (*80*). When 5-Me_3-tpa is used as the tripodal ligand, the green intermediate species can be isolated as a solid upon standing overnight at −40°C. Spectroscopic and labeling experiments point toward the presence of a $Fe_2(\mu\text{-}O)_2$ core in $[Fe_2(\mu\text{-}O)_2(5\text{-}Me_3\text{-}tpa)_2](ClO_4)_3$, which thus represents the first well-characterized non-heme iron complex with a formal $Fe^{III}Fe^{IV}$ oxidation state and it makes the complex relevant as a model for the dioxygen activation chemistry of MMO (*81*). Lippard and co-workers have also proposed the presence of an $Fe^{III}Fe^{IV}$ species which is derived from the (μ-peroxo)diiron(III) intermediate (*82*).

The incorporation of amido functions into tetradentate ligands has been investigated in a few model systems in order to improve their reactivity and stability of the iron complexes. The complexes $\{[Fe(opba)]_2O\}$ and $\{[Fe(tpca)(H_2O]_2O\}(ClO4)_2$ show modest to good catalytic activity in the oxidation of cyclohexane and adamantane (Table IV) (*83,84*). The compound

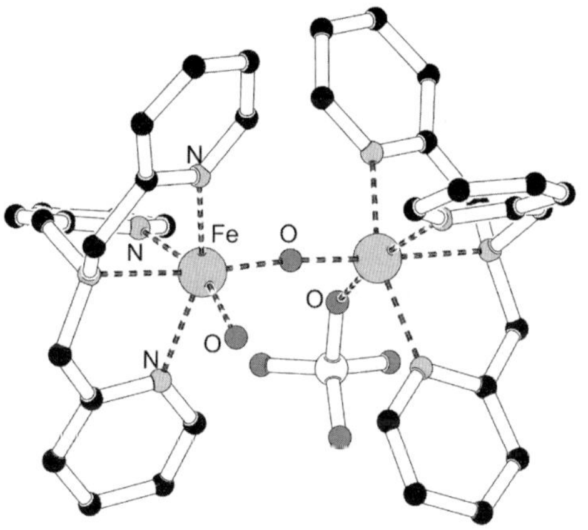

FIG. 14. Molecular structure of the cation $[Fe_2O(tpa)(H_2O)(ClO_4)]^{3+}$.

$\{[Fe(opba)]_2O\}$ catalyzes the oxidation of cyclohexane and adamantane with total yields up to 5% based on the oxidant after 24 h with a ratio A/K of 0.9 and a ratio 3°/2° of 3 (*84*). These results are in line with a mechanism involving hydroxyl radicals as transient intermediates. In the oxidation of cyclohexane catalyzed by $\{[Fe(tpca)(H_2O]_2O\}(ClO_4)_2$, the average alcohol-to-ketone ratio is equal to 7, which is much higher than the value characteristic of radical chain oxidation (*54*). Also, the stereospecificity observed in the oxidation of *cis*- and *trans*-1,2-dimethylcyclohexane (up to 97% retention of configuration) points toward the involvement of a metal-based oxidant (*83*). A few other complexes showing low to moderate catalytic activity in the oxidation of cyclohexane and/or adamantane are listed in Table IV.

B. *cis*-Dihydroxylation vs. Olefin Epoxidation

In the last few years, several non-heme iron complexes have been identified as functional models for non-heme iron dioxygenases (*85–88*). These model complexes are able to catalyze the *cis*-dihydroxylation of olefins as well as the epoxidation of olefins using H_2O_2 as the primary oxidant. Table V presents the results of olefin oxidation by some representative mononuclear and dinuclear non-heme iron complexes in combination with H_2O_2.

Kimura and co-workers reported the first iron-catalyzed olefin epoxidation with H_2O_2 as oxidant and $Fe(acac)_3$ as the catalyst (*89*). The oxidation of *cis*- and *trans*-stilbene afforded *trans*-epoxide as the major product with a product yield of 96%. Similar stereochemical features were observed in the epoxidation of other substrates, such as methyl oleate and *cis*- or *trans*-9-octadecen-1-ol (*89*).

Several iron complexes of tetradentate ligands containing pyridine arms have been reported to be able of carrying out olefin *cis*-hydroxylation and/or epoxidation in combination with H_2O_2 (*90*). In general, the *cis*-diol/epoxide ratio can be tuned by the nature of the metal coordination environment. Mononuclear iron(II) complexes of tetradentate ligands typically lead to

TABLE V

CATALYSIS OF OLEFIN OXIDATION BY SOME REPRESENTATIVE MONONUCLEAR AND DINUCLEAR NON-HEME IRON COMPLEXES IN USING H_2O_2 AS OXIDANT

Catalyst	Substrate	% diol	% epoxide	% by-products	Ref.
$[Fe(tpa)(CH_3CN)_2](ClO_4)_2$	Cyclooctene	2.6	2.3	–	*(88)*
$[Fe(5\text{-}Metpa)(CH_3CN)_2](ClO_4)_2$	Cyclohexene	45	30	–	*(85)*
	Oct-1-ene	67	18	–	*(85)*
$[Fe(bph)(CH_3CN)_2](ClO_4)_2$	Cyclooctene	0	2.5	–	*(85)*
$[Fe(cyclam)(CH_3CN)_2](ClO_4)_2$	Cyclohexene	–	20	–	*(91)*
$[Fe(N4py)(CH_3CN)](ClO_4)$	Cyclooctene	0	0.6	–	*(88)*
$\{[Fe(mebpa)Cl]_2O\}(ClO_4)_2$	Cyclooctene	0	20	2.5	*(85)*
$(Me_4N)[Fe_2(L_3)(OAc)_2]$	Cyclohexene	0	10	~10	*(97)*

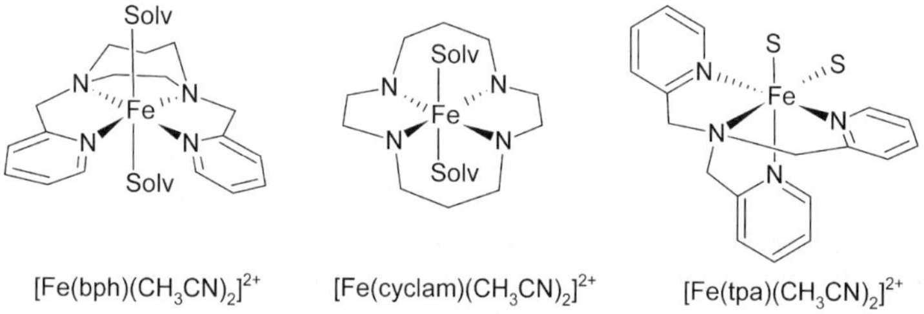

FIG. 15. Non-heme iron complexes as models for naphthalene 1,2-dioxygenase *(88)*.

hexacoordinate metal complexes with two labile ligands on the iron(II) center; upon treatment with H_2O_2 a conversion to the iron(III) active form is observed. Depending on whether these labile sites are located *cis* or *trans* to each other, different types of selectivity can be observed. Complexes with *trans*-open coordination sites (Fig. 15) like $[Fe(cyclam)(CH_3CN)_2](OTf)_2$ *(91)* and $[Fe(bph)(CH_3CN)_2](ClO_4)_2$ *(88)*, catalyze the epoxidation of olefins with H_2O_2 as terminal oxidant. Complexes with two *cis*-open coordination sites (Fig. 15) like $[Fe(tpa)(CH_3CN)_2](ClO_4)_2$, $[Fe(tpa)(OTf)_2]$, $[Fe(6\text{-}Me_3\text{-}tpa)(CH_3CN)_2](ClO_4)_2$, $[Fe(6\text{-}Me_3\text{-}tpa)(OTf)_2]$, $[Fe(bpmen)(CH_3CN)_2](ClO_4)_2$, $[Fe(6\text{-}Me_2\text{-}bpmen)(OTf)_2]$, and $[Fe(6\text{-}Me_2\text{-}bpmen)(OTf)_2]$, not only catalyze the epoxidation of olefins but also the *cis*-dihydroxylation reaction *(85–88)*. Employing the complex $[Fe(6\text{-}Me_3\text{-}tpa)(CH_3CN)_2](ClO_4)_2$, which contains two *cis*-coordinated acetonitrile molecules, the *cis*-diol was observed as the major product in the oxidation of cyclooctene *(87)*.

The catalytic behavior associated with these catalysts arises from the spin state of Fe^{III}–OOH intermediates that react with the substrate in distinct ways. As indicated by spectroscopic and labeling studies, the complexes $[Fe(tpa)(CH_3CN)_2](ClO_4)_2$ and $[Fe(tpa)(OTf)_2]$ form low-spin Fe^{III}–OOH intermediates, which give rise to *cis*-diol products with one oxygen derived from H_2O_2 and the other from H_2O (Fig. 16) *(87)*. For this type of catalysts,

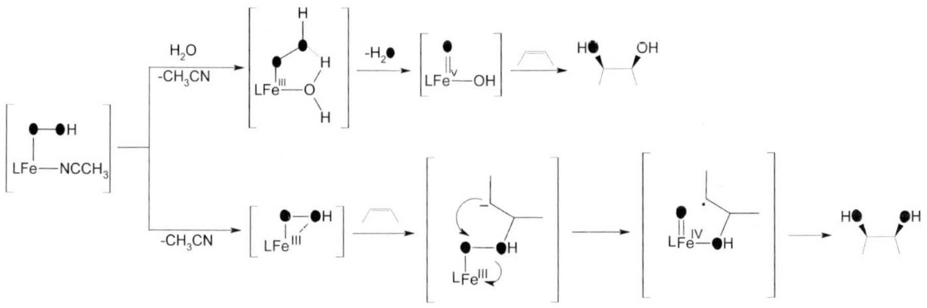

FIG. 16. Proposed mechanism for *cis*-dihydroxylations catalyzed by non-heme iron(II) catalysts (*200*).

epoxidation represents a significant fraction (40–90 %) of the olefin oxidation reactivity. The introduction of 6-methyl substituents into the tpa framework favors the high-spin state for the iron center due to the steric effects of the 6-methyl group that restrict the extent to which the metal-ion cavity can shrink. Thus, in the case of [Fe(6-Me_3-tpa)(CH_3CN)$_2$](ClO_4)$_2$ and [Fe(6-Me_2-bpmcn)(OTf)$_2$], from ^{18}O labeling studies it was concluded that the oxygen atoms into the *cis*-diol products are exclusively delivered from the oxidant H_2O_2 (Fig. 16), making the catalysts functional models for the bacterial Rieske dioxygenases (*87*). The mechanism that was proposed for the model complexes of these enzymes involves the nucleophilic attack of the Fe^{III} η^2–OOH species on the olefin, followed by reductive O–O bond homolysis.

Iron complexes of cyclam and related ligands have shown to be effective catalysts for olefin epoxidations by aqueous H_2O_2 in acetonitrile or methanol as solvent (*91,92*). In all cases, moderate turnovers based on the iron complex (up to 20), moderate yields based on H_2O_2 (up to 40 %), moderate stereospecifity, and small amounts of allylic oxidation products have been obtained. Particularly significant was the case of the substrate *cis*-stilbene, which resulted in a 26 % yield of *cis*-stilbene oxide and less than 2 % *trans*-stilbene oxide (*91*). White and co-workers have reported that the epoxide (40 % yield) is the major product in the oxidation of 1-decene catalyzed by the compound [Fe(bpmen)(CH_3CN)$_2$](ClO_4)$_2$ (*93*). An improvement to 82 % selectivity for epoxide formation was observed in the presence of one equiv. of acetic acid. In the latter case, it was suggested that the active species is a (μ-oxo)(μ-carboxylate) diiron(III) complex which is generated by self-assembly under the conditions of epoxidation (*93*). However, this result is in disagreement with those reported by the group of Que, who attribute the lack of catalytic activity of (μ-oxo)(μ-carboxylate) diiron(III) species to the presence of the carboxylate bridge, which hampers the facile coordination of H_2O_2 (*44,87*).

The importance of two labile coordination sites positioned *cis* to each other in order to achieve stereospecific olefin epoxidation has been further confirmed by the catalytic activity of the compound [Fe(N4Py)(CH_3CN)](ClO_4)$_2$ in the oxidation of olefins, in which the epoxide is obtained as minor product and the allylic oxidation products are formed with moderate yields (*60*). This complex contains a pentadentate ligand that is closely related to the tpa-family, but with an additional pyridine group (Fig. 6), the resulting complex thus possessing only one labile coordination site. However, different results are obtained with the catalyst $[Fe(pma)(CH_3CN)](ClO_4)_2$ that also contains a pentadentate ligand, but an iron(III) metal center (*94*). In the latter case, effective oxo-transfer to a variety of substrates has been observed. Thus, the catalytic oxidation of *cis*-stilbene generates the *cis*-epoxide (stereochemistry retained), while rotation around the C–C bond in the transition state results in the formation of only a small amount of *trans*-epoxide; *trans*-stilbene affords only the *trans*-epoxide (*94*). Using an iron(III) complex of a chiral derivative ligand of Hpma, the optically active *cis*-β-epoxide was obtained in about 45 % *ee* in the oxidation of *cis*-β-methylstyrene, whereas the epoxidation of *trans*-β-methylstyrene resulted in the formation of the racemic product (*95*).

Interesting results in the oxidation of olefins catalyzed by iron complexes of the tetradentate ligands N3Py-Me and N3py-Ph have been reported by Feringa and co-workers (*62*). Using acetonitrile as the solvent, olefins are oxidized to the corresponding *cis*-epoxides and *cis*-diols. In acetone as the solvent, a different reactivity pathway was found, resulting in the *trans*-dihydroxylation of *cis*-olefins. In the case of the complex $[Fe(N3Py\text{-}Me)(CH_3CN)_2](ClO_4)_2$, with cyclooctene as substrate and acetonitrile as solvent, 14 turnovers for *cis*-epoxide and 22 turnovers for *meso*-diol were obtained; a small preference for *cis*-diol was also observed. In acetone, epoxide and 1,2-diols were obtained with the *rac*-1,2-cyclooctanediol as main product (selectivity 95 %). These differences in the catalytic activity have been ascribed to different intermediates that might be formed in acetonitrile and acetone (*62*). Recently, Que and co-workers have shown that the reaction of $[Fe(tpa)(OTf)_2]$ with H_2O_2 leads to different intermediates in acetonitrile and acetone (*96*). Thus, the reaction of $[Fe(tpa)(OTf)_2]$ with dihydrogen peroxide using acetonitrile as solvent results in the formation of Fe^{III}–OOH, as indicated by UV-VIS and resonance Raman isotope-labeling experiments. When the reaction is performed in acetone the intermediate adduct of H_2O_2 and acetone is formed, which is most likely also the case for $[Fe(N3Py\text{-}Me)(CH_3CN)_2](ClO_4)_2$ in acetone (Fig. 17) (*96*).

A few examples of olefin epoxidation catalyzed by dinuclear iron(III) complexes have been reported (*79,97,98*). The complex $(Me_4N)[Fe_2(L3)(OAc)_2]/H_2O_2$ (L3 = N,N′-(2-hydroxy-5-methyl-1,3-xylene)bis(N-carboxymethylglycine), (Fig. 18) in combination with H_2O_2 catalyzes the oxidation of cyclohexene, styrene, and *cis*-stilbene to the corresponding epoxides, under inert atmosphere and in DMF as solvent (*97*). In all cases, the epoxide constitutes

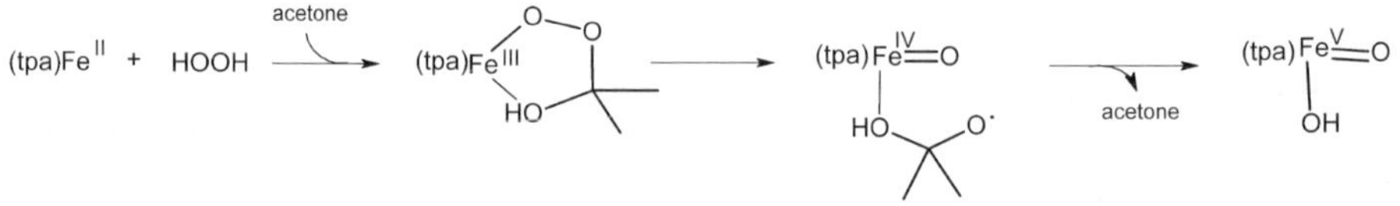

FIG. 17. Proposed intermediates in the reaction of $[Fe(tpa)(OTf)_2]$ with H_2O_2 in acetone (*96*).

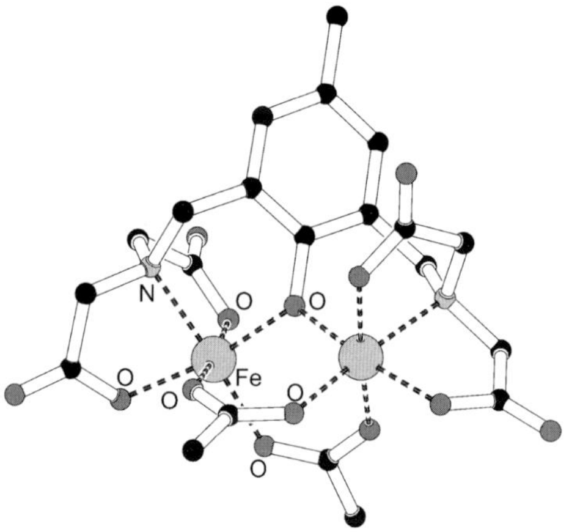

FIG. 18. Molecular structure of the $[Fe(L3)(OAc)_2]^-$ anion.

10 % of the H_2O_2 consumed in the reaction, but by-products were also observed in low concentrations. The epoxidation of *cis*-stilbene results in the formation of 95 % *trans*- and 5 % *cis*-epoxide. The compound $\{[Fe(L2)(NO_3)]_2O\}(NO_3)_2$ efficiently catalyzes the epoxidation of styrene, cyclohexene, 2-methyl-2-butene, and *cis*- and *trans*-2-heptene with 51–84 % conversion and high selectivity (71–99 %) for epoxide products (*98*). The stereospecific epoxidation as well as the high degree of retention of configuration in the oxidation of *cis*- and *trans*-2-heptene suggest that a non-radical species is involved in the oxidation of these substrates, but detailed mechanistic studies have not been undertaken. The behavior of the complex $\{[Fe(mebpa)Cl]_2O\}(ClO_4)_2$ in combination with H_2O_2 towards catalytic oxidation of cyclohexene, cyclooctene, and styrene has also been studied (*79*). In all cases, both epoxidation and allylic oxidation have been observed with a substrate conversion of 39–84 %.

C. AROMATIC HYDROXYLATION

Several model systems have been developed to contribute to the understanding of the mechanism of action of iron-containing aromatic hydroxylases and eventually to mimic their activities. These studies include iron-catalyzed oxygenations of coordinated phenolates and arene-ligand substituents by molecular oxygen in the presence of reducing agents (*99–102*). Alkyl hydroperoxide-driven reactions have also been reported (*103–105*). Some aromatic hydroxylations with iron(III) salts and H_2O_2 in the presence of reducing agents have been reported (*106,107*). Also, a few model complexes have been reported to perform aromatic hydroxylation in the presence of H_2O_2 as oxidant in the absence of reducing agents (*51,108*).

The dinuclear iron(III) complex $[Fe_2O(L1)_2(OAc)_2]^-$ (L1 = N,N′-bis(3,4,5-trimethoxybenzyl)ethylenediamine-N,N′-diacetate) (Fig. 6), in a reaction with 3 equiv. of H_2O_2 reacts to form a new deep-blue species with a maximum absorption at 560 nm. As indicated by the spectroscopic studies, the new

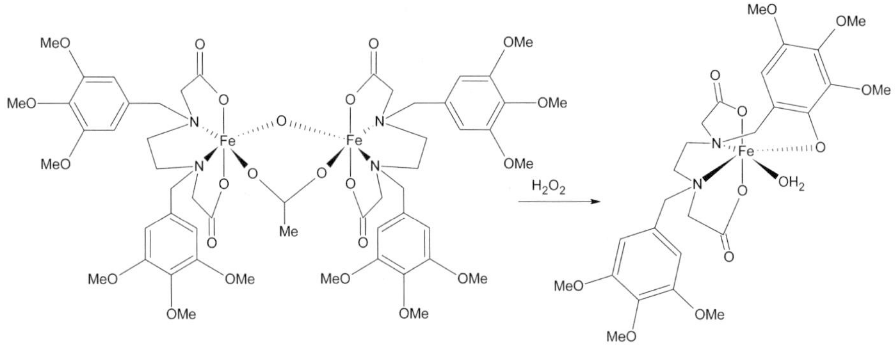

FIG. 19. Aromatic hydroxylation catalyzed by $[Fe_2O(L_1)_2(CH_3COO)_2)]^-$ (*108*).

species is a mononuclear iron(III) complex in which the ligand L1 is hydroxylated at an *ortho* position of one of the phenyl groups (Fig. 19) (*108*). Thus, this system mimics in an elegant manner, the catalytic function of ribonucleotide reductase where the hydroxylated reaction product remains bound to the iron center. The dinuclear compound $[Fe_2O(L1)_2(OAc)_2]^-$ may be attacked by the nucleophilic, electron-rich phenyl ring, in agreement with the experiments carried out in the presence of $^{18}O_2$, for which the full incorporation of labeled oxygen was observed (*100*). The dinuclear species is stabilized only in the presence of the acetate anion with the two Fe^{III} ions doubly bridged by an oxo and an acetato group, thus providing phenyl groups adjacent to the iron center as substrates for oxidation. Because of the lability of the acetato bridge, the reaction of $[Fe_2O(L1)_2(OAc)_2]^-$ with dihydrogen peroxide leads to aromatic hydroxylation, with the formation of a mononuclear product (*108*). The mononuclear structure of the final product is a result of the good electron-donating properties of the phenolate ligand, which decreases the Lewis acidity of the iron(III) ion and may prevent it from forming a dinuclear oxo-bridged complex.

Girerd and co-workers have reported a few other non-heme iron complexes of hexa-, penta-, or tetra-azadentate ligands, which are able of catalytic hydroxylation of various aromatic substrates (such as anisole, toluene, ethylbenzene, benzene, and naphthalene) in the presence of hydroquinones or tetrahydropterins as reducing agents (*51,109*). The yield of the oxidation reactions ranges from 18 up to 69%, based on dihydrogen peroxide. Labeling experiments have shown that the transfer of an oxygen atom from $H_2^{18}O_2$ into anisole substrate occurred (*51*). Also, it has been observed that with toluene and ethylbenzene for which the aromatic hydroxylation is in competition with the benzylic oxidation, the increased amounts of the aromatic hydroxylation product in the presence of a reducing agent were always accompanied by an increase of the aromatic/benzylic hydroxylation ratio.

D. ROLE OF ADDITIVES

The use of organic additives in the iron-catalyzed alkane hydroxylation has been demonstrated to play an important role in product selectivity.

Addition of 2-pyrazine carboxylic acid in cyclohexane oxidation catalyzed by $[Fe_2(Hptb)(\mu\text{-}OH)(NO_3)_2](NO_3)_2 \cdot CH_3OH \cdot 2H_2O$ (Hptb = N,N,N′,N′-tetrakis (2-benzimidazolylmethyl)-2-hydroxo-1,3-diaminopropane) affords cyclohexyl hydroperoxide as the main reaction product (TON = 140), but the oxidation reaction proceeds non-stereoselectively and high concentrations (up to 25 equiv. relative to the catalyst) of additive are required (*110*). A likely explanation of the increased concentration of cyclohexyl hydroperoxide is that the 2-pyrazine carboxylic acid (Hpca) facilitates the proton transfer between the coordinated H_2O_2 molecule and the ligands at the iron centers to produce the Fe^{III}–OOH intermediate (*110*). Similar catalytic action was observed for the iron(III) complex with the ligand 1,4,7-triazacyclononane (H_3tacn), $[Fe_2O(H_3tacn)_2(OAc)_2]I_2 \cdot 0.5NaI \cdot 3H_2O$, and the experimental results suggest the involvement of hydroxyl radicals as main oxidizing species (*111*). However, recent studies have shown that the role of Hpca in catalytic oxidations is strongly dependent on the reaction conditions. In the oxidation of alkanes catalyzed by $\{[Fe(mebpa)Cl]_2O\}(ClO_4)_2$ using 10 equiv. of H_2O_2, the presence of Hpca in various concentrations results in complete inhibition of the oxidation reactions (*79*). Using 250 equiv. of H_2O_2, the oxidation of cyclohexane catalyzed by $\{[Fe(mebpa)Cl]_2O\}(ClO_4)_2$ in the presence of 2 equiv. of Hpca leads to 4.6 equiv. of cyclohexanol and 1.7 equiv. cyclohexanone (*79*). These differences can only be accounted for by the presence of different reaction intermediates and further studies are needed.

IV. Biomimetic Manganese Catalysts

A. Alkane Oxidation

Only one manganese complex has been reported to catalyze the oxidation of alkanes in the presence of H_2O_2 as the primary oxidant (*111–115*). Shul'pin and co-workers have found that the complex $[Mn_2(\mu\text{-}O)_3(Me_3tacn)_2](PF_6)_2$ (Fig. 20) catalyzes the oxidation of saturated hydrocarbons with H_2O_2 if small amounts of a carboxylic acid are added to the reaction solution (*112*). In the absence of a carboxylic acid, the compound catalyzes the decomposition of dihydrogen peroxide to molecular oxygen in acetonitrile solution and at room temperature. In the early stages of the reaction, higher alkanes are

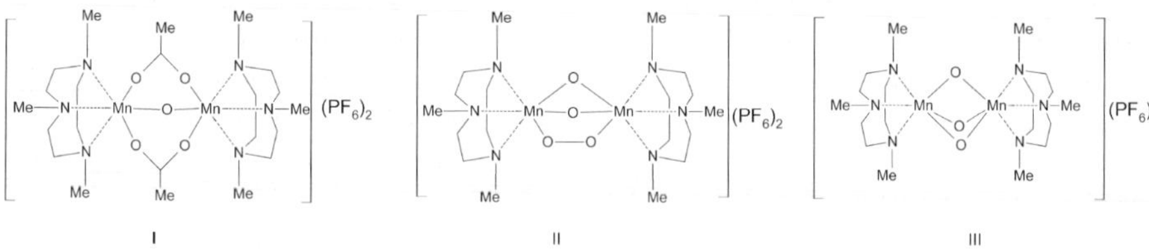

Fig. 20. Schematic structure of dinuclear manganese complexes that can be formed with the ligand Me_3tacn ligand under different synthetic conditions (*141*).

oxidized to the corresponding alkyl hydroperoxides, which are transformed gradually into carbonyl compounds and alcohols (*113,116*). Acetic acid was found to be the most convenient co-catalyst, although other carboxylic acids such as propionic acid and trifluoroacetic acid can be used (*116*). The proposed mechanism for the catalytic cycle includes the formation of dinuclear hydroperoxo derivatives that can be transformed under the action of acetic acid to form an $O{=}Mn^{V}{-}Mn^{IV}{-}OOH$ species. The substrate RH reacts with the manganese-oxo group resulting in the formation of $R^{\bullet}$ radicals, that further react with the $Mn^{IV}{-}OOH$ species to give the alkyl hydroperoxide as the main primary product of the oxidation process (*117*).

B. Olefin Epoxidations

B.1. Mn Catalysts Containing Salen Ligands

Mn-salen-type complexes have been extensively investigated as catalysts for the epoxidation of olefins, with special attention for asymmetric catalysis (*118*). Although effective stereochemical communication between substrate and catalyst is essential for attaining high enantioselectivities in asymmetric catalytic reactions, olefin epoxidation catalyzed by the Mn-salen complexes is the most useful and widely applicable reaction that does not involve substrate precoordination with an appropriate choice of ligand and reaction conditions (*119*). In 1990, a breakthrough was achieved for the highly selective asymmetric epoxidation of unfunctionalized alkenes. The groups of Jacobsen and Katsuki independently introduced a chiral diamine group in the salen ligand (*120,121*). The formation of a saturated (octahedral) inactive manganese complex is disadvantageous to catalysis. The introduction of bulky substituents near axial coordination sites can eliminate this problem and enhance the catalytic performance of the Mn-salen complexes. With regard to the origin of the enantioselectivity, it is generally assumed that it is dominated by the steric interactions between the chiral Mn-salen complex and the two prochiral faces of the olefin. However, the electronic effects, in particular the nature of the substituents in the 5,5′-positions of the phenolato ring of the ligand, also have a strong influence on the selectivity.

The most commonly used oxidants in the asymmetric epoxidation reactions are iodosylbenzene and sodium hypochlorite. Dihydrogen peroxide has been used in only a limited number of studies with chiral Mn-salen complexes and all turnover numbers reported are lower than those obtained with the porphyrin complexes (*122*).

In the olefin epoxidation, the mechanistic scheme commonly proposed for the oxygen-transfer reaction consists of a two-step catalytic cycle (Fig. 21) (*123*). In the first step, an oxygen atom is transferred from the primary oxidant to the Mn^{III}-salen catalyst, which in the second step carries the activated oxygen to the olefinic double bond. The main problem in Mn-salen catalyzed epoxidation with H_2O_2 was the formation of $HO^{\bullet}$ radicals by the homolytic cleavage of the weak O–O bond, leading to indiscriminate oxidation (*124*). Addition of Lewis bases, such as imidazole, pyridine, or

2,6-lutidine, was found to favor the heterolytic bond cleavage, leading to reactive manganese-oxo species (*125*). An alternative approach to avoid the formation of hydroxyl radicals or catalase activity is to make use of anhydrous H_2O_2 adducts like $Ph_3P(O){\cdot}H_2O_2$ in combination with maleic anhydride, which produces peracids in solution (*126*). Also, the adduct urea·H_2O_2 has been used as oxidant in combination with ammonium acetate as additive (*126*).

Berkessel and co-workers reported the first asymmetric epoxidation of unfuctionalized 1,2-dihydronaphthalene using H_2O_2 as terminal oxidant and acetonitrile as the solvent or a two-phase system of water-dichloromethane (*127*). In this method, oxygen transfer to the substrate olefin is effected by a manganese complex of a chiral pentadentate dihydrosalen ligand with a covalently attached imidazole group (Fig. 22). The (1R, 2S)-epoxide was predominantly formed in an enantiomeric excess (*ee*) up to 64 % based on substrate and a product yield up 77 % based on substrate. When the substrate olefin was changed to 6,8-dimetyl-1,2-dihydronaphthalene, a similar *ee* was obtained (63 %), but at significantly lower yield. When the *tert*-butyl group in the 5-position of the ligand is omitted (Fig. 23), the manganese complex catalyzes the epoxidation of 1,2-dihydronaphthalene to the corresponding epoxide under two-phase conditions, in even slightly higher *ee* (66 %) and a product yield up 47 % (*128*). The urea clathrate of H_2O_2 has also been tested as the primary oxidant, resulting in enantiomeric excesses up to 48 % in the oxidation of 1,2-dihydronaphthalene (product yield

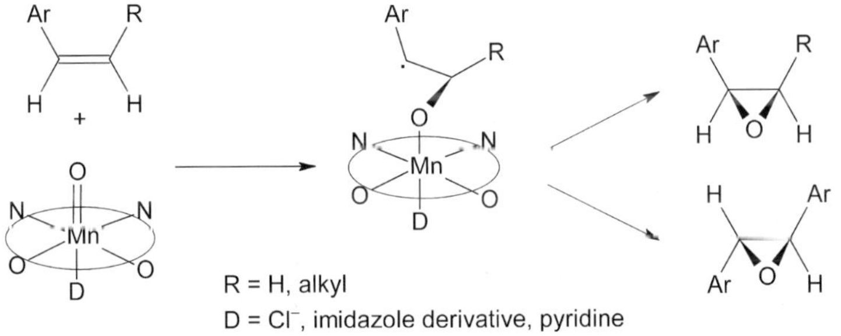

FIG. 21. Proposed mechanism for manganese-catalyzed olefin epoxidations (*123*).

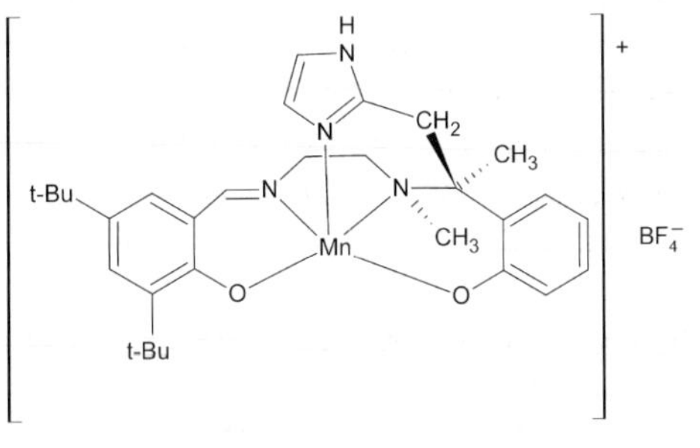

FIG. 22. Pentadentate dihydrosalen complex of manganese as peroxidase model.

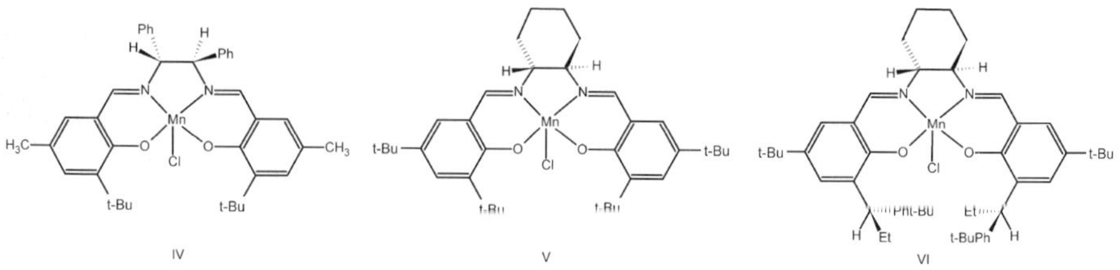

FIG. 23. Jacobsen-type catalysts.

up 70 %). In all catalytic systems, the limited stability of the catalyst was the major weakness with the epoxide yields, reflecting the kinetic competition between the oxygen transfer to a substrate and the oxidative degradation of the catalyst.

An alternative approach for the epoxidation of simple olefins has been developed by Pietikainen and co-workers and Katsuki and co-workers by using the Jacobsen-type catalysts (Fig. 23) and *in situ* addition of imidazole or N-methylimidazole (*122,129*). The highest *ee*-values obtained were 60 % for 1,2-dihydronaphthalene oxide and 47 % for trans-β-methylstyrene oxide in the presence of N-methylimidazole (*129*). Under similar conditions, 6-acetamido-2,2-dimethyl-7-nitrochromene was converted into the corresponding epoxide of 95 % *ee* in 98 % yield based on substrate (*122*).

Asymmetric epoxidation of olefins with dihydrogen peroxide and using carboxylate salts as co-catalysts has been reported by Pietikainen and co-workers (*118,126*). Several simple soluble salts (acetates, formates, benzoates) were studied, all giving good yields of epoxides with moderate to excellent enantioselectivity. The addition of NH_4OAc in the oxidation of 1,2-dihydronaphthalene catalyzed by the manganese complex **I** (Fig. 23) has resulted in 66 % *ee* with a 73 % yield. A particularly interesting result was obtained with the substrate spiro[chromen-2,1′-cyclohexane]; the chromene oxide was produced in 90 % yield and showed *ee*-values exceeding 90 % (*126*). Electron-donating substituents at the 5 and 5′ positions of the salicylide ligand in the catalyst **I** attenuated the reactivity of the catalyst resulting in a longer reaction time. On the other hand, introduction of an electron-withdrawing Br-substituent in the salicylide ligand accelerated the epoxidation reaction (*126*).

With *in situ* generated peroxycarboxylic acids from urea·H_2O_2, carboxylic acid anhydrides and N-methylmorpholine N-oxide as an additive, chiral derivatives of Mn-salen complex **I** were found to oxidize effectively dihydronaphthalene and indene (*126*). The combination employing maleic anhydride has led to good results, up to 87 % *ee* and 71 % yield. Generally, lowering the temperature from 2 to −18°C had marginal positive effect on the yield and enantiomeric excess of the epoxides.

Dinuclear or polynuclear manganese complexes of salen-type ligands were also tested in the epoxidation of olefins with H_2O_2 as oxidant. Enantioselective epoxidation of several olefins with urea·H_2O_2 and a dinuclear Mn-salen type of complex has been reported by Kureshy and co-workers (*130*). Conversions of more than 99 % *ee* were obtained with chromenes and

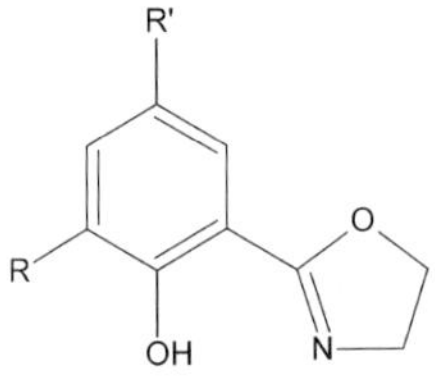

Ligand	*R*	*R'*
Hphox	H	H
H-5'-NO_2phox	H	NO_2
H-3',5'-dClphox	Cl	Cl
H-5'-Clphox	H	Cl
H-5'-Mephox	H	CH_3
H-3'-Mephox	CH_3	H
H-5'-MeOphox	H	OCH_3

FIG. 24. Structure of the H-phox derivatives.

indene, using ammonium acetate as co-catalyst. Only marginal improvement in *ee* values was observed in the presence of coordinating co-catalysts, while both the conversion and the *ee* decreased in the presence of nitrogen-containing, coordinating co-catalysts. Vancheesan and co-workers have found that polynuclear manganese complexes of 2,5-dihydroxyterephthalaldehyde are moderate catalysts for olefin epoxidation in the presence of imidazole (*131*). In particular, the oxidation of cyclohexene afforded up to 49% conversion and up to 92% selectivity toward epoxide.

B.2. Mn Catalysts Containing Phenol-oxazoline Ligands

The family of manganese catalysts based on 2-(2′-hydroxyphenyl)oxazoline (Hphox) and its derivatives (Fig. 24) has been developed by Reedijk and co-workers (*132–135*). As compared with salen-based ligands, these ligands have been reported to have the advantage of being more stable against hydrolysis (*136*). A series of manganese(III) complexes of the type [Mn(phox)$_3$] were synthesized and tested in the epoxidation of styrene and *cis*-stilbene using acetone as solvent and N-methylimidazole as additive (Table VI). In the oxidation of styrene, up to 220 turnover numbers were obtained towards styrene oxide, while in the oxidation of *cis*-stilbene both *trans*- and *cis*-stilbene oxides were formed, along with small amounts of the cleavage product benzaldehyde (*135*). As shown in Table VI the catalytic activity of the [Mn(phox)$_3$] complexes is affected significantly by the substituents on the phenol ring (*135*). For the

TABLE VI

OXIDATION OF STYRENE AND *CIS*-STILBENE BY MN-PHOX COMPLEXES (*135*)[a]

Substrate	Styrene			*cis*-Stilbene		
	Styrene oxide	Benzaldehyde	Diol	*cis*-Oxide	*trans*-Oxide	Benzaldehyde
[Mn(5′-NO_2phox)$_3$]	207	5	18	32	64	2
[Mn(3′,5′-dClphox)$_3$]	108	4	5	22	37	2
[Mn(5′-Clphox)$_3$]	192	4	12	18	50	2
[Mn(phox)$_3$]	220	5	11	17	50	2
[Mn(5′-Mephox)$_3$]	181	4	9	13	46	1
[Mn(3′-Mephox)$_3$]	80	3	3	9	31	2
[Mn(5′-MeOphox)$_3$]	76	2	5	4	19	1

[a]Catalytic activity reported in TON (turnover numbers) = mol product/mol catalyst.

reactions carried out in methanol, all Mn-phox complexes gave turnover numbers between 23 and 34 towards styrene oxide after 15 min, which is an efficiency relative to dihydrogen peroxide of only 1.5 to 2.3 %. The oxidation reactions in acetone show a dramatic increase in turnover numbers compared to the reactions in methanol, possibly related to the formation of the acetone·H_2O_2 adduct (*135*). The efficiency relative to the oxidant has simultaneously increased, ranging from 5.1 to 14.7 %. Electrochemistry experiments have clearly shown that the ligand substituents have a dramatic effect on the Mn^{III}/Mn^{II} and Mn^{III}/Mn^{IV} redox couples (*135*). The oxidation potentials of the complexes generally increase with a more electron-donating nature of the substituents (*135*). The use of Mn-phox complexes with the more electron-donating ligands results in a significant decrease in efficiency compared to $[Mn(phox)_3]$, while the use of complexes with electron-withdrawing ligands only results in a slightly lower activity (*135*). The catalytic reactions with most of the Mn-phox complexes proceed very fast during the first few minutes, and no clear differences concerning the rate of the reactions can be observed. After 15 min no catalytic activity can be observed anymore, except for the $[Mn(5'\text{-}NO_2phox)_3]$ catalyst, which remains active up to one hour. The prolonged activity of $[Mn(5'\text{-}NO_2phox)_3]$ probably arises from a higher stability of the catalyst, which is in accordance with the slower disappearance of the color of the complex compared to the other reactions (*135*). An increase in the stability of complexes upon introduction of electron-withdrawing substituents has also been reported for Mn-salen catalysts (*137*).

The relatively low catalytic activity of $[Mn(phox)_3]$ and its derivatives have been first attributed to the hexa-coordination of the metal center which might prevent the substrate to access the manganese coordination sites (Fig. 25a). As a result, attempts to isolate complexes with labile coordination sites have been made (Fig. 25b) (*133*). However, detailed studies have shown that ligand degradation appears to be the limiting factor of the oxidation catalysis, because the catalytic activity of the complexes appeared to be related to the number of ligands present in the complexes (*133*). This was further supported by the addition of fresh amounts of Hphox ligand to the reaction mixtures, which regenerated the catalytic activity (*133*). Mechanistic studies indicated that $[Mn(phox)_2]^+$ and $[Mn(phox)_2(Nmiz)]^+$ and related species are present in the catalytic reaction mixtures and that the oxazoline group is where decomposition takes place (*134*).

B.3. Mn Catalysts with 1,4,7-trimethyl-1,4,7-triazacyclononane and Derived Ligands

Dinuclear manganese complexes of 1,4,7-trimethyl-1,4,7-triazacyclononane (Me_3tacn) were first described by Wieghardt and co-workers as synthetic models for enzymes like manganese-containing superoxide dismutase and manganese catalase (*138–143*). The triazacyclononane ligand imitates the coordination sphere of the three histidines that are often encountered in metalloproteins and therefore metal complexes of tacn are of interest as structural and functional models. Furthermore, the tacn ligands appeared to be unique in keeping the manganese complex stable under alkaline and

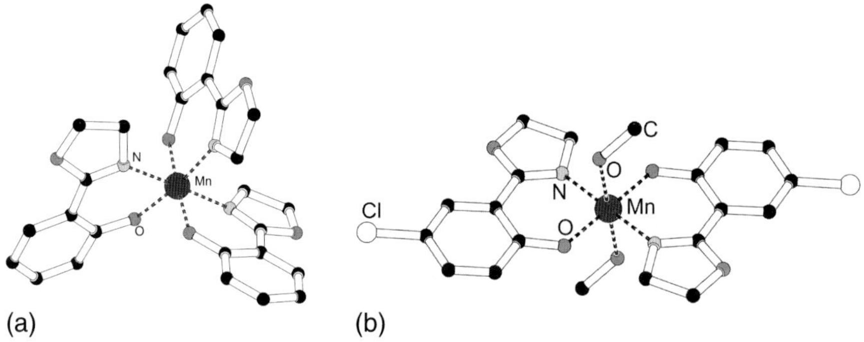

FIG. 25. Molecular structures of the catalysts [Mn(phox)$_3$] (a) and [Mn(5′-Clphox)$_2$(CH$_3$OH)$_2$]$^+$ (b).

oxidizing conditions. A number of tacn-derived ligands have been prepared and aspects of their coordination chemistry have been examined (Fig. 27) (*144–149*). Various dinuclear complexes can be formed, depending on the synthesis conditions (Fig. 20).

The application of Mn-Me$_3$tacn complexes as promising oxidation catalysts engenders from the work at Unilever on low-temperature bleaching additives for washing detergents (*141,150*). However, under the conditions reported, the epoxidation reaction was seriously plagued by the catalytic disproportionation of dihydrogen peroxide (*150*). In the following years, the Mn-Me$_3$tacn system was further improved and selective epoxidation of olefins with H_2O_2 was reported by Bein and co-workers (*151*). In acetone and at ambient temperature, less-reactive olefins were oxidized to the corresponding epoxides with selectivity up to 99% and turnovers up to 1000. The epoxide selectivity has been found to increase in the order *trans*-hept-3-ene (81%) < *trans*-hex-2-ene < cyclohexane < *cis*-hex-2-ene (>98%), reflecting that especially *cis*-olefins are smoothly epoxidized. Replacing Me$_3$tacn with the non-methylated H$_3$tacn resulted in rapid peroxide decomposition, without any product formation. The oxidation profile of the Mn-Me$_3$tacn catalyst in acetone was explained by a mechanism that involves the nucleophilic addition of dihydrogen peroxide to acetone, resulting in the formation of 2-hydroperoxy-2-hydroxypropane (as shown in Fig. 17) (*151*). Nevertheless, acetone is not acceptable for large-scale applications with dihydrogen peroxide because of the risk of formation of explosive cyclic peroxides. Rather simultaneously, Kerschner and co-workers have reported that the mononuclear complex [Mn(Me$_3$tacn)(OMe)$_3$](PF$_6$) (Fig. 26) activates dihydrogen peroxide effectively in water at room temperature and catalytically oxidizes several water-soluble olefins, such as 4-vinylbenzoic acid and styrylacetic acid (Table VII) (*152*). In the oxidation of 4-vinylbenzoic acid a conversion of 82% was attained and the epoxide was the main product, whereas several products were

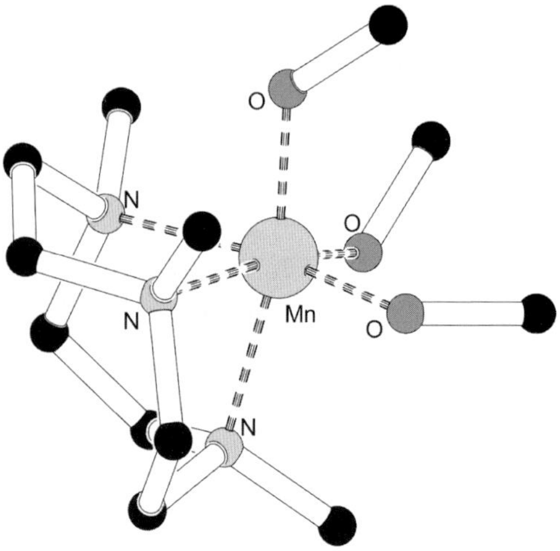

FIG. 26. Molecular structure of the complex $[Mn(Me_3tacn)(OCH_3)](PF_6)$.

TABLE VII

OXIDATION OF WATER-SOLUBLE OLEFINS WITH H_2O_2 CATALYZED BY $[Mn(Me_3tacn)(OMe)_3](PF_6)$ (*152*)

Substrate	Structure	% Conversion[a]	Time (h)
4-vinylbenzoic acid	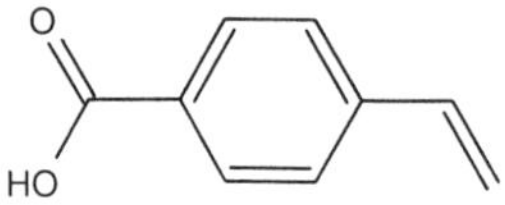	82 92	1.0 1.5
styrylacetic acid	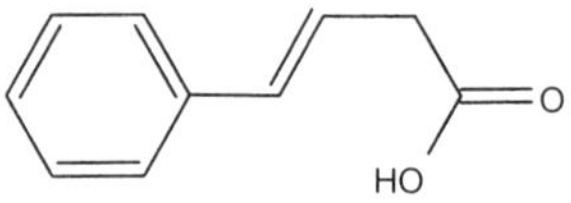	69 98	1.5 3.0

[a]conversion based on the substrate.

identified during the oxidation of styrylacetic acid. The epoxide of styrylacetic acid was found to be labile under the reaction conditions and it hydrolyzes to the corresponding diol or undergoes an intramolecular ring-opening reaction affording 3-hydroxy-4-phenylbutyrolactone (*152*). However, the catalytic species has been shown to achieve more than 400 turnovers without any noticeable catalyst degradation (*152*).

A catalytic amount of an oxalate/oxalic acid buffer strongly enhances the catalytic properties of the Mn-Me$_3$tacn complexes and yields for the epoxides of allyl acetate or 1-hexene reach up to 99 % and 65 % on olefin and peroxide basis, respectively (*153*). It has been suggested that the addition of a catalytic amount of the didentate oxalate ligand prevents the formation of μ-peroxo-bridged dinuclear species (Fig. 20) and as a result the catalase type decomposition of H_2O_2 often associated with dinuclear complexes is suppressed (*148*). The addition of ascorbic acid as co-catalyst allowed for further improvement of olefin epoxidation catalyzed by Mn-Me_3tacn complexes (*154*). The most remarkable result was obtained in the epoxidation of methyl acrylate: 0.03 mol % catalyst was sufficient for full conversion of the substrate (more than 3,000 catalyst turnovers) and the epoxidation was complete after *ca.* 2 h. The oxidation of 1-octene afforded the corresponding epoxide with a yield of 83 %. The exact role of ascorbic acid is still unclear, but this catalytic system developed by Berkessel and co-workers is the most effective one reported so far (*154*).

High catalytic efficiency was reported by Feringa and co-workers in the oxidation of olefins catalyzed by $[Mn_2O_3(Me_3tacn)_2](PF_6)_2$ using acetonitrile as solvent and gmha as co-catalyst (gmha = glyoxylic acid methylester methyl hemiacetal) (*155*). In particular, styrene afforded a high yield of styrene oxide (97 %) in the presence of gmha, whereas the use of oxalate resulted in only modest conversion of styrene (44 %). Under the reaction conditions, the presence of gmha as co-catalyst enables high conversions with only 30 % excess of oxidant with respect to substrate and the H_2O_2 conversion is dramatically improved as compared with the previously described Mn-Me_3tacn systems. In the group of Shul'pin it was reported that the complex $[Mn_2O_3(Me_3tacn)_2](PF_6)_2$ efficiently catalyzes the epoxidation of sterically accessible olefins, including natural terpenes, if a small amount of carboxylic acid is present in the solution (*156*). It has been found that acetic acid acts with the highest efficiency and the overall yield of epoxides was up to 72 % based on the substrate (*156*).

Attempts to mediate asymmetric epoxidation using manganese complexes of chiral tacn derivatives have also been carried out. Bolm and co-workers reported the first asymmetric epoxidation catalyzed by an *in situ* prepared manganese complex of the ligand **5** depicted in Fig. 27 (*157*). Styrene was converted to styrene epoxide in 15 % yield and with an *ee* of 43 %. Replacing the methyl groups with isopropyl groups in the ligand **5**, the *ee* was found in the range 13–38 % for styrene and chromene. The catalytic activity of the dinuclear manganese complex of the enantiopure proline-derived triazacyclononane ligand **6** has been tested in the epoxidation of vinylarenes (*146*). With H_2O_2 as oxidant and 2 mol % of catalyst in acetone at −25°, approximately 28 % conversion of styrene was observed after 2 h giving the styrene epoxide with 24 % *ee*. Extending the reaction time to 4 h increased the conversion (88 %) of the olefin, but reduced the enantioselectivity (15 % *ee*). Styrene derivatives, such as 3-nitrostyrene and 4-chlorostyrene were oxidized to the corresponding epoxides with an enantiomeric excess up to 26 %.

Various successful attempts to fine-tune the catalyst selectivity have been made. Epoxidation reactions using manganese complexes of tacn derivatives bearing a lipophilic alkyl chain on one of the ring carbons were reported by Koek and co-workers (*147*). These lipophilic complexes permit them to be used

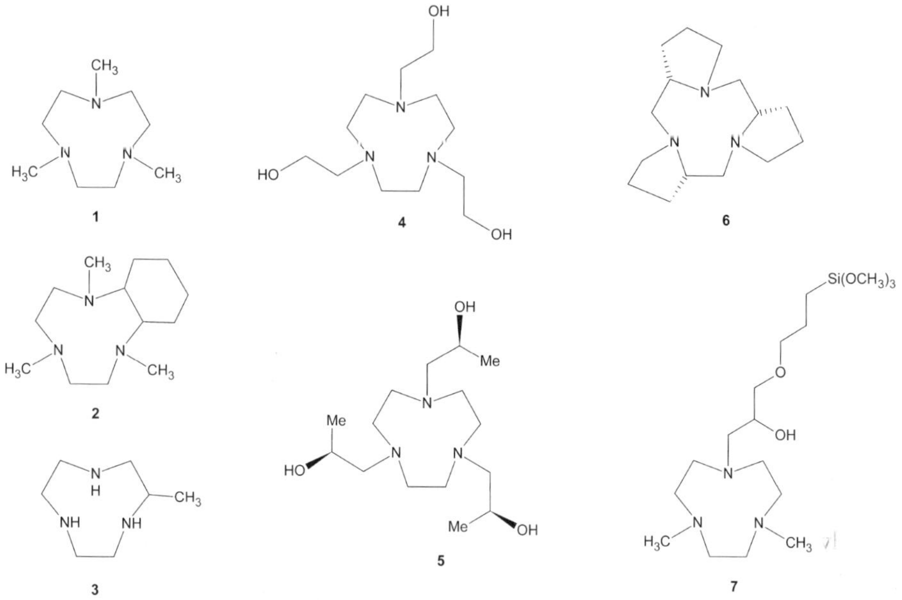

FIG. 27. Some tacn derivatives used for the synthesis of manganese epoxidation catalysts.

as homogeneous catalysts in two-phase systems in the epoxidation of lipophilic substrates like styrene and dodecene with conversions towards epoxide up to 21 %. Encapsulation of the Mn-Me_3tacn complex in zeolites increased the epoxidation selectivity (*151*). By immobilization of the triazacyclononane ligand **7** (Fig. 27) on an inorganic support a new group of active heterogeneous manganese epoxidation catalysts were introduced (*148*). Improved selectivities were found but the conversions obtained were lower than with the homogenous catalysts.

B.4. Miscellaneous Catalysts

The tedious preparation of triazacyclononane derivatives as well as the sensitivity of the corresponding metal complexes to the changes in the Me_3tacn structure has led to the development of new manganese catalysts based on polypyridine ligands (*158*). Feringa and co-workers have designed the hexadentate dinucleating ligand tptn (tptn = N,N,N′,N′-tetrakis(2-pyridylmethyl)propane-1,3-diamine) featuring a three nitrogen donor set for each manganese site (Fig. 28) (*158*). The dinuclear manganese complex of this ligand is an effective catalyst for the epoxidation of olefins using acetone as solvent and at reaction temperatures of 0°C (*158*). Thus, styrene is oxidized to the corresponding epoxide with up to 270 turnovers using 8 equiv. of H_2O_2 with respect to the substrate. High selectivity is observed in the epoxidation reaction of cyclic alkenes, and besides epoxides no allylic oxidation products were found. In the oxidation of *cis*-β-methyl-styrene, both *cis*- and *trans*-epoxides were formed in up to 150 turnovers (*158*).

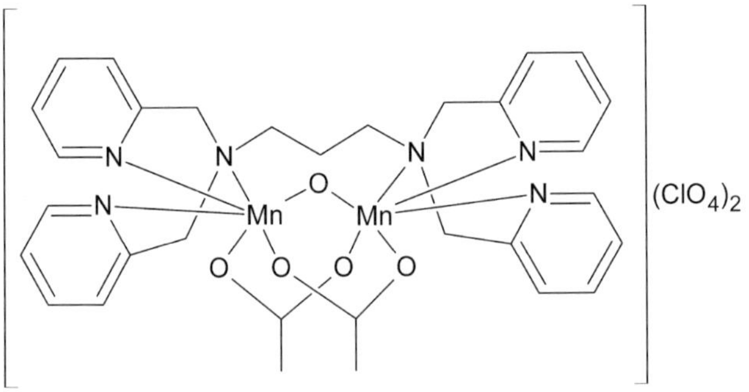

FIG. 28. Schematic structure of the dinuclear manganese complex $[Mn_2O(OAc)_2(tptn)](ClO_4)_2$.

A very simple catalytic system of $MnSO_4$ and a carbonate buffer for the epoxidation of olefins has been reported by Burgess and co-workers (*159,160*). Catalytic experiments were performed in DMF or *tert*-butyl alcohol as solvents and with the slow addition of a mixture of 30 % dihydrogen peroxide and aqueous 0.2 M sodium hydrogen carbonate buffer (*159*). A variety of aryl-substituted, cyclic and tri-alkyl-substituted olefins were epoxidized in these conditions using 10 equiv. of H_2O_2 and in all cases a product yield up to 99 % has been reported (*159,160*). In the reaction mixture, the combination of dihydrogen peroxide and hydrogen carbonate results in an equilibrium to produce peroxycarbonate (Fig. 29). This equilibrium is established in a few minutes, but the epoxidation reactions that rely on this species alone require reaction times of 1–2 days (*160*). The deprotonation of hydrogen carbonate

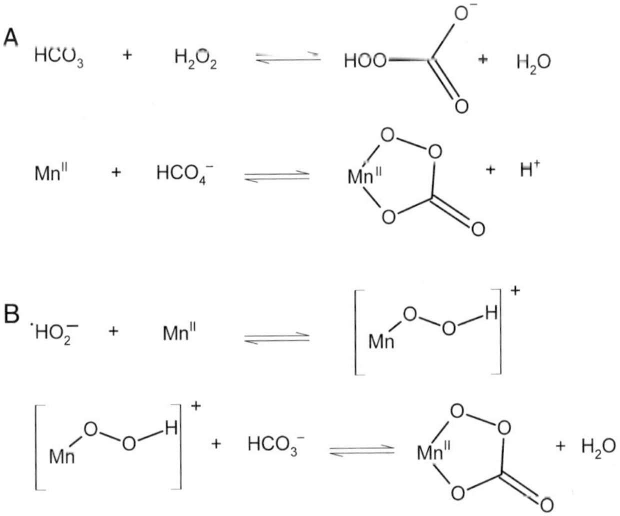

FIG. 29. Formation of the manganese peroxycarbonate complex (A) by the direct reaction of peroxycarbonate or (B) by the reaction of the peroxy complex with carbonate.

($pK_a = 10.3$) becomes significant at pH values above 8~9. Consequently, epoxide yields in the manganese-catalyzed reactions decrease from around 95% at pH values lower than 8.2 to less than 50% at pH 9.0, producing evidence for the key role of the peroxycarbonate species in the process (*160*). EPR studies show that the Mn^{II} salt is initially consumed in the catalytic reaction but is regenerated towards the end of the process when presumably the dihydrogen peroxide is spent. The mechanistic pathway is proposed to involve a manganese η^2-peroxycarbonate complex (*160*). In Fig. 29 routes are given by which such a complex can be formed. One possibility involves the generation of an equilibrium concentration of peroxycarbonate, followed by the coordination of this species to Mn^{II}. Alternatively, the coordination of the hydroperoxide anion to Mn^{II}, deprotonation, and the reaction of this species with hydrogen carbonate would lead to the same intermediate, as is indicated in Fig. 29 (*160*).

C. *cis*-Dihydroxylation

Whereas several catalytic methods are currently available for manganese-catalyzed epoxidation with aqueous H_2O_2, high turnover numbers for *cis*-dihydroxylation reactions so far have only been achieved with osmium compounds (*161–165*). However, manganese-catalyzed oxidation reactions have a few inherent advantages, such as the low price of the manganese salts and complexes and their non-toxic nature.

Recently, it has been shown that selective oxidation of olefins with H_2O_2 and a heterogenized manganese catalyst might open new perspectives for the synthesis of *cis*-diols (*148,166*). Heterogenized Mn-tacn catalysts can be obtained by synthesis of cationic $[Mn(Me_3tacn)]^{2+}$ in zeolite Y or by covalent attachment of the related ligand 1,4-dimethyl-1,4,7-triazacyclononane (HMe_2tacn) to a functionalized silica material (Fig. 30) (*166*). The heterogenization procedure of the tacn ligand employs the conversion

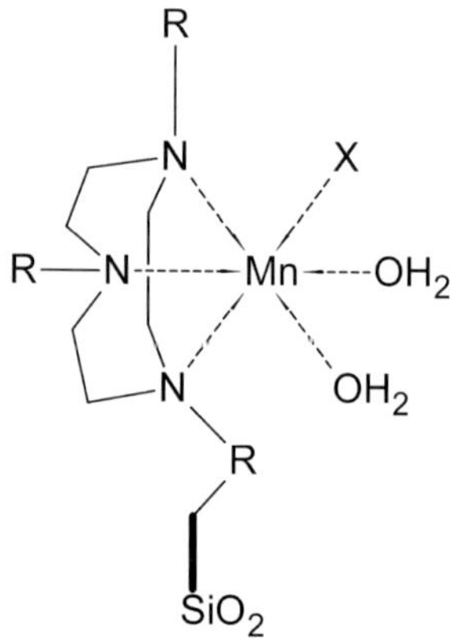

Fig. 30. The proposed epoxidation/*cis*-dihydroxylation active structure for the heterogenized Mn-Me_2tacn catalyst.

of HMe_2tacn to a silylated compound 3-(glycidyloxy)propyltrimethoxysilane followed by immobilization on a SiO_2 surface and subsequent complexation with $MnSO_4{\cdot}H_2O$ (*148*). Remarkably, during olefin oxidation with this new heterogenized catalyst substantial amounts of *cis*-diol were formed apart from the expected epoxide. However, the catalyst activity with respect to *cis*-diol formation is still modest (10–60 mol *cis*-diol/mol Mn). For cyclohexene and *trans*-2-hexene also a small amount of products due to allylic oxidation was observed. The selectivity for *cis*-diol formation ranges from 6 % for geraniol up to 25 % in the case of linalool. The diol formation was not observed in the presence of 1,2-epoxyethane or epoxycyclohexane as substrates, proving that epoxide hydrolysis is not the source of the observed diols. In the oxidation of *cis*-2-hexene, stereoretention for the diol was slightly higher than for the epoxide (94 % vs. 91 %). The authors suggested that epoxide and diol are formed from a single intermediate that is sufficiently long-lived to allow either some bond rotation or a *cis* insertion of a second manganese-coordinated oxygen atom to form the *cis*-diol (*148*). The results of catalytic oxidation of various substrates have shown that electronic effects are important for the regioselectivity of Mn-catalyzed *cis*-dihydroxylation, as the most substituted double bonds react fastest (*166*). *Trans*-dihydroxylation products, originating from epoxide hydrolysis, typically amount to less than 1–2 % of the total products, which is in agreement with the low acidity of the reaction medium. Control experiments with homogeneous HMe_2tacn complexes resulted in strong peroxide decomposition and no oxidation products were obtained. A single, sufficiently long-lived species (Fig. 31) was postulated as

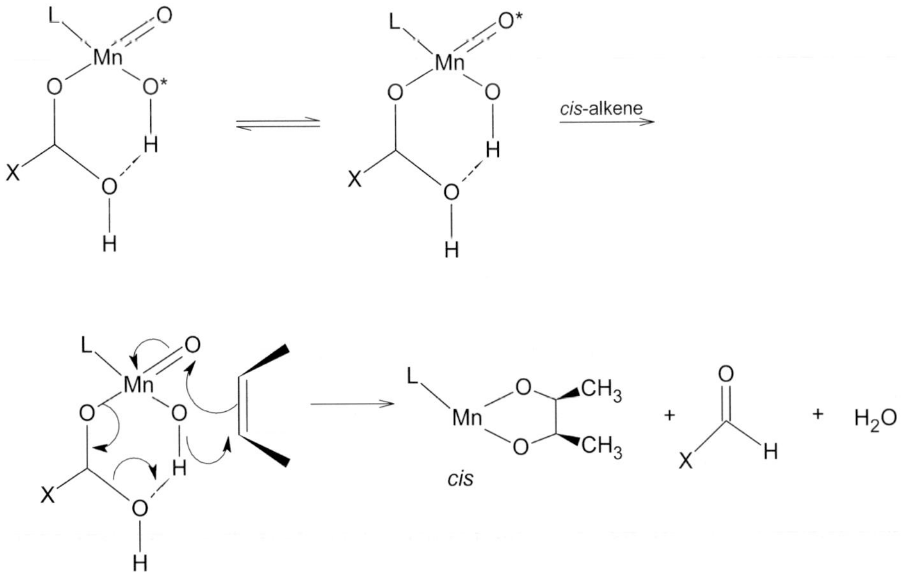

FIG. 31. Proposed *cis*-dihydroxylation mechanism for the oxidation of olefins catalyzed by $[Mn_2O_3(Me_3tacn)_2](PF_6)_2$/gmha (*155*).

the active intermediate for both epoxidation and *cis*-dihydroxylation. This intermediate contains two labile coordination sites e.g. H_2O next to X, the activated oxygen, which might be inserted into the olefin substrate (*148*).

More recently, the use of $[Mn_2O_3(Me_3tacn)_2](PF_6)_2$ in combination with gmha has been reported to result in a highly active system which is able to effect *cis*-dihydroxylation with H_2O_2 and with high turnovers up to 420 (*155*). The highest selectivity to *cis*-diol was found for cyclooctene, which afforded the *cis*-diol as the main product (42 %) beside the epoxide (36 %). The ring size of the cycloolefin has a major influence on the epoxide/*cis*-diol ratio, and no *trans*-diol could be detected in the oxidation of cyclic olefins (*155*). Yields of diol are significantly lower for *trans*-2-hexene than for *cis*-2-hexene, but the epoxide/*cis*-diol ratio was found to be similar in both cases. Attempts to increase the *cis*-diol selectivity by addition of acid or base resulted in a dramatic decrease of conversion. It has been proposed that the *cis*-diol formation involves the reaction of the olefin with a mononuclear manganese oxo-hydroxo species with a coordinated hydrated carbonyl ligand, induced through a hydrogen bonded 6-membered ring transition state (Fig. 31). Reoxidation of the manganese center by H_2O_2, release of the diol from manganese and hydration of the carbonyl compound closes the catalytic cycle (*155*).

V. Concluding Remarks

The present review has outlined the efforts to develop biomimetic non-heme iron and manganese catalysts for alkane hydroxylation, olefin epoxidation, and *cis*-dihydroxylation reactions. However, the examples reviewed here are mostly presented as reported in the literature, since the various reaction conditions involved in the catalytic oxidations hamper a direct comparison and critical evaluation of the data. The survey has not only illustrated a rich variety of iron and manganese complexes that lead to the successful structural modeling of important non-heme iron and manganese enzymes, but also significant features of the oxidation reactions catalyzed by these complexes in combination with dihydrogen peroxide.

With regard to the biomimetic non-heme iron complexes, the work devoted to develop catalysts that perform catalytic alkane hydroxylation has resulted in a large number of iron complexes, which generate $Fe^{IV}{=}O$ iron-oxo species characterized by different spectroscopic techniques. There is now direct evidence that the involvement of high-valent iron-oxo species leads to stereospecific alkane hydroxylation, while hydroxyl radicals contribute to non-selective oxidations. The impressive work performed by Que and co-workers has demonstrated that olefin epoxidation and *cis*-dihydroxylation are different facets of the reactivity of a common Fe^{III}–OOH intermediate, whose spin state can be modulated by the electronic and steric properties of the ligand environment. Labeled $H_2{}^{18}O$ and $H_2{}^{18}O_2$ experiments have shown that the highly reactive $Fe^{IV}{=}O$ species exchange their oxygen atom with $H_2{}^{18}O$ and $H_2{}^{18}O_2$. In spite of significant progress, only a few biomimetic iron catalysts assure that biomimetic hydrocarbon oxidations can be actually achieved. Further improvement of these catalysts as well as the rational design of other structural and functional models for the non-heme iron enzymes might contribute to the development of new catalysts for practical use.

Various biomimetic manganese complexes have been identified to perform catalytic hydrocarbon oxidation in combination with dihydrogen peroxide. Highly selective asymmetric epoxidations of unfunctionalized olefins can be achieved using Mn-salen catalysts. The introduction of a chiral diamine ligand in the salen ligand has resulted in chiral epoxides with significantly high ee values. The addition of ligands that can coordinate axially and favor the formation of oxomanganese(V) intermediates, which are presumed to be the actual oxidizing species, lead to improvement of the selectivity and catalyst stability. The catalyst degradation observed in the olefin epoxidations with Mn-phox complexes has indicated that the instability of the catalyst remains a major problem in the oxidation reactions performed by manganese complexes containing C=N bonds. For non-asymmetric epoxidation reactions, complexes with more stable ligands, such as Me_3tacn and the pyridine-derived ligands are more suitable. These complexes are very stable, versatile, and tunable catalysts and are therefore attractive in an economic sense.

The future development of new catalysts is likely to be boosted by the application of combinatorial chemistry. By using this method, a new iron complex as catalyst for epoxidations with dihydrogen peroxide has been reported (*167*). However, the ultimate aim of the biomimetic catalysis remains the development of iron and manganese catalysts that use dioxygen as primary oxidant, which is nature's way to carry out most of the oxidations. Several studies concerning the activation of dioxygen by non-heme iron, or manganese complexes have shown that there are many intrinsic problems that should be overcome to achieve control of dioxygen activation by metal complexes (*168–173*). It has been demonstrated that the solvent system as well as the structure of the complexes are important factors that control the formation of oxometal(IV) intermediates by activating O_2 (*168,169*). Also, some reports reinforce the idea that reactivity of metal complexes may be regulated by H-bonds (*172*). Finally, it is essential to explore possible intermediates in the reaction pathway, and to distinguish intra- from intermolecular mechanisms.

Acknowledgments

This work was financially supported by the Dutch Economy, Ecology, Technology (EET) programme, a joint programme of the Ministry of Economic Affairs, the Ministry of Education, Culture and Science, and the Ministry of Housing, Spatial Planning and the Environment.

References

1. Parshall, G. W.; Ittel, S. D. "*Homogeneous Catalysis. The Applications and Chemistry of Catalysis by Soluble Transition Metal Complexes*"; 2nd edn., Wiley-Interscience: New York, **1992**.
2. Sheldon, R. A. *Top. Curr. Chem.* **1993**, *164*, 21.
3. Ojima, I. "*Catalytic Asymmetric Synthesis*"; VCH: Weinheim, **1993**.
4. Costas, M.; Mehn, M. P.; Jensen, M. P.; Que, L. *Chem. Rev.* **2004**, *104*, 939.
5. Wu, A. J.; Penner-Hahn, J. E.; Pecoraro, V. L. *Chem. Rev.* **2004**, *104*, 903.
6. Reedijk, J.; Bouwman, E. "*Bioinorganic Catalysis*"; Marcel Dekker: New York, **1999**.
7. Schuchardt, U.; Cardoso, D.; Sercheli, R.; Pereira, R.; de Cruz, R. S.; Guerreiro, M. C.; Mandelli, D.; Spinace, E. V.; Fires, E. L. *Appl. Catal. A-Gen.* **2001**, *211*, 1.
8. Thomas, J. M.; Raja, R.; Sankar, G.; Bell, R. G. *Acc. Chem. Res.* **2001**, *34*, 191.
9. Berkessel, A.; Frauenkron, M.; Schwenkreis, T.; Steinmetz, A. *J. Mol. Catal. A-Chem.* **1997**, *117*, 339.
10. Costas, M.; Chen, K.; Que, L. *Coord. Chem. Rev.* **2000**, *200*, 517.
11. Shul'pin, G. B. *C. R. Chim.* **2003**, *6*, 163.
12. Nakamura, I.; Yamamoto, Y. *Chem. Rev.* **2004**, *104*, 2127.
13. Trost, B. M.; Crawley, M. L. *Chem. Rev.* **2003**, *103*, 2921.
14. Lautens, M.; Fagnou, K.; Hiebert, S. *Acc. Chem. Res.* **2003**, *36*, 48.
15. Groves, J. T.; Meyers, R. S. *J. Am. Chem. Soc.* **1983**, *105*, 5791.
16. Collman, J. P.; Lee, V. J.; Kellenyuen, C. J.; Zhang, X. M.; Ibers, J. A.; Brauman, J. I. *J. Am. Chem. Soc.* **1995**, *117*, 692.
17. Naruta, Y.; Ishihara, N.; Tani, F.; Maruyama, K. *Bull. Chem. Soc. Jpn.* **1993**, *66*, 158.
18. Collman, J. P.; Zhang, X. M.; Lee, V. J.; Uffelman, E. S.; Brauman, J. I. *Science* **1993**, *261*, 1404.
19. Campbell, L. A.; Kodadek, T. *J. Mol. Catal. A-Chem.* **1996**, *113*, 293.
20. Hill, C. L.; Prosser-McCartha, C. M. *Coord. Chem. Rev.* **1995**, *143*, 407.
21. Lippard, S. J.; Berg, J. M. "*Principles of Bioinorganic Chemistry*"; University Science Books: Mill Valley, **1994**.
22. Hendrick, M. J.; May, M. T.; Phliska, M. J.; Robinson, K. D. "*Metals in Biological Systems*"; Ellis Horwood: New York, **1992**.
23. Sono, M. *Chem. Rev.* **1996**, *96*, 2841.
24. Law, N. A.; Caudle, M. T.; Pecoraro, V. L. *Adv. Inorg. Chem.* **1999**, *46*, 305.
25. Lehnert, N.; George, S. D.; Solomon, E. I. *Curr. Opin. Chem. Biol.* **2001**, *5*, 176.
26. Solomon, E. I.; Brunold, T. C.; Davis, M. I.; Kemsley, J. N.; Lee, S. K.; Lehnert, N.; Neese, F.; Skulan, A. J.; Yang, Y. S.; Zhou, J. *Chem. Rev.* **2000**, *100*, 235.
27. Solomon, E. I.; Randall, D. W.; Glaser, T. *Coord. Chem. Rev.* **2000**, *200*, 595.
28. Nanda, S.; Yadav, J. S. *J. Mol. Catal. B-Enzym.* **2003**, *26*, 3.
29. Kauppi, B.; Lee, K.; Carredano, E.; Parales, R. E.; Gibson, D. T.; Eklund, H.; Ramaswamy, S. *Struct. Fold. Des.* **1998**, *6*, 571.
30. Stubbe, J.; Kozovich, J. W.; Wu, W.; Vanderwall, D. E. *Chem. Rev.* **1996**, *96*, 322.
31. Baik, M. H.; Newcomb, M.; Friesner, R. A.; Lippard, S. J. *Chem. Rev.* **2003**, *103*, 2385.
32. Wallar, B. J.; Lipscomb, J. D. *Chem. Rev.* **1996**, *96*, 2625.
33. Tshuva, E. Y.; Lippard, S. J. *Chem. Rev.* **2004**, *104*, 987.
34. Pecoraro, V. L.; Hsieh, W. Y. The use of model complexes to elucidate the structure and function of manganese redox enzymes. In: "*Metal Ions in Biological Systems*"; vol. 37, Marcel Dekker: New York, **2000**, p. 429.
35. Yocum, C. F.; Pecoraro, V. L. *Curr. Opin. Chem. Biol.* **1999**, *3*, 182.
36. Iwata, S.; Barber, J. *Curr. Opin. Struct. Biol.* **2004**, *14*, 447.
37. Ferreira, K. N.; Iverson, T. M.; Maghlaoui, K.; Barber, J.; Iwata, S. *Science* **2004**, *303*, 1831.
38. Mukhopadhyay, S.; Mandal, S. K.; Bhaduri, S.; Armstrong, W. H. *Chem. Rev.* **2004**, *104*, 3981.

39. Pecoraro, V. L.; Gelasco, A.; Baldwin, M. J. Reactivity and mechanism of manganese enzymes – A modeling approach. In: "*Mechanistic Bioinorganic Chemistry*"; vol. 246, ACS: Washington, **1995**, p. 265.
40. Boelrijk, A. E. M.; Dismukes, G. C. *Inorg. Chem.* **2000**, *39*, 3020.
41. Boelrijk, A. E. M.; Khangulov, S. V.; Dismukes, G. C. *Inorg. Chem.* **2000**, *39*, 3009.
42. Rowland, J. M.; Olmstead, M.; Mascharak, P. K. *Inorg. Chem.* **2001**, *40*, 2810.
43. Lehnert, N.; Ho, R. Y. N.; Que, L.; Solomon, E. I. *J. Am. Chem. Soc.* **2001**, *123*, 8271.
44. Chen, K.; Que, L. *J. Am. Chem. Soc.* **2001**, *123*, 6327.
45. Chen, K.; Que, L.; *Chem. Commun.* **1999**, 1375.
46. Gozzo, F. *J. Mol. Catal. A-Chem.* **2001**, *171*, 1.
47. Sheu, C.; Richert, S. A.; Cofre, P.; Ross, B.; Sobkowiak, A.; Sawyer, D. T.; Kanofsky, J. R. *J. Am. Chem. Soc.* **1990**, *112*, 1936.
48. Milas, N. A.; Golubovic, A. *J. Am. Chem. Soc.* **1959**, *81*, 6461.
49. Girerd, J. J.; Banse, F.; Simaan, A. J. Characterization and properties of non-heme iron peroxo complexes. In: "*Metal-Oxo and Metal-Peroxo Species in Catalytic Oxidations"; Structure Bonding* vol. 97, Springer-Verlag, **2000**, p. 145.
50. Balland, V.; Banse, F.; Anxolabehere-Mallart, E.; Nierlich, M.; Girerd, J. J. *Eur. J. Inorg. Chem.* **2003**, 2529.
51. Balland, V.; Mathieu, D.; Pons-Y-Moll, N.; Bartoli, J. F.; Banse, F.; Battioni, P.; Girerd, J. J.; Mansuy, D. *J. Mol. Catal. A-Chem.* **2004**, *215*, 81.
52. Kim, J.; Harrison, R. G.; Kim, C.; Que, L. *J. Am. Chem. Soc.* **1996**, *118*, 4373.
53. Menage, S.; Vincent, J. M.; Lambeaux, C.; Fontecave, M. *J. Chem. Soc., Dalton Trans.* **1994**, 2081.
54. Ingold, K. U.; MacFaul, P. A. "*Biomimietic Oxidations Catalyzed by Transition Metal Complexes*"; Ed. Meunier, B., World Scientific Publishing and Imperial College Press: London, **2000**.
55. Buxton, G. V.; Greenstock, C. L.; Helman, W. P.; Ross, A. B. *J. Phys. Chem.* **1988**, *17*, 513.
56. Groves, J. T.; Nemo, T. E. *J. Am. Chem. Soc.* **1983**, *105*, 6243.
57. Kim, C.; Chen, K.; Kim, J. H.; Que, L. *J. Am. Chem. Soc.* **1997**, *119*, 5964.
58. Roelfes, G.; Lubben, M.; Chen, K.; Ho, R. Y. N.; Meetsma, A.; Genseberger, S.; Hermant, R. M.; Hage, R.; Mandal, S. K.; Young, V. G.; Zang, Y.; Kooijman, H.; Spek, A. L.; Que, L.; Feringa, B. L. *Inorg. Chem.* **1999**, *38*, 1929.
59. Lubben, M.; Meetsma, A.; Wilkinson, E. C.; Feringa, B.; Que, L. *Angew. Chem. Int. Edit. Engl.* **1995**, *34*, 1512.
60. Roelfes, G.; Lubben, M.; Hage, R.; Que, L.; Feringa, B. L. *Chem. Eur. J.* **2000**, *6*, 2152.
61. Lehnert, N.; Neese, F.; Ho, R. Y. N.; Que, L.; Solomon, E. I. *J. Am. Chem. Soc.* **2002**, *124*, 10810.
62. Klopstra, M.; Roelfes, G.; Hage, R.; Kellogg, R. M.; Feringa, B. L. *Eur. J. Inorg. Chem.* **2004**, 846.
63. Wada, A.; Ogo, S.; Nagatomo, S.; Kitagawa, T.; Watanabe, Y.; Jitsukawa, K.; Masuda, H. *Inorg. Chem.* **2002**, *41*, 616.
64. Horwitz, C. P.; Fooksman, D. R.; Vuocolo, L. D.; Gordon-Wylie, S. W.; Cox, N. J.; Collins, J. T. *J. Am. Chem. Soc.* **1998**, *120*, 4867.
65. Collins, J. T. *Acc. Chem. Res.* **2001**, *35*, 782.
66. Bartos, M. J.; Gordon-Wylie, S. W.; Fox, B. G.; Wright, L. J.; Weintraub, S. T.; Kauffmann, K. E.; Munck, E.; Kostka, K. L.; Uffelman, E. S.; Rickard, C. E. F.; Noon, K. R.; Collins, T. *J. Coord. Chem. Rev.* **1998**, *174*, 361.
67. Ghosh, A.; de Oliveira, F. T.; Yano, T.; Nishioka, T.; Beach, E. S.; Kinoshita, I.; Munck, E.; Ryabv, A. D.; Horwitz, C. P.; Collins, J. T.; *J. Am. Chem. Soc.* **2005**, *127*, 2505.
68. Bukowski, M. R.; Zhu, S. R.; Koehntop, K. D.; Brennessel, W. W.; Que, L. *J. Biol. Inorg. Chem.* **2004**, *9*, 39.
69. Fish, R. H.; Konings, M. S.; Oberhausen, K. J.; Fong, R. H.; Yu, W. M.; Christou, G.; Vincent, J. B.; Coggin, D. K.; Buchanan, R. M. *Inorg. Chem.* **1991**, *30*, 3002.
70. Kulikova, V. S.; Gritsenko, O. N.; Shteinman, A. A. *Mendeleev Commun.* **1996**, 119.
71. Duboc-Toia, C.; Menage, S.; Lambeaux, C.; Fontecave, M. *Tetrahedron Lett.* **1997**, *38*, 3727.

72. Mekmouche, Y.; Duboc-Toia, C.; Menage, S.; Lambeaux, C.; Fontecave, M. *J. Mol. Cat. A-Chem.* **2000**, *156*, 85.
73. Fontecave, M.; Menage, S.; Duboc-Toia, C. *Coord. Chem. Rev.* **1998**, *178–180*, 1555.
74. Ito, S.; Okuno, T.; Matsushima, H.; Tokii, T.; Nishida, Y. *J. Chem. Soc. Dalton Trans.* **1996**, 4479.
75. Ito, S.; Okuno, T.; Matsushima, H.; Tokii, T.; Nishida, Y. *J. Chem. Soc. Dalton Trans.* **1996**, 4037.
76. Ito, S.; Suzuki, M.; Kobayashi, T.; Itoh, H.; Harada, A.; Ohba, S.; Nishida, Y. *J. Chem. Soc. Dalton Trans.* **1996**, 2579.
77. Ito, S.; Okuno, T.; Itoh, H.; Ohba, S.; Matsushima, H.; Tokii, T.; Nishida, Y. *Z. Naturforsch. (B)* **1997**, *52*, 719.
78. Nishida, Y.; Okuno, T.; Ito, S.; Harada, A.; Ohba, S.; Matsushima, H.; Tokii, T. *Chem. Lett.* **1995**, 885.
79. Tanase, S.; Foltz, C.; de Gelder, R.; Hage, R.; Bouwman, E.; Reedijk, J. *J. Mol. Cat. A-Chem.* **2005**, *225*, 161.
80. Leising, R. A.; Brennan, B. A.; Que, L.; Munck, E. *J. Am. Chem. Soc.* **1991**, *113*, 3988.
81. Dong, Y. H.; Fujii, H.; Hendrich, M. P.; Leising, R. A.; Pan, G. F.; Randall, C. R.; Wilkinson, E. C.; Zang, Y.; Que, L.; Fox, B. G.; Kauffmann, K.; Munck, E. *J. Am. Chem. Soc.* **1995**, *117*, 2778.
82. Feig, A. L.; Lippard, S. J. *Chem. Rev.* **1994**, *94*, 759.
83. Gutkina, E. A.; Rubtsova, T. B.; Shteinman, A. A. *Kinet. Catal.* **2003**, *44*, 106.
84. Pardo, E.; Lloret, F.; Carrasco, R.; Munoz, M. C.; Temporal-Sanchez, T.; Ruiz-Garcia, R. *Inorg. Chim. Acta* **2004**, *357*, 2713.
85. Ryu, J. Y.; Kim, J.; Costas, M.; Chen, K.; Nam, W.; Que, L. *Chem. Commun.* **2002**, 1288.
86. Costas, M.; Tipton, A. K.; Chen, K.; Jo, D. H.; Que, L. *J. Am. Chem. Soc.* **2001**, *123*, 6722.
87. Chen, K.; Costas, M.; Kim, J. H.; Tipton, A. K.; Que, L. *J. Am. Chem. Soc.* **2002**, *124*, 3026.
88. Chen, K.; Que, L. *Angew. Chem. Int. Edit.* **1999**, *38*, 2227.
89. Yamamoto, T.; Kimura, M.; *J. Chem. Soc. Chem. Commun.* **1977**, 948.
90. Chen, K.; Costas, M.; Que, L.; *J. Chem. Soc. Dalton Trans.* **2002**, 672.
91. Nam, W.; Ho, R.; Valentine, J. S. *J. Am. Chem. Soc.* **1991**, *113*, 7052.
92. Kist, L. T.; Trujillo, M. J. F.; Szpoganicz, B.; Manez, M. A.; Basallote, M. G. *Polyhedron* **1997**, *16*, 3827.
93. White, M. C.; Doyle, A. G.; Jacobsen, E. N. *J. Am. Chem. Soc.* **2001**, *123*, 7194.
94. Guajardo, R. J.; Hudson, S. E.; Brown, S. J.; Mascharak, P. K. *J. Am. Chem. Soc.* **1993**, *115*, 7971.
95. Kaku, Y.; Otsuka, M.; Ohno, M. *Chem. Lett.* **1989**, 611.
96. Payeras, A. M.; Ho, R. Y. N.; Fujita, M.; Que, L. *Angew. Chem. Int. Ed. Engl.* **2004**, *20*, 4944.
97. Murch, B. P.; Bradley, F. C.; Que, L. *J. Am. Chem. Soc.* **1986**, *108*, 5027.
98. Wang, X. M.; Wang, S. X.; Li, L. J.; Sundberg, E. B.; Gacho, G. P. *Inorg. Chem.* **2003**, *42*, 7799.
99. Mathieu, D.; Bartoli, J. F.; Battioni, P.; Mansuy, D. *Tetrahedron* **2004**, *60*, 3855.
100. Menage, S.; Galey, J. B.; Dumats, J.; Hussler, G.; Seite, M.; Luneau, I. G.; Chottard, G.; Fontecave, M. *J. Am. Chem. Soc.* **1998**, *120*, 13370.
101. Duprat, A. F.; Capdevielle, P.; Maumy, M. *J. Chem. Soc. Chem. Commun.* **1991**, 464.
102. Smith, J. R. L.; Shaw, B. A. J.; Foulkes, D. M.; Jeffrey, A. M.; Jerina, D. M. *J. Chem. Soc. Perkin Trans. 2* **1977**, 1583.
103. Jensen, M. P.; Lange, S. J.; Mehn, M. P.; Que, E. L.; Que, L. *J. Am. Chem. Soc.* **2003**, *125*, 2113.
104. Avenier, F.; Dubois, L.; Latour, J. M. *New J. Chem.* **2004**, *28*, 782.
105. Shul'pin, G. B.; Stoeckli-Evans, H.; Mandelli, D.; Kozlov, Y. N.; Vallina, A. T.; Woitiski, C. B.; Jimenez, R. S.; Carvalhoc, W. A. *J. Mol. Catal. A-Chem.* **2004**, *219*, 255.
106. Tamagaki, S.; Sasaki, M.; Tagaki, W. *Bull. Chem. Soc. Jpn.* **1989**, *62*, 153.
107. Tamagaki, S.; Suzuki, K.; Tagaki, W. *Bull. Chem. Soc. Jpn.* **1989**, *62*, 148.
108. Menage, S.; Galey, J. B.; Hussler, G.; Seite, M.; Fontecave, M. *Angew. Chem., Int. Engl. Ed.* **1996**, *35*, 2353.

109. Bartoli, J. F.; Lambert, F.; Morgenstern-Badarau, I.; Battioni, P.; Mansuy, D. *C. R. Chim.* **2002**, *5*, 263.
110. Nizova, G. V.; Krebs, B.; Suss-Fink, G.; Schindler, S.; Westerheide, L.; Cuervo, L. G.; Shul'pin, G. B. *Tetrahedron* **2002**, *58*, 9231.
111. Nizova, G. V.; Bolm, C.; Ceccarelli, S.; Pavan, C.; Shul'pin, G. B. *Adv. Synth. Catal.* **2002**, *344*, 899.
112. Shul'pin, G. B.; Lindsay-Smith, J. R. *Russ. Chem. Bull.* **1998**, *47*, 2379.
113. Shul'pin, G. B.; Suss-Fink, G.; Shul'pina, L. S. *J. Mol. Catal. A-Chem.* **2001**, *170*, 17.
114. Ryu, J. Y.; Kim, S. O.; Nam, W.; Heo, S.; Kim, J. *Bull. Korean Chem. Soc.* **2003**, *24*, 1835.
115. Kozlov, Y. N.; Mandelli, D.; Woitiski, C. B.; Shul'pin, G. B. *Russ. J. Phys. Chem.* **2004**, *78*, 370.
116. Smith, J. R. L.; Shul'pin, G. B. *Tetrahedron Lett.* **1998**, *39*, 4909.
117. Shul'pin, G. B.; Nizova, G. V.; Kozlov, Y. N.; Pechenkina, I. G. *New J. Chem.* **2002**, *26*, 1238.
118. Pietikainen, P. *Tetrahedron* **1998**, *54*, 4319.
119. Jacobsen, E. N. Asymmetric aziridination. In: "*Catalytic Asymmetric Synthesis*," Ed. Ojima, I., VCH: New York, **1993**.
120. Irie, R.; Noda, K.; Ito, Y.; Matsumoto, N.; Katsuki, T. *Tetrahedron Lett.* **1990**, *31*, 7345.
121. Zhang, W.; Loebach, J. L.; Wilson, S. R.; Jacobsen, E. N. *J. Am. Chem. Soc.* **1990**, *112*, 2801.
122. Irie, R.; Hosoya, N.; Katsuki, T. *Synlett.* **1994**, 255.
123. Palucki, M.; Finney, N. S.; Pospisil, P. J.; Guler, M. L.; Ishida, T.; Jacobsen, E. N. *J. Am. Chem. Soc.* **1998**, *120*, 948.
124. Cavallo, L.; Jacobsen, H. *Eur. J. Inorg. Chem.* **2003**, 892.
125. Meunier, B. *Chem. Rev.* **1992**, *92*, 1411.
126. Pietikainen, P. *J. Mol. Catal. A-Chem.* **2001**, *165*, 73.
127. Schwenkreis, T.; Berkessel, A. *Tetrahedron Lett.* **1993**, *34*, 4785.
128. Berkessel, A.; Frauenkron, M.; Schwenkreis, T.; Steinmetz, A.; Baum, G.; Fenske, D. *J. Mol. Catal. A-Chem.* **1996**, *113*, 321.
129. Pietikainen, P. *Tetrahedron Lett.* **1994**, *35*, 941.
130. Kureshy, R. I.; Khan, N. H.; Abdi, S. H. R.; Patel, S. T.; Jasra, R. V. *Tetrahedron: Asymm.* **2001**, *12*, 433.
131. Krishnan, R.; Vancheesan, S. *J. Mol. Catal. A-Chem.* **2000**, *157*, 15.
132. Hoogenraad, M., Ramkisoensing, K., Kooijman, H.; Spek, A. L.; Bouwman, E.; Haasnoot, J. G.; Reedijk, J. *Inorg. Chim. Acta* **1998**, *279*, 217.
133. Hoogenraad, M.; Ramkisoensing, K.; Driessen, W. L.; Kooijman, H.; Spek, A. L.; Bouwman, E.; Haasnoot, J. G.; Reedijk, J. *Inorg. Chim. Acta* **2001**, *320*, 117.
134. Hoogenraad, M.; Kooijman, H.; Spek, A. L.; Bouwman, E.; Haasnoot, J. G.; Reedijk, J. *Eur. J. Inorg. Chem.* **2002**, 2897.
135. Hoogenraad, M.; Ramkisoensing, K.; Gorter, S.; Driessen, W. L.; Bouwman, E.; Haasnoot, J. G.; Reedijk, J.; Mahabiersing, T.; Hartl, F. *Eur. J. Inorg. Chem.* **2002**, 377.
136. Hoveyda, H. R.; Karunaratne, V.; Rettig, S. J.; Orvig, C. *Inorg. Chem.* **1992**, *31*, 5408.
137. Kureshy, R. I.; Khan, N. H.; Abdi, S. H. R.; Bhatt, A. K. *J. Mol. Catal. A-Chem.* **1996**, *110*, 33.
138. Bossek, U.; Hummel, H.; Weyhermuller, T.; Wieghardt, K.; Russell, S.; van der Wolf, L.; Kolb, U. *Angew. Chem. Int. Edit. Engl.* **1996**, *35*, 1552.
139. Bossek, U.; Wieghardt, K.; Nuber, B.; Weiss, J. *Inorg. Chim. Acta* **1989**, *165*, 123.
140. Wieghardt, K.; Bossek, U.; Bonvoisin, J.; Beauvillain, P.; Girerd, J. J.; Nuber, B.; Weiss, J.; Heinze, J. *Angew. Chem. Int. Edit. Engl.* **1986**, *25*, 1030.
141. Hage, R. *Recl. Trav. Chim. Pays-Bas-J. Roy. Neth. Chem. Soc.* **1996**, *115*, 385.
142. Wieghardt, K.; Bossek, U.; Nuber, B.; Weiss, J.; Bonvoisin, J.; Corbella, M.; Vitols, S. E.; Girerd, J. J. *J. Am. Chem. Soc.* **1988**, *110*, 7398.
143. Bossek, U.; Weyhermuller, T.; Wieghardt, K.; Nuber, B.; Weiss, J. *J. Am. Chem. Soc.* **1990**, *112*, 6387.
144. Scheuermann, J. E. W.; Ronketti, F.; Motevalli, M.; Griffiths, D. V.; Watkinson, M. *New J. Chem.* **2002**, *26*, 1054.

145. Scheuermann, J. E. W.; Sibbons, K. F.; Benoit, D. M.; Motevalli, M.; Watkinson, M. *Org. Biomol. Chem.* **2004**, *2*, 2664.
146. Bolm, C.; Meyer, N.; Raabe, G.; Weyhermuller, T.; Bothe, E. *Chem. Commun.* **2000**, 2435.
147. Koek, J. H.; Kohlen, E.; Russell, S. W.; van der Wolf, L.; ter Steeg, P. F.; Hellemons, J. C. *Inorg. Chim. Acta* **1999**, *295*, 189.
148. de Vos, D. E.; de Wildeman, S.; Sels, B. F.; Grobet, P. J.; Jacobs, P. A. *Angew. Chem. Int. Edit.* **1999**, *38*, 980.
149. Golding, S. W.; Hambley, T. W.; Lawrance, G. A.; Luther, S. M.; Maeder, M.; Turner, P. *J. Chem. Soc. Dalton Trans.* **1999**, 1975.
150. Hage, R.; Iburg, J. E.; Kerschner, J.; Koek, J. H.; Lempers, E. L. M.; Martens, R. J.; Racherla, U. S.; Russell, S. W.; Swarthoff, T.; Vanvliet, M. R. P.; Warnaar, J. B.; van der wolf, L.; Krijnen, B. *Nature* **1994**, *369*, 637.
151. de Vos, D.; Bein, T. *Chem. Commun.* **1996**, 917.
152. Quee Smith, V. C.; DelPizzo, L.; Jureller, S. H.; Kerschner, J. L.; Hage, R. *Inorg. Chem.* **1996**, *35*, 6461.
153. de Vos, D. E.; Sels, B. F.; Reynaers, M.; Rao, Y. V. S.; Jacobs, P. A. **1998**, *39*, 3221.
154. Berkessel, A.; Sklorz, C. A. *Tetrahedron Lett.* **1999**, *40*, 7965.
155. Brinksma, J.; Schmieder, L.; van Vliet, G.; Boaron, R.; Hage, R.; de Vos, D. E.; Alsters, P. L.; Feringa, B. L. *Tetrahedron Lett.* **2002**, *43*, 2619.
156. Woitiski, C. B.; Kozlov, Y. N.; Mandelli, D.; Nizova, G. V.; Schuchardt, U.; Shul'pin, G. B. *J. Mol. Cat. A-Chem.* **2004**, *222*, 103.
157. Bolm, C.; Luong, T. K. K.; Schlingloff, G. *Synlett* **1997**, 1151.
158. Brinksma, J.; Hage, R.; Kerschner, J.; Feringa, B. L. *Chem. Commun.* **2000**, 537.
159. Lane, B. S.; Burgess, K. *J. Am. Chem. Soc.* **2001**, *123*, 2933.
160. Lane, B. S.; Vogt, M.; DeRose, V. J.; Burgess, K. *J. Am. Chem. Soc.* **2002**, *124*, 11946.
161. Severeyns, A.; de Vos, D. E.; Fiermans, L.; Verpoort, F.; Grobet, P. J.; Jacobs, P. A. *Angew. Chem. Int. Edit.* **2001**, *40*, 586.
162. Rosini, C.; Tanturli, R.; Pertici, P.; Salvadori, P. *Tetrahedron Asymm.* **1996**, *7*, 2971.
163. Nakajima, M.; Tomioka, K.; Iitaka, Y.; Koga, K. *Tetrahedron* **1993**, *49*, 10793.
164. Halterman, R. L.; McEvoy, M. A. *J. Am. Chem. Soc.* **1992**, *114*, 980.
165. Erdik, E.; Kahya, D.; Daskapan, T. *Synth. Commun.* **1998**, *28*, 1.
166. Sels, B. F.; Villa, A. L.; Hoegaerts, D.; de Vos, D. E.; Jacobs, P. A. *Top. Catal.* **2000**, *13*, 223.
167. Francis, M. B.; Jacobsen, E. N. *Angew. Chem. Int. Ed. Engl.* **1999**, *38*, 937.
168. Carson, E. C.; Lippard, S. J. *J. Am. Chem. Soc.* **2004**, *126*, 3412.
169. Moreira, R. F.; Tshuva, E. Y.; Lippard, S. J. *Inorg. Chem.* **2004**, *43*, 4427.
170. MacBeth, C. E.; Golombek, A. P.; Young, V. G.; Yang, C.; Kuczera, K.; Hendrich, M. P.; Borovik, A. S. *Science* **2000**, *289*, 938.
171. MacBeth, C. E.; Golobek, A. P.; Young, V. G.; Yang, C.; Kuckzera, K.; Hendrich, M. P.; Borovik, A. S. *Science* **2000**, *289*, 928.
172. MacBeth, C. E.; Gupta, R.; Mitchell-Koch, K. R.; Young, V. G.; Lushington, G. H.; Thompson, W. H.; Hendrich, M. P.; Borovik, A. S. *J. Am. Chem. Soc.* **2004**, *126*, 2556.
173. Kim, S. O.; Sastri, C. V.; Seo, M. S.; Kim, J.; Nam, W. *J. Am. Chem. Soc.* **2005**, *127*, 4178.
174. Hufton, S. E.; Jennings, I. G.; Cotton, R. G. H. *Biochem. J.* **1995**, *311*, 353.
175. Ohlendorf, D. H.; Orville, A. M.; Lipscomb, J. D. *J. Mol. Biol.* **1994**, *244*, 586.
176. Lim, J. H.; Yu, Y. G.; Han, Y. S.; Cho, S.; Ahn, B. Y.; Kim, S. H.; Cho, Y. *J. Mol. Biol.* **1997**, *270*, 259.
177. Roach, P. L.; Clifton, I. J.; Fülöp, V.; Harlos, K.; Barton, G. J.; Hajdu, J.; Andersson, I.; Schofield, C. J.; Baldwin, J. E. *Nature* **1995**, *375*, 700.
178. Minor, W.; Steczko, J.; Stec, B.; Otwinowski, Z.; Bolin, J. T.; Walter, R.; Axelrod, B. *Biochemistry* **1996**, *35*, 10687.
179. Huang, W.; Jia, J.; Cummings, J.; Nelson, M. J.; Schneider, G.; Lindqvist, Y. *Structure* **1997**, *5*, 691.
180. Eklund, H.; Uhlin, U.; Farnegardh, M.; Logan, D. T.; Nordlund, P. *Prog. Biophys. Mol. Biol.* **2001**, *77*, 177.
181. Rosenzweig, A. C.; Frederick, C. A.; Lippard, S. J.; Nordlund, P. *Nature* **1993**, *366*, 537.

182. Guddat, L. W.; McAlpine, A. S.; Hume, D.; Hamilton, S.; de Jersey, J.; Martin, J. L. *Struct. Fold. Des.* **1999**, *7*, 757.
183. Lindqvist, Y.; Huang, W.; Schneider, G.; Shanklin, J. *EMBO J.* **1996**, 15.
184. Pikus, J. D.; Studts, J. M.; Achim, C.; Kauffmann, K. E.; Munck, E.; Steffan, R. J.; McClay, K.; Fox, B. G. *Biochemistry* **1996**, *35*, 9106.
185. Cadieux, E.; Vrajmasu, V.; Achim, C.; Powlowski, J.; Munck, E. *Biochemistry* **2002**, *41*, 10680.
186. Small, F. J.; Ensign, S. A. *J. Biol. Chem.* **1997**, *272*, 24913.
187. Sluis, M. K.; Sayavedra-Soto, L. A.; Arp, D. J. *Microbiology* **2002**, *148*, 3617.
188. Hsu, H. F.; Que, L.; Shanklin, J. *J. Inorg. Biochem.* **1999**, *74*, 168.
189. Rea, S. *FEBS Lett.* **2001**, *509*, 389.
190. Stallings, W. C.; Pattridge, K. A.; Strong, R. K.; Ludwig, M. L. *J. Biol. Chem.* **1984**, *259*, 695.
191. Que, L.; Widom, J.; Crawford, R. L. *J. Biol. Chem.* **1981**, *256*, 941.
192. Beyer, W. F.; Fridovich, I. *Biochemistry* **1985**, *24*, 6460.
193. Hornsten, L.; Su, C.; Osbourn, A. E.; Hellman, U.; Oliw, E. H. *Eur. J. Biochem.* **2002**, *269*, 2690.
194. Tanase, S.; Marques-Gallego, P.; Bouwman, E.; Hage, R.; Long, G. J.; Shahin, A.; de Gelder, R.; Mutikainen, I.; Turpeinen, U.; Reedijk, J., manuscript in preparation.
195. Tung, H. C.; Kang, C.; Sawyer, D. T. *J. Am. Chem. Soc.* **1992**, *114*, 3445.
196. Sheu, C. S.; Sawyer, D. T. *J. Am. Chem. Soc.* **1990**, *112*, 8212.
197. Mekmouche, Y.; Menage, S.; Duboc-Toia, C.; Fontecave, M.; Galey, J. B.; Lebrun, C.; Pecaut, J. *Angew. Chem. Int. Ed. Engl.* **1999**, *50*, 949.
198. Balogh-Hergovich, E.; Speier, G.; Reglier, M.; Giorgi, M.; Kuzmann, E.; Vertes, A. *Eur. J. Inorg. Chem.* **2003**, 1735.
199. Okuno, T.; Ito, S.; Ohba, S.; Nishida, Y. *J. Chem. Soc. Dalton Trans.* **1997**, 3547.
200. Fujita, M.; Costas, M.; Que, L. *J. Am. Chem. Soc.* **2003**, *125*, 9912.

DNA OXIDATION BY COPPER AND MANGANESE COMPLEXES

MARGUERITE PITIÉ, CHRISTOPHE BOLDRON and GENEVIÈVE PRATVIEL

Laboratoire de Chimie de Coordination du CNRS, 205 route de Narbonne, 31077 Toulouse Cedex 4, France

I. Introduction

Although DNA proves to be highly resistant to hydrolysis, its sensitivity toward oxidation is now well documented. DNA appears to be the critical target for many oxidation processes due to reactive oxygen species, ionizing radiations, UV light, photosensitizers, or dyes. It is therefore implicated in cancer, aging, and neurological disorders. Oxidative DNA damage is also at the basis of the mechanism of action of some antitumoral drugs and among them iron bleomycin.

Redox-active metal complexes have been known for decades to be able to mediate oxidative degradation of DNA. Because of the large literature on iron complexes, the most famous representatives of which are Fe-bleomycin and Fe-edta, this review will be restricted on copper and manganese complexes. It will focus particularly on $Cu(1,10\text{-phenanthroline})_2$ and the manganese cationic porphyrin, Mn^{III}-bis(aqua)*meso*-tetrakis(4-*N*-methyl-pyridiniumyl)-porphyrin, Mn-TMPyP. Surprisingly these two complexes of relatively simple structure exhibit high reactivity toward DNA and have been used as tools to study the mechanisms of oxidative DNA damage at a molecular level.

ADVANCES IN INORGANIC CHEMISTRY
VOLUME 58 ISSN 0898-8838 / DOI: 10.1016/S0898-8838(05)58003-6

II. Oxidative DNA Cleavage by Copper Complexes

A. Copper Phenanthroline

The DNA cleavage activity of Cu^I complexes of 1,10-phenanthroline (phen, Fig. 1) with H_2O_2 as coreactant was initially observed by Sigman *et al.* in 1979 (*1*). It is probably one of the best-studied DNA oxidizing agent able to mediate principally single-strand cleavage on DNA through sugar oxidation. Double-strand scission of DNA by activated $Cu(phen)_2$ results from two successive single-strand nicks closely located on both complementary strands. Two types of complexes are able to oxidize nucleic acids: $Cu^I(phen)$ and $Cu^I(phen)_2$, but the last species is clearly the more efficient and, for this reason, the determination of double-stranded DNA oxidation mechanisms has been mainly studied with $Cu^I(phen)_2$ (*2–8*).

A.1. Nature of the Active Species

Formation of $Cu^I(phen)_2$ is described by equilibriums summarized in Eqs. (1), (2) in Fig. 2 (*9,10*). Phen binds also Cu^{II} ions with equilibrium constants $\log K_1$ and $\log K_2$ of 8.8 and 6.6, respectively (*10*). Analysis by X-ray diffraction of monocrystals showed that the geometrical environment of the metal ion in $Cu^I(phen)_2$ is tetrahedral (*11*), but the $Cu^{II}(phen)_2$ complex adopts different coordination geometries as trigonal bipyramid or octahedron with one or two additional ligand(s) as H_2O or chloride (*12*). The complex was often

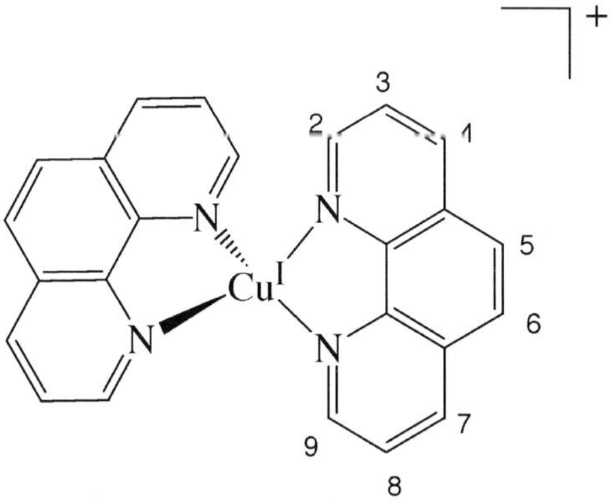

Fig. 1. Cu^I complex of 1,10-phenanthroline: $Cu^I(phen)_2$.

$$Cu^I + phen \rightleftharpoons Cu^I(phen) \qquad \log K_1 = 10.3 \qquad (1)$$

$$Cu^I(phen) + phen \rightleftharpoons Cu^I(phen)_2 \qquad \log K_2 = 5.5 \qquad (2)$$

Fig. 2. Cu^I complexation by phen.

used at micromolar concentration and it results from the low value observed for K_2 that DNA oxidations were generally conducted in the presence of an excess of up to 10 equivalents of phen for one equivalent of copper in order to favor the 2 phen/1 copper stoichiometry. However, free phenanthroline is a weak competitive inhibitor of $Cu^I(phen)_2$ binding to nucleic acids (*13,14*).

Cu^I complexes of phenanthroline are air sensitive and generate reduced oxygen species ("ROS"). Consequently DNA oxidation reactions have been rather studied with stable Cu^{II} complexes of phenanthroline, in the presence of a reductant and air. Series of experiments allowed to determine the reactants involved (*15*): (i) the inhibition of the reaction by catalase showed the role of H_2O_2; (ii) the DNA cleavage reaction proceeds as quickly under anaerobic conditions as under aerobic conditions in the presence of H_2O_2 and $Cu^I(phen)_2$; (iii) in anaerobic conditions, the DNA cleavage reaction by $Cu^I(phen)_2$ is markedly slower in the absence of added H_2O_2; (iv) DNA cleavage reaction by $Cu^{II}(phen)_2$ in aerobic conditions was potentiated with the superoxide generator constituted of xanthine and xanthine oxidase, but no cleavage was observed in the absence of the copper complex; (v) superoxide dismutase altered the rate of the DNA cleavage in a manner which is consistent with superoxide affecting the steady-state levels of $Cu^{II}(phen)_2$ and H_2O_2, but not with $O_2^{\bullet -}$ playing a direct role in the chemistry of the cleavage; (vi) addition of scavengers of $HO^\bullet$ as alcohol does not inhibit cleavage events.

According to these results, succession of redox events has been proposed for the generation of the active species (Fig. 3). The Cu^{II} complex is first reduced to the Cu^I complex [Eq. (3)], which reacts with dioxygen to produce superoxide anion [Eq. (4)]. The Cu^I complex reduced then superoxide anion to H_2O_2 [Eq. (5)]. Finally $Cu^I(phen)_2$ reacts with H_2O_2 and forms the active species that is able to cleave DNA [Eq. (6)]: only these last two reactants are involved during DNA oxidation, the other ones allowing the generation of H_2O_2.

Despite the frequent use of $Cu(phen)_2$ as a cleaving agent, the exact nature of the reactive species involved in this DNA cleavage is still unknown.

$$Cu^{II}(phen)_2 + e^- \longrightarrow Cu^I(phen)_2 \qquad (3)$$

$$Cu^I(phen)_2 + O_2 \rightleftharpoons Cu^{II}(phen)_2 + O_2^{\bullet -} \qquad (4)$$

$$Cu^I(phen)_2 + O_2^{\bullet -} + 2\,H^+ \longrightarrow Cu^{II}(phen)_2 + H_2O_2 \qquad (5)$$

$$Cu^I(phen)_2 + H_2O_2 \longrightarrow \text{reactive species} \qquad (6)$$

FIG. 3. Succession of redox events involved to form a reactive species able to cleave DNA with $Cu^{II}(phen)_2$ in the presence of a reductant and dioxygen.

A popular view is that the hydrophobic cationic $Cu^I(phen)_2$ complex binds to DNA and forms an intermediate that reacts with H_2O_2 to generate a hydroxyl radical ($HO^\bullet$). However, according to reports on the detailed chemistry of the cleavage reaction, freely diffusible hydroxyl radicals are not responsible for the strand scission (*15,16*). It is more likely a copper bound oxidant, such as Cu^{III}-hydroxo (which is equivalent to a hydroxyl radical coordinated to a Cu^{II} ion) or Cu^I-hydroperoxide, but Cu^{III}-oxo species was also proposed (see below for a discussion about these different reactive species). However, Williams *et al.* (*17*) concluded, from cleavage of DNA duplexes including a bulge, that $Cu(phen)_2$ cleaves DNA *via* a freely diffusible intermediate as $HO^\bullet$ that is generated proximal to the site of attack. Thus steric exclusion and close proximity between sugar residues and the copper center could explain that hydroxyl trapping agents could not inhibit the $Cu(phen)_2$ mediated reaction. However, damage to DNA bases by activated $Cu(phen)_2$ is typical of hydroxyl radical-induced products, suggesting that at least a minor reaction pathway involves free hydroxyl radicals (*18*).

The one-electron reduction of the starting complex $Cu^{II}(phen)_2$ can be accomplished by addition of exogenous reducing agents [ascorbate (*13,14,18*), thiols like mercaptopropionic acid (*14,15,19–23*) or 2-mercaptoethanol (*14,18*), NADH (*24*) or by the use of superoxide generators as xanthine/xanthine oxidase (*15*), or cobalt-60 γ-radiolysis (*16*)]. One-electron reductants such as superoxide anion and thiol accomplish directly the reduction of the Cu^{II}-phenanthroline complex, but 2-electron reductants such as NADH are involved in a multistep reaction with superoxide playing a central role (*24*). H_2O_2 can be added exogenously in order to increase the efficiency of the system.

However, reductants (always in large excess in the experiments) used for the activation of the Cu^{II}, e.g. thiols, are susceptible to compete with phen itself for copper coordination site and so decrease the quantity of $Cu^I(phen)_2$ available for DNA cleavage (*14*).

A.2. Interactions with Nucleic Acids

Kinetic studies demonstrated that the nuclease activity of $Cu^{II}(phen)_2$ proceeds by an ordered mechanism: the freely diffusing $Cu^{II}(phen)_2$ is first reduced to the Cu^I complex $Cu^I(phen)_2$ which binds reversibly to DNA. This non-covalently bound Cu^I complex is then oxidized by H_2O_2 to generate the reactive species responsible for DNA strand scission (*19*). The association constant for DNA binding of $Cu^I(phen)_2$ was determined to be $4.7 \times 10^4\ M^{-1}$ (DNA base pairs) (*13*). However, $Cu^{II}(phen)_2$ also binds to DNA, but the direct reduction of the complex $Cu^{II}(phen)_2$---DNA by $O_2^{\bullet -}$ is very slow and can be neglected (*25*).

During footprint experiments, the protection of a zone of cleavage by netropsin (a minor groove binder) or the lack of cleavage protection by EcoRI (which interacts with the major groove of the DNA) suggest a minor groove interaction of $Cu(phen)_2$ (*21*). The reaction mechanisms involving the hydrogen atoms at C1′, C4′, and C5′ of the deoxyriboses (which are located within the minor groove) as the main targets of the oxidative attack also support this positioning of the cleaver.

Initial observation of the inhibition of cleavage by intercalators suggested that the cleavage reaction might proceed through an intercalative complex (*15*). DNA binding by intercalation was also strongly suggested by: (i) DNA-induced changes in the $Cu^{I}(phen)_2$ visible absorption spectrum and the induction of a strong $Cu^{I}(phen)_2$ circular dichroism spectrum; (ii) significant increase in DNA solution viscosity upon introduction of $Cu^{I}(phen)_2$ and (iii) a binding-site size consistent with the neighbor-exclusion parameter $n \approx 2$ base pairs (*13*).

$Cu^{I}(phen)_2$ is precluded by its tetrahedral geometry from full intercalation, but a modelization of its interaction with double-stranded DNA has been performed from crystallographic values of $[Cu^{I}(phen)_2]ClO_4$. In this plausible model, one phen ligand partially intercalates at a TA step and the angle of 50° between the two phen entities of the complex allows the other phen to favorably interact in the minor groove of the DNA (*3,13,14,26–28*). Since $Cu(phen)_2$ attacks DNA from a binding site within the minor groove, the geometry of this structural domain must govern the cleavage specificity by the coordination complex.

The DNA cleavage activity of $Cu(phen)_2$ is sequence-dependent, but not nucleotide specific (*26,29*). The unsubstituted aromatic phenanthroline ligands of $Cu(phen)_2$ preclude hydrogen-bonding interactions with nucleobases as determinants of specificity. The only observed selectivity is related to a preference for minor groove binding at 5′-TAT triplets, which were cleaved intensively at the deoxyadenosine sugar ring. The related sequences TGT, TAAT, TAGPyr, and CAGT were less preferred while CAT and TAC triplets, polypurine and polypyrimidine sequences were cleaved at a lower frequency (*26*). The $Cu(phen)_2$ preference for TAT may be related, in part, to sequence-dependent variations of the B conformation of DNA. Stacking interactions are favored with a "propeller-twisted" base pair. Changes in roll and twist angles open the minor groove of DNA at the TA step and may facilitate the helix penetration by a phen ring in either a partial intercalation or groove-binding mode.

Mutational changes on double-stranded DNA can generate additional sites detectable by the coordination complex when, for example, DNase I does not detect substantial changes accompanying these base substitutions (*30*).

On another way, single-stranded DNA is a poor substrate (*15,31*) and the $Cu(phen)_2/H_2O_2$ system cleaves A, B, and Z forms of DNA at different rates (*32*). The B structure is the most easily cleaved DNA. A DNA is less efficiently cleaved presumably because of fewer favorable contacts between the complex and the widened minor groove of the A form double helix. The A structure, formed by RNA–DNA hybrids, is cleaved on both strands at roughly one-third of the rate for B DNA under comparable conditions. In contrast, the left-handed Z structure, with its deep narrow minor groove, is completely resistant to $Cu(phen)_2$ degradation.

RNA is also a substrate for the nuclease activity of $Cu(phen)_2$. The copper complex shows a preference for single-stranded loops relative to double-stranded A-structures present in RNA, but the detailed mechanism of cleavage is still unknown (*33*).

A.3. DNA Oxidation Mechanisms

As expected for a minor groove binder, C–H bonds at C1′, C4′, and C5′ of deoxyriboses, three sugar C–H bonds accessible from the minor groove, are the main targets of activated $Cu^I(phen)_2$ (*4,5,7*).

Oxidation of polydA-dT allowed first the characterization of free nucleobases release and, (*20*) after a heating step or by freezing, an oxidized sugar residue: 5-methylene furanone (5-MF), specific of C1′ oxidation was observed by HPLC and GC/MS (*34*). Using ^{18}O-enriched H_2O_2 and water, the carbonyl oxygen of 5-MF was found to derive from H_2O (*35*). Control experiments in the presence of $Cu^I(phen)_2$, H_2O_2, and a reductant under anaerobic conditions showed that O_2 is not involved in the reaction. 3′- and 5′-phosphate ends were characterized enzymatically and confirmed by PAGE analysis of cleavage products of 5′- or 3′-^{32}P-labeled double-stranded oligonucleotides or restriction DNA fragments (*21*). During these PAGE analysis, direct strand-cleavages were observed and a metastable intermediate at the 3′-end was trapped which was proposed to be an α,β-unsaturated lactone intermediate.

To accommodate these observations, the mechanism summarized in Fig. 4 was proposed. Activated $Cu(phen)_2$ abstracts H′1 of deoxyribose then the initial 1′-deoxyribosyl radical **1** is converted to a carbocation **2** by one-electron oxidation probably mediated by the copper complex. Then nucleophilic attack by water at C1′ releases the attached nucleobase allowing the formation of a deoxyribonolactone abasic site **3**. A first β-elimination liberates a 5′-phosphate end (the single-strand break event) and a metastable α,β-unsaturated lactone intermediate **4** still attached at the 3′-end of the DNA strand. 5-MF **5** is released from its precursor by a thermal step (90°C for 1 h) or a piperidine treatment necessary to facilitate the second β-elimination allowing the recovery of the 3′-phosphorylated-ending DNA fragments (*21,34*). A key feature of this pathway is that a strand break is observed at room temperature; heat and alkaline treatment are not necessary for strand cleavage although a half-life of 32–54 h has been estimated for a deoxyribonolactone abasic site in duplex DNA under physiological conditions (*36*). Use of mononucleotide model led Chen and Greenberg to propose that $Cu(phen)_2$ itself catalyses the first β-elimination producing strand breakage (*37*). It was confirmed in the case of an activated Cu(3-Clip-Phen) derivative which performs also direct strand-cleavage from C1′ oxidation of DNA (*38*).

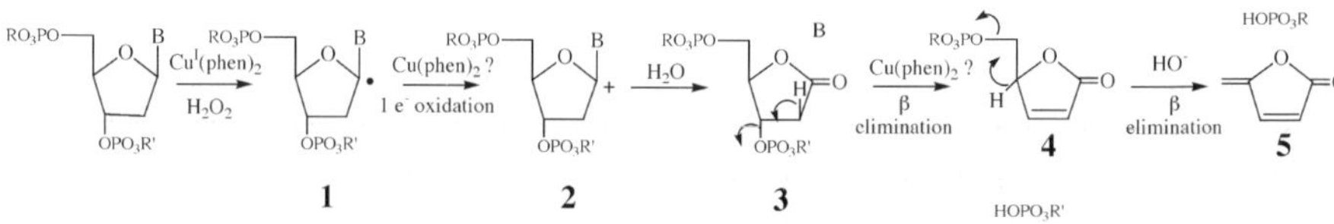

FIG. 4. Mechanism of C1′ oxidation of DNA by activated $Cu(phen)_2$.

Deoxyribonolactone abasic sites were nevertheless observed by Oyushi and Sugiyama during LC/ESI-MS analysis of oxidation products of a hexanucleotide duplex generated by activated $Cu(phen)_2$ (*23*). In this case, the resistance of the oxidized abasic site to the $Cu(phen)_2$ promoted direct cleavage and was associated to the melting of the short duplex after its oxidation since the presence of deoxyribonolactone decreases the melting temperature of duplex DNA (*38,39*).

C1′ oxidation has been observed with many other DNA oxidizing agents such as γ-radiolysis, neocarzinostatin, Mn-TMPyP (*4,5,7*), copper complexes of benzosulfonamides (*40*), Cu-kanamycin (*41*), or for copper complexed to guanine of DNA (*42*), but it was not associated to direct strand-cleavage events except for metal complexes like Cu(3-Clip-Phen) (*43*), or Ru(terpyridine)(bipyridine) complexes (*44*).

Other sugar oxidation mechanisms were less studied. The frequent observation of 3′-phosphoglycolate fragments, migrating faster than 3′-phosphate fragments during PAGE analysis of oxidation products induced on DNA duplex by activated $Cu(phen)_2$, is characteristic of a C4′ oxidation of deoxyribose (*21*). Although no mechanistic study was performed with $Cu(phen)_2$, this C4′ oxidation can be observed with many other agents able to abstract a hydrogen atom on deoxyribose and is particularly well documented in the case of Fe-bleomycin or neocarzinostatin (which, as $Cu(phen)_2$, oxidize DNA in the presence of reductant and O_2), but also for γ-radiolysis that generates $HO^•$. Molecular oxygen is always involved (*4,5,7*). One of the processes shown in Fig. 5 is probably involved in the formation of 3′-phosphoglycolate fragments with activated $Cu(phen)_2$.

In the pathway A, proposed in the case of Fe-bleomycin (*4,5,7,45,46*), the metal complex abstracts H4′ on deoxyribose to form an initial C4′ radical **6** that combines with dioxygen to form a peroxyl radical **7** which is then reduced to hydroperoxide **8** (since the reaction medium in which 3′-phosphoglycolate fragments were observed contains reductant). Then the peroxide **8** undergoes a Criegee rearrangement resulting in oxygen insertion between the C4′ and C3′ carbon of the target nucleotide and the formation of a C4′ cation **9**. Criegee rearrangement is generally an acid catalyzed mechanism, but it has been proposed in the case of Fe-bleomycin that the metal complex can act as a Lewis acid (*4*). The intermediate **9** formed in the above reaction promotes the specific elimination of the 2′-pro-R-proton and the C1′-O bond rupture giving rise to a new intermediate **10**. Subsequently, DNA strand scission occurs with the release of a new DNA terminus bearing a 3′-phosphoglycolate moiety **11**. The remaining nucleotide fragment **12** connected to the new 5′-terminus formed at the site of strand scission is then hydrolyzed through the addition of solvent to the 3′-carbon leading to the release of a base propenal **13** and a new 5′-phosphorylated DNA terminus.

The pathway B begins as the pathway A, but the C4′ radical **6** breaks the secondary C–O bond of the phosphate linkage at the 3′ position in a heterolytic reaction. This releases a 5′-phosphorylated DNA fragment and a radical cation **14**. This radical cation reacts with H_2O to generate a new 4′-DNA radical **15**, which forms a hydroperoxide **16** by reaction with O_2. Then the

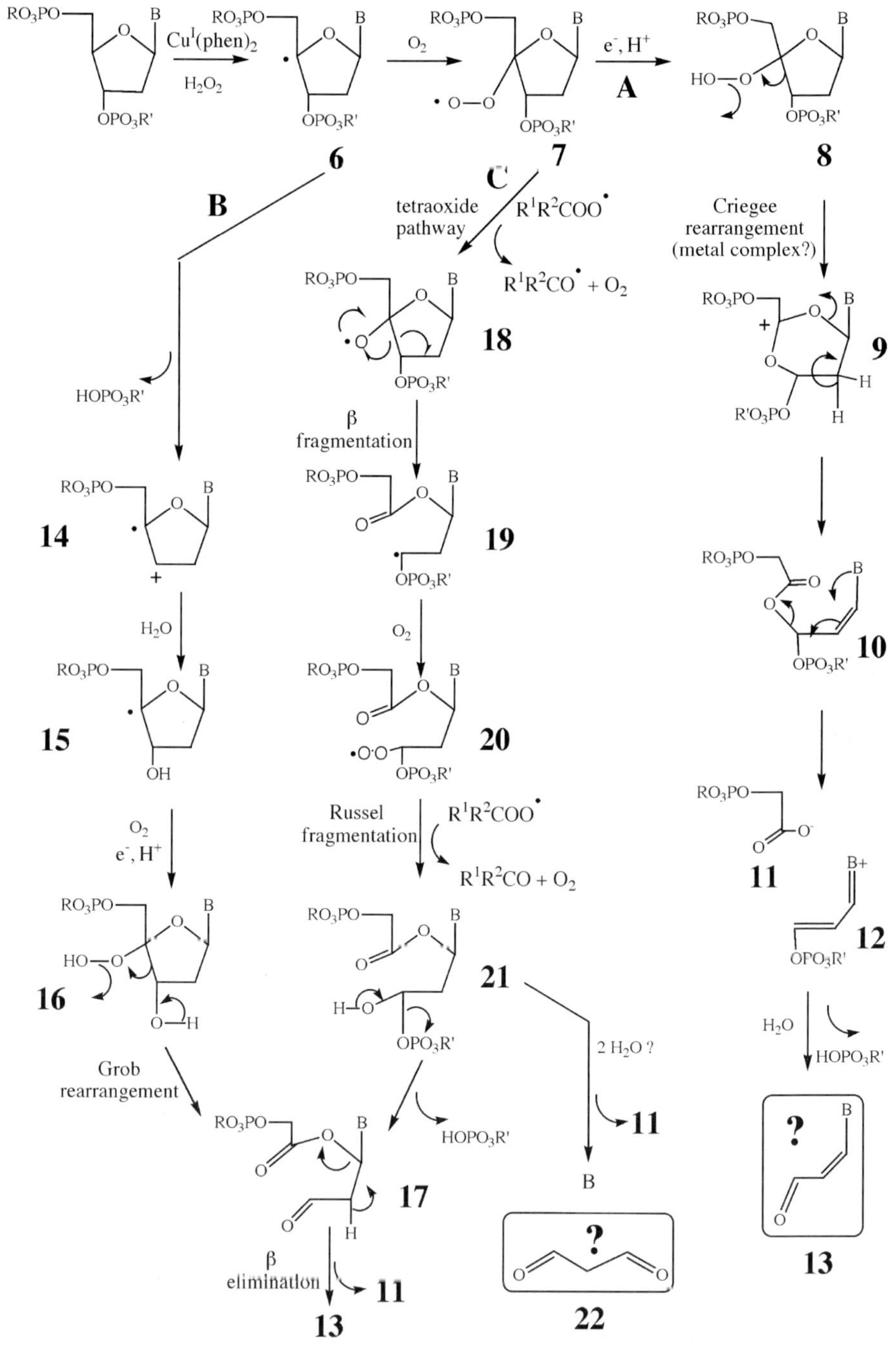

FIG. 5. Propositions of mechanisms of C4′ oxidation of DNA by activated $Cu(phen)_2$ producing 3′-phosphoglycolate fragments **11**.

fragments for sequencing, cloning, or chromosomal mapping (*6,8*). Conjugates with an intercalator (*77*) or a minor groove binder like Hoechst 33258, or different peptides have also been prepared (*67*).

All these conjugates contain only one phen, so the coordination sphere of the chelated copper ion is probably completed by unknown ligands (water molecules, basic sites of DNA, or an excess of the conjugated molecule?).

B. Copper Clip-Phen

B.1. Design of Clip-Phen

A solution to favor the 2 phen/Cu stoichiometry, even in conjugates, is to link the two phen entities by a bridge. Unfortunately, first attempts, reported by Hélène *et al.* who prepared two phen entities linked by a very long 15 atoms tether on the C5 of the phen cycle (*78*), or by Ogawa *et al.* who synthesized the rigid SMC ligand with two phen doubly bridged by two sulfur atoms on the C2 of phen (*79*), failed to increase DNA cleavage activity of their copper complexes. The Clip-Phen strategy was more successful (Fig. 9). It consists of the covalent linkage of two phen according to different criteria: (i) the bridge between the two phen must be fixed on the C2 or C3 of phen, near the nitrogen atoms chelating copper, in order to be of smaller size; (ii) the length of this bridge must be sufficient to allow the flexibility necessary for the conformational changes occurring during the reduction of Cu^{II} (a trigonal bipyramid or octahedron complex in the case of $Cu^{II}(phen)_2$) to Cu^{I} (a tetrahedral complex for $Cu^{I}(phen)_2$); (iii) this bridge must be functionalized in order to facilitate the attachment of different vectors without modification of the phenanthroline skeleton and such functionalization of the bridge must be possible without inhibiting the ability of the phenanthroline groups to chelate copper or to interact with nucleic acids; (iv) phen must be monofunctionalized in order to limit steric constraints induced by the presence of substituents which might hamper the interaction with DNA; (v) the C9 of phen must be without functionalization to avoid a "neocuproine-like" ligand which does not form a redox active 2:1 complex due to steric clash between

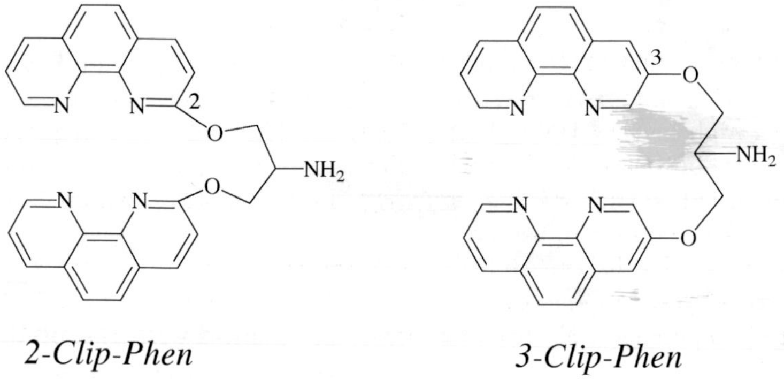

Fig. 9. Structure of 2- and 3-Clip-Phen.

the *ortho* methyl groups (which inhibits redox cycling through the square planar Cu^{II} complex) (*3,80*); (vi) the new ligand must be as simple as possible and easy to prepare in large quantity if needed, so therefore must be symmetric.

The first bridge tested was serinol and allows the preparation of 2- and 3-Clip-Phen where phen are bridged between their C2 (*81*), and C3 (*82*), respectively. After their metallation by one equivalent of Cu^{II}, their ability to cleave ΦX174 supercoiled DNA in the presence of reductant (mercaptopropionic acid or ascorbate) and air was compared to Cu(phen) and $Cu(phen)_2$. All the activated complexes performed single-strand cleavages on DNA, but DNA cleavage activity was increased by a factor 2 for Cu(2-Clip-Phen) and by a factor 60 for Cu(3-Clip-Phen) when compared to $Cu(phen)_2$. Moreover, Cu(3-Clip-Phen) was active at nanomolar concentration and gave the same number of single-strand breaks when metallated by 1 equivalent of copper salt (1 Cu for 2 phen entities, corresponding to a Cu(3-Clip-Phen) complex) or in the presence of 2 equivalents of copper salt (1 Cu per phen entity, corresponding to a putative Cu_2(3-Clip-Phen) compound) as expected for a ligand able to chelate the same copper ion with its two phen subunits.

Quantifications of the number of single-strand cleavages induced by different concentrations of complexes were in accordance with systems having no or poor DNA sequence selectivity as it was confirmed from the analysis by PAGE of oligonucleotides cleaved by activated Cu(2-Clip-Phen) or Cu(3-Clip-Phen). Analysis of DNA oxidation products showed that Cu(3-Clip-Phen) interacts with the minor groove of the double-stranded DNA (*43*). As observed for $Cu(phen)_2$, addition of H_2O_2 to a sample containing Cu(Clip-Phen) and a reductant increased DNA cleavage meaning that their mode of activation is probably similar (*83*).

In order to optimize further their DNA cleavage activity, two series of new Clip-Phen ligands including different bridges have been synthesized (*84*). For a same bridge, activated copper complexes of 3-Clip-Phen derivatives were always more active than their isomers with a junction at C2 of phen. Differences between 2- and 3-Clip-Phen series also clearly appeared during electrochemical analyses of the different copper complexes that suggested that the lower DNA cleavage efficiency of copper complexes of 2-Clip-Phen derivatives could be correlated to their higher stabilization of Cu^{I} species. Moreover, copper complexes with a bridge containing three methylene units appeared to give the best results in DNA cleavage experiments. Thus serinol, chosen for reasons previously mentioned, appears to have an optimal length to favor a high level of DNA cleavage activity. This activity was increased in the case of Cu(3-Clip-Phen) by the presence of the primary amine function included in the serinol bridge, but this amine function, which is probably coordinated to copper in the case of Cu(2-Clip-Phen) as observed on monocrystals, decreased its reactivity toward the DNA.

Unfortunately, copper complexes of Clip-Phen seemed to be able to adopt different geometries and stoichiometries in solution with the predominant presence of monomeric LCu species. Consequently their physico-chemical characterization was found difficult.

B.2. Conjugates of Clip-Phen

Nuclease activity of activated Cu(Clip-Phen) was modulated by their conjugation to different minor groove binders. The conjugates with 3-Clip-Phen were always the more active compared to the 2-Clip-Phen derivatives.

So, conjugates of Cu(Clip-Phen) with the natural polyamine spermine have shown an enhanced nuclease activity after their activation in the presence of a reductant and air (*83*). Interestingly, an analog containing only one phen conjugated to spermine appeared to be poorly active. Therefore, it can be concluded that the high activity of Cu(Clip-Phen-spermine) was not only due to the presence of spermine, which increases the affinity of the Clip-Phen entity for DNA, but was effectively due to the presence of the two bridged phen entities, as expected from the Clip-Phen strategy.

Cu(2-Clip-Phen) or Cu(3-Clip-Phen) conjugates with intercalators such as acridine derivatives have also shown an enhanced nuclease activity that can be modulated by the length of the arm of junction with the intercalator (*85,86*). UV-visible spectral data on the complexes in the presence of calf-thymus DNA were consistent with the expected intercalative mode of binding of the acridine part.

In order to increase the sequence selectivity of DNA cleavage by Cu(Clip-Phen) derivatives on a succession of a minimum of 5 AT base pairs, four conjugates with the distamycin analogue Py-Py-Py-β-Dp have been prepared with phen functionalized on the C2, 2-Clip-Phen, 3-Clip-Phen, or 2-Clip-Phen containing a long tether (*87*). After metallation by $CuCl_2$, the DNA cleavage activities of the different conjugates in the presence of ascorbate and air were compared on a restriction fragment by PAGE. Cleavage patterns showed that the polyamide moiety of the conjugates directs the cleavage activity in the vicinity of AT tracts, but the precise cleavage selectivity of these conjugates was dependent on the type of phenanthroline residue linked to the poly-*N*-methylpyrrole entity. The conjugate with 3-Clip-Phen (Fig. 10) was the more active and cleaved always on the 5′- and 3′-sides of a seven AT base pairs box when the preferential cleavage site for the copper complex of the conjugate with 2-Clip-Phen was inside this seven AT box. As previously observed for spermine conjugates, the "distamycin" conjugate with only one phen was again less active than the one including Clip-Phen and the one with large tether induced more diffuse cleavage sites. No modification of

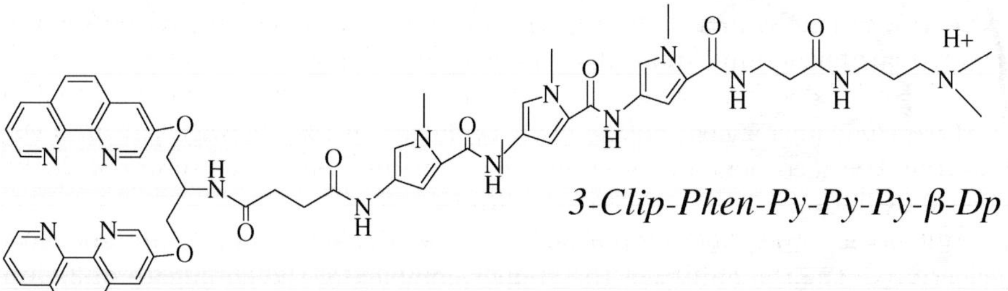

Fig. 10. Structure of 3-Clip-Phen-Py-Py-Py-β-Dp.

the cleavage pattern was observed after piperidine treatment, indicating that these conjugates performed direct DNA breaks, probably due to deoxyribose oxidation, as previously observed for copper complexes of phenanthroline.

B.3. Mechanism of DNA Cleavage

Mechanisms of DNA oxidation by copper complexes of 3-Clip-Phen and its conjugate with a distamycin analogue, in the presence of a reductant and air, were studied (*43*). HPLC characterization of the direct release of nucleobases associated, after a heating step at pH = 8, to the production of 5-MF **5** indicated that these copper complexes oxidized the C1′ position of deoxyribose (Fig. 4). The oxidation at C1′ was the major degradation route. The release, after the heating step, of nucleobases and of the characteristic sugar fragment furfural **27** (issued from two β-elimination steps from the DNA fragment containing a 5′-aldehyde end) showed that the systems performed also C5′ oxidation (Fig. 8). The facility of production of free nucleobases by activated Cu(3-Clip-Phen) was A ≈ C > G ≈ T while calf thymus DNA used for these experiments contains 39 % GC. Remarkably, quantification of released nucleobases showed that the system was catalytic. Comparison of DNA cleavage activities showed also that the conjugation of Cu(3-Clip-Phen) with the distamycin analogue decreased the rate of the reaction.

Digestion of the DNA oxidation products by P1 nuclease and bacterial alkaline phosphatase allowed the characterization of glycolic acid residues coming from 3′-phosphoglycolate fragments **11** and indicating that these copper complexes performed also C4′ oxidation (Fig. 5), but this pathway was not associated with base-propenal release although malondialdehyde (their hydrolysis product) was probably characterized through the chromophore that it forms with thiobarbituric acid. These results are in agreement with the work of von Sonntag *et al.* on γ-radiolysis of DNA (known to produce DNA fragments with 3′-phosphoglycolate ends). No base-propenals **13** were detected although malondialdehyde was observed, indicating that the "glycolate-pathway" does not always lead to the formation of base propenals (*49*).

PAGE analysis of cleavage fragments of a 35-mer double-stranded oligonucleotide confirmed these data and showed that the systems, similarly to activated $Cu(phen)_2$, performed direct strand cleavages. 5′-Phosphate, 3′-phosphate, and 3′-phosphoglycolate ends were observed, but also metastable 3′-end products resulting probably of C1′ oxidation (probably fragments with a 3′-ribonolactone end) that produced a smear on PAGE which was easily transformed in 3′-phosphate ends by a heating step at pH = 8.

In order to detect nucleobase oxidation if any, a heating step in piperidine 1 M was also performed, but no clear evolution of the cleavage patterns was observed in accordance to systems realizing probably preferentially deoxyribose oxidation.

Since activated Cu(3-Clip-Phen) cleaves DNA without clear sequence selectivity, it is the ability of the copper complex of the conjugate of 3-Clip-Phen with the distamycin analogue to produce, after activation, sequence selective DNA cleavage that allowed easier confirmation of these mechanisms of DNA oxidation by PAGE. Scissions on both strands were shifted by two base pairs toward the 3′-ends confirming that the chemical nuclease oxidized sugar residues within the minor groove (an attack in the major groove would have given a 5′-shift) (*88*). Remarkably, the evolution of a 5′-aldehyde-fragment **25** was followed by its reduction with $NaBH_4$ or its transformation in a 5′-phosphate fragment with alkaline heating.

Kinetic comparison with a DNA duplex including a deoxyribonolactone abasic site **3** at the specific cleavage site allowed Greenberg *et al.* to show that the system performs direct strand breakage from C1′ oxidation as proposed for activated $Cu(phen)_2$ (*38*).

Clean cleavage patterns observed were also in accordance with the fact that the copper complex forms a non-diffusible active species responsible of the oxidation of duplex DNA.

C. Generalities upon Copper Complexes Able to Cleave DNA

Many copper complexes are known to cleave DNA through hydrolysis and/or to oxidize it through photooxidation and/or through oxido/reduction chemistry. This last point is probably the major phenomenon and is the scope of the present survey. However, it is important to notice that excellent papers and reviews discuss the other two modes of DNA degradation by copper complexes (*89–91*).

A non-exhaustive list of copper complexes or copper systems able to oxidize DNA, through oxido/reduction chemistry, can be proposed in order to show their large diversity. It includes (i) inorganic copper salts (*19,42,59,92*); (ii) synthetic ligands as aromatic nitrogenous ligands [phenanthroline (*3,84,93*), polypyridine (*94–96*)], aromatic sulfonamide (*40,97*), and their derivatives, salens (*98–100*), NSO Schiff bases (*101*), thiosemicarbazones (*102*), polyphenols (*103*), orthoquinones (*104*), peptides (*105–108*), but also aliphatic ligands as polyamines (*109*), or aminoacids (*110*); (iii) antibiotic or antitumoral natural ligands as polypyrrole (bilirubine, prodigiosine, tambjanine) (*111–114*), aminoglycoside (kanamycin) (*41,115,116*), anthracycline (doxorubicin) (*117*), peptidic derivatives (bleomycin) (*45,46,118*), polyphenols (catechin, resorsinol) (*119–121*) or methoxyphenols (syringaresinol, syringaldehyde) (*119*), aminoquinone (mitomycine C) (*122*); (iv) pollutants and their metabolites as benz[a]anthracene for which it can be proposed that the nucleic acids oxidation in the presence of copper could be involved in their toxicity (*123,124*).

Although copper complexes are currently used in order to oxidize and cleave DNA, DNA oxidation mechanisms have been studied for only a minor part of them.

Cu^{II}/Cu^{I} redox potential couples were generally observed between +0.2 and +0.8 V (*125,126*). These values are easily accessible in biological conditions and allow, as summarized in Fig. 11, Cu^{I} species, derived from the reduction of stable Cu^{II} species [Eq. (7)], to induce the formation of reactive oxygen species from dioxygen. Superoxide anion ($O_2^{\bullet-}$) results of an electron transfer between O_2 and Cu^{I} [Eq. (8)]. H_2O_2 is formed by reduction of $O_2^{\bullet-}$ by the Cu^{I} complex [Eq. (9)] or by dismutation of $O_2^{\bullet-}$ [Eq. (10)]. A Fenton

$$Cu^{II} + e^- \rightleftharpoons Cu^I \quad (7)$$

$$Cu^I + O_2 \rightleftharpoons Cu^{II} + O_2^- \quad (8)$$

$$Cu^I + O_2^{\cdot -} + 2H^+ \longrightarrow Cu^{II} + H_2O_2 \quad (9)$$

$$2O_2^{\cdot -} + 2\,H^+ \longrightarrow H_2O_2 + O_2 \quad (10)$$

$$Cu^I + H_2O_2 \longrightarrow Cu^{II} + OH^- + HO^{\cdot} \quad (11)\ \text{Fenton}$$

$$H_2O_2 + O_2^{\cdot -} \longrightarrow OH^- + HO^{\cdot} + O_2 \quad (12)\ \text{Haber–Weiss}$$

$$RH + HO^{\cdot} \longrightarrow R^{\cdot} + H_2O \quad (13)$$

FIG. 11. Redox events involved in the formation of diffusible reactive oxygen species by Cu^{II} complexes in the presence of reductant and dioxygen. DNA radical formation by $HO^{\bullet}$.

reaction Eq. (11), or a Haber–Weiss reaction [Eq. (12)] produce hydroxyl radical ($HO^{\bullet}$) (*127,128*). This diffusible species is in fact the most reactive of the three reduced species of dioxygen. The hydroxyl radical is able to perform oxidative damage on DNA as hydrogen atom abstraction [Eq. (13)] that generates a radical on the nucleic acid like those involved in the initial events of deoxyribose oxidation. Details on DNA degradation products were given in the previous paragraphs.

Non-diffusible active species generated by Cu^I complexes in the presence of H_2O_2 have been also proposed to oxidize DNA. It can be an encaged hydroxyl radical: formally a hydroxyl radical generated in a confined zone in the close vicinity of the DNA target as summarized in Fig. 12, Eq. (14) (*3,16,129*). It can abstract a hydrogen atom from DNA conducting to a radical [Eq. (15)]. This radical could then eventually be oxidized to a cation by an electron transfer to the Cu^{II}-hydroxo species [Eq. (16)] that regenerates the Cu^I species.

These proposed non-diffusible active species generated by Cu^I complex in the presence of H_2O_2 can also be centered on the metal.

It can be a Cu^I-hydroperoxo species formed as in Eq. (17) of Fig. 13. A homolysis of the O–O bound could allow the abstraction of a hydrogen atom from DNA that gives rise to a DNA radical and a new copper complex (Cu^I–$O^{\bullet}$ or its canonical form Cu^{II}=O), Eq. (18). This last intermediate may

$$Cu^{I} + H_2O_2 \longrightarrow \underset{\text{encaged species}}{\{Cu^{II}\text{-OH} + HO^{\cdot}\}} \quad (14)$$

$$RH + \{Cu^{II}\text{-OH} + HO^{\cdot}\} \longrightarrow R^{\cdot} + H_2O + Cu^{II}\text{-OH} \quad (15)$$

$$R^{\cdot} + Cu^{II}\text{-OH} \xrightarrow{e^{-}\ \text{transfer}} R^{+} + Cu^{I}\text{-OH} \xrightarrow{H^{+}} Cu^{I} + H_2O \quad (16)$$

FIG. 12. Proposed redox events for the oxidation of DNA deoxyribose by an "encaged" hydroxyl radical, a proposed non-diffusible active species formed in the presence of a Cu^{I} complex and H_2O_2.

$$Cu^{I} + H_2O_2 \longrightarrow Cu^{I}\text{-OOH} + H^{+} \quad (17)$$

$$RH + Cu^{I}\text{-OOH} \xrightarrow[\text{homolysis}]{\text{O-O}} R^{\cdot} + H_2O + Cu^{I}\text{-O}^{\cdot}\ (\text{or } Cu^{II}{=}O) \quad (18)$$

$$R^{\cdot} + Cu^{I}\text{-O}^{\cdot} \xrightarrow[e^{-}\ \text{transfer}]{H^{+}} R^{+} + Cu^{I}\text{-OH} \xrightarrow{H^{+}} Cu^{I} + H_2O \quad (19)$$

FIG. 13. Proposed redox events for the oxidation of DNA deoxyribose by a Cu^{I}-hydroperoxo non-diffusible active species formed in the presence of a Cu^{I} complex and H_2O_2.

$$Cu^{I} + H_2O_2 \longrightarrow Cu^{III}\text{-OH} + OH^{-} \quad (20)$$

$$RH + Cu^{III}\text{-OH} \longrightarrow R^{\cdot} + H_2O + Cu^{II} \quad (21)$$

$$R^{\cdot} + Cu^{II} \xrightarrow{e^{-}\ \text{transfer}} R^{+} + Cu^{I} \quad (22)$$

FIG. 14. Proposed redox events for the oxidation of DNA deoxyribose by a Cu^{III}-hydroxo non-diffusible active species formed in the presence of a Cu^{I} complex and H_2O_2.

eventually oxidize the DNA radical through an electron transfer to form a DNA cation and a Cu^{I} species as in Eq. (19) (*59,127,130–133*).

A putative Cu^{III}-hydroxo species, generated as in Eq. (20) of Fig. 14, could also be able to oxidize DNA as in Eqs. (21,22), since this trivalent state of copper has been observed when it is stabilized by a ligand having strong σ donor effects as hydroxyl or deprotonated amides: it was involved in biological processes like in the case of the galactose oxidase reaction (*134–136*). In this pathway, the Cu^{III}-hydroxo species abstracts a hydrogen atom on DNA to form a radical and a Cu^{II} complex [Eq. (21)]. Then an electron transfer

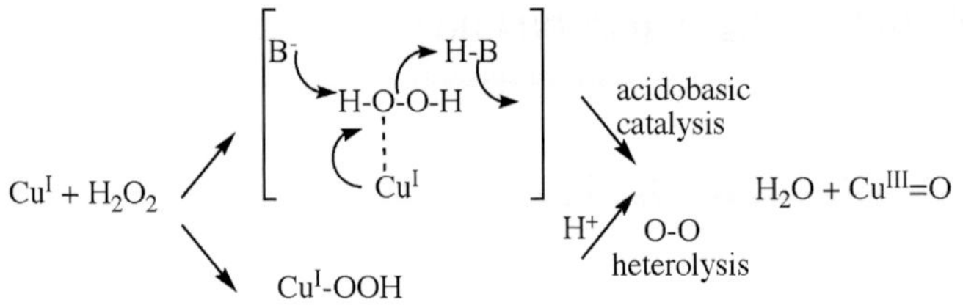

(23)

$$RH + Cu^{III}{=}O \longrightarrow R^{\cdot} + Cu^{II}\text{-}OH \quad (24)$$

$$R^{\cdot} + Cu^{II}\text{-}OH \xrightarrow[e^- \text{ transfer}]{H^+} R^+ + Cu^I + H_2O \quad (25)$$

FIG. 15. Proposed redox events for the oxidation of DNA deoxyribose by a Cu^{III}-oxo non-diffusible active species formed in the presence of a Cu^I complex and H_2O_2.

from the radical to the Cu^{II} complex may eventually oxidize the radical to form a DNA cation and regenerate the Cu^I complex [Eq. (22)].

A putative Cu^{III}-oxo species, formed from an acido-basic catalysis or from the heterolysis of the O–O bound of a Cu^I-hydroperoxo intermediate as in Eq. (23) of Fig. 15, was also proposed although its orbital populating seems unfavorable (*137*). Eqs. (24), (25) summarize the pathway that could be involved with this species during DNA oxidation events. A hydrogen atom abstraction on the DNA by the Cu^{III}-oxo species would produce a radical on DNA and a Cu^{II}-hydroxo species [Eq. (24)]. Then an eventual electron transfer between them may allow the oxidation of the radical to a cation and the regeneration of the initial Cu^I complex.

Other non-diffusible active species centered on the metal and generated by Cu^I complexes in the presence of dioxygen can be also proposed.

Thus a non-diffusible species such as Cu^{II}-hydroperoxo species can be generated as in Eq. (26) in Fig. 16 by Cu^I complexes in the presence of O_2 and an electron (by first fixation of O_2 on the metal center then reduction or by reaction with a beforehand-formed superoxide anion). This intermediate

$$Cu^I + e^- + O_2 \ (\text{or } Cu^I + O_2^{\cdot -}) \longrightarrow Cu^{II}\text{-}OO^- \xrightarrow{H^+} Cu^{II}\text{-}OOH \quad (26)$$

$$RH + Cu^{II}\text{-}OOH \xrightarrow[\text{homolysis}]{\text{O-O}} R^{\cdot} + H_2O + Cu^{II}\text{-}O^{\cdot} \ (\text{or } Cu^{III}{=}O) \quad (27)$$

$$R^{\cdot} + Cu^{II}\text{-}O^{\cdot} \xrightarrow[e^- \text{ transfer}]{H^+} R^+ + Cu^{II}\text{-}OH \xrightarrow{H^+} Cu^{II} + H_2O \quad (28)$$

FIG. 16. Proposed redox events for the oxidation of DNA deoxyribose by a Cu^{II}-hydroperoxo non-diffusible active species formed in the presence of a Cu^I complex and dioxygen.

(a) "Cu^{II}-O-O-Cu^{II} pathway" of dicuprous complex:

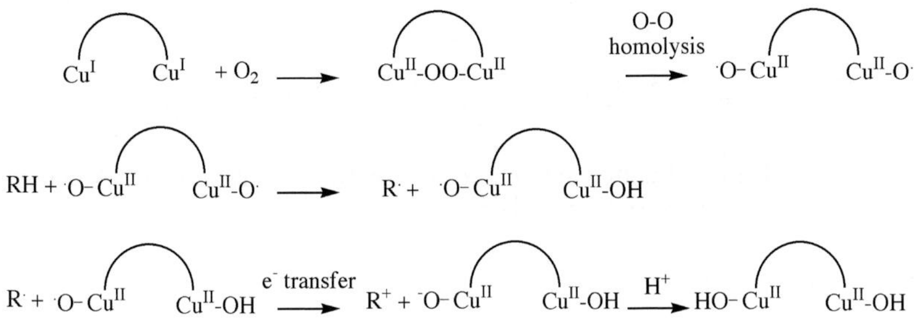

(29)

(30)

(31)

(b) Initiation of the "Cu^{II}-hydroperoxo pathway" for dicuprous complex:

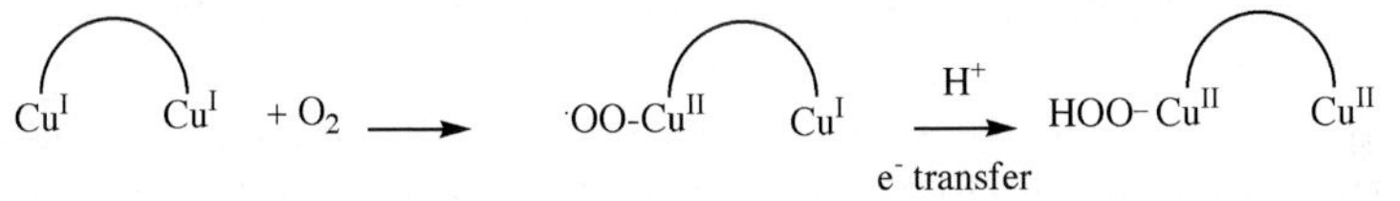

(32)

FIG. 17. Proposed redox events involving non-diffusible active species for the oxidation of DNA deoxyribose by a dinuclear Cu^{I} complex in the presence of dioxygen. (a) The "Cu^{II}–O–O–Cu^{II}" pathway. (b) Initiation of the "Cu^{II}-hydroperoxo" pathway. The Cu^{II}-hydroperoxo species can then react as in Eqs. (27), (28) in Fig. 16.

was proposed for the hydrogen atom abstraction catalyzed by copper enzymes (as dopamine β-monoxygenase or peptidylglycine monoxygenase) (*138–141*). Cu^{II}-hydroperoxo could oxidize DNA as in Eqs. (27), (28). A O–O homolysis of the Cu^{II}–OOH allows the hydrogen atom abstraction [Eq. (27)] that forms a DNA radical and a new copper complex (Cu^{II}–$O^{\bullet}$ or its canonical form Cu^{III}=O) which might be able to oxidize the DNA radical to a cation by electron transfer [Eq. (28)], this last reaction releasing a Cu^{II} complex.

Dinuclear Cu^{I} complexes can also generate Cu^{II}–O–O–Cu^{II} species in the presence of dioxygen. This peroxy-bridged complex can be able to perform oxidation reactions through O–O homolysis (*142*) generating two Cu^{II}–$O^{\bullet}$ reactive centers as in Eq. (29) in Fig. 17. One of the center may abstract a hydrogen atom from DNA, Eq. (30) and thus generate a radical that may be oxidized into a cation through an electron transfer to the other Cu^{II}–$O^{\bullet}$ center.

However, dinuclear Cu^{I} complexes can also generate Cu^{II}-hydroperoxo species in the presence of dioxygen [Eq. (32)] which can then react with DNA as in Eqs. (27), (28) in Fig. 16 (*94–96,138,139*).

In another way, Cu^{II} complexes can also react with H_2O_2. The main reactive species proposed to explain the DNA oxidation by Cu^{II} complexes in the presence of H_2O_2 is Cu^{II}-hydroperoxo species formed as Eq. (33) in Fig. 18 (*138,143*), but the formation of Cu^{III}–OH or Cu^{III} species associated to the release of $HO^{\bullet}$ as in Eq. (34) has been also proposed (*127,136,144,145*).

$$Cu^{II} + H_2O_2 \longrightarrow Cu^{II}\text{-}OOH + H^+ \quad (33)$$

$$Cu^{II} + H_2O_2 \longrightarrow Cu^{III} + HO^- + HO^{\cdot} \quad (\text{or } Cu^{III}\text{-}OH + HO^{\cdot}) \quad (34)$$

FIG. 18. Proposed non-diffusible active species for the oxidation of DNA deoxyribose by a Cu^{II} complex in the presence of H_2O_2.

The different pathways involving non-diffusible active species like "encaged $HO^{\bullet}$" species [Eqs. (15), (16)], Cu^{I}-hydroperoxo [Eqs. (18), (19)], Cu^{III}-hydroxo [Eqs. (22), (23)], Cu^{III}-oxo [Eqs. (24), (25)], Cu^{II}-hydroperoxo [Eqs. (27), (28)], or dinuclear copper complexes [Eqs. (30), (31)] are working hypotheses. Our mechanism proposals were done in accordance to hydrogen abstractions, but also electron transfers observed during the deoxyribose oxidation of double-stranded DNA (Figs 4, 5, 7, and 8). The mechanisms of DNA cleavage showed that the active species abstracts a hydrogen atom from deoxyribose to generate, on the DNA sugar, a radical that reacts then with O_2 or a carbocation that reacts with H_2O. The radical results of a one-electron oxidation event. The formation of the carbocation is equivalent to a two-electron oxidation of the substrate which can be done in two steps with the active species proposed above. However, when the activated species can only remove a hydrogen atom to give rise to a carbon centered radical, it must be proposed that a second equivalent of these active species, dioxygen or another oxidant present in the medium, should be involved during the electron transfer reaction oxidizing this radical to a carbocation.

Some of these activated species like $HO^{\bullet}$, Cu^{I}-hydroperoxo, or Cu^{III}-hydroxo have been also proposed in the case of the oxidations of the DNA nucleobases (*55*). Various mechanisms like $HO^{\bullet}$ addition on a double-bound, hydrogen abstraction on the methyl groups or electron transfer induce nucleobases oxidations and copper complexes are oxidant enough to perform them, but, in the presence of excess of reductants, such as in the conditions often used during DNA oxidation by copper complexes, oxidized nucleobases (base radicals and radical cations) may be reduced back to undamaged species. Thus the ability of copper complexes to oxidize nucleobases could be underestimated.

The nature of the active species involved during DNA oxidation depends on copper ligands that more or less stabilize the different activated species proposed (*134–146*). The ability of some of these copper complexes to undergo autoactivation (without any added reductant) was also related to the nature of the ligand (*100,111,112*). The ligand allows also (or no) a more or less selective interaction with double-stranded DNA and thus modulates the selectivity of the oxidation events observed.

Different experiments were generally used to analyze the involvement of reactive oxygen species during DNA oxidation by copper complexes. H_2O_2 necessity can be revealed from the inhibition of the DNA oxidation by catalase (which catalyses the dismutation of H_2O_2 to H_2O and O_2). Decrease of DNA oxidation by the use of superoxide dismutase (which catalyses the dismutation of $O_2^{\bullet-}$ to H_2O_2 and O_2) proves that superoxide anion is involved. No reaction is observed under anaerobic condition if the mechanism involves dioxygen. Diffusible hydroxyl radical can be trapped by alcohols (as ethanol, mannitol, or glycerol) or by DMSO or Rhodamine B. NaN_3 is a classical trap of singlet dioxygen (the reactive form of dioxygen), but this species is more involved during photooxidation events. 5,5′-dimethylpyrolline (DMPO) forms adducts with radicals like $HO^{\bullet}$, $HOO^{\bullet}$, $H^{\bullet}$, or $CH_3^{\bullet}$, which give characteristic signals in EPR analysis.

The contribution of the copper during the DNA degradation can be revealed by the addition of neocuproine or bathocuproine, which trap copper to form non-redox active Cu^{I} complexes unable to oxidize DNA.

III. Oxidative DNA Damage by Manganese Complexes

The few examples of manganese complexes able to damage DNA by oxidative processes came from the intensive work on metal complexes used as catalysts for oxidation reactions like epoxidation of olefins and hydroxylation of alkanes. These metal-complexes were initially designed to mimic the catalytic cycle of heme enzymes, cytochrome P450 (*147*) or peroxidase (*148*). For these enzymes, a high-valent iron-oxo intermediate is known to be the key intermediate in the oxidation catalysis. Although manganese porphyrins are not found in the active sites of biological enzymes they have been shown to have unusually high reactivity toward olefin epoxidation and alkane hydroxylation. The oxidation catalysts were based not only on the porphyrin ligand, which is reminiscent of the natural heme (protoporphyrin-IX), but also on the Schiff base motif, an easily synthesized ligand compared to porphyrins. However, salen complexes were less robust than metalloporphyrins (*147–158*).

Additionally, the mechanistic similarities between bleomycin and cytochrome P450 (*45,159–162*) led to the test of manganese bleomycin for DNA strand scission (*163,164*).

Among the manganese complexes tested for oxidative DNA damage, the cationic manganese porphyrin, Mn^{III}-bis(aqua)*meso*-tetrakis(4-*N*-methylpyridiniumyl)-porphyrin, Mn-TMPyP, (Fig. 19) proved to be the most efficient and will be described in more detail.

A. High-valent Manganese-oxo Species

Mn^{III} complexes are not strong oxidants and must be activated in order to become good oxidation catalysts. To mimic the catalytic cycle of heme enzymes, for which the active iron-oxo species is generated from Fe^{III} and H_2O_2 (peroxidase) or from Fe^{II}, a source of electrons and O_2 (cytochrome P450), the activation of manganese complexes into high-valent

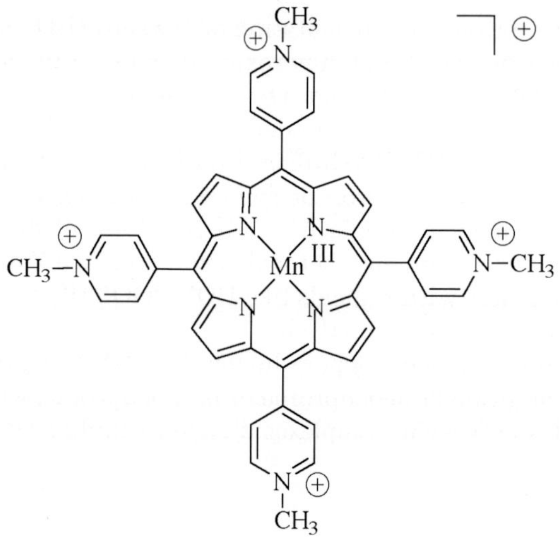

Mn-TMPyP

FIG. 19. Structure of the cationic manganese porphyrin used for oxidative DNA damage: Mn^{III}-bis(aqua)*meso*-tetrakis(4-*N*-methylpyridiniumyl)-porphyrin (Mn-TMPyP). This complex has overall five positive charges. Axial ligands (two water molecules) were omitted for clarity.

manganese-oxo species is performed with an "oxygen atom donor" (Fig. 20A). The coordination of one oxygen atom from a dissymmetric peroxide bond (or an oxygen–halogen bond) on the Mn^{III} followed by the heterolytic cleavage of the oxygen–oxygen (oxygen–halogen) bond leads to the formation of a Mn^{V}-oxo entity. This heterolytic cleavage is crucial and depends on the presence of a good leaving group. Within the enzymatic proteins, the heterolytic cleavage is favored by the protonation of the second oxygen atom. The good leaving group is, in that case, a water molecule (Fig. 20A). Different "oxygen-atom donors" were used for inducing DNA damage: potassium monopersulfate ($KHSO_5$), magnesium monoperoxyphthalate (MMPP), peroxynitrite ($ONOO^-$), and iodosylbenzene (PhIO) (Fig. 20B). For a review on the use of these different activating agents with metalloporphyrins and other metal complexes in oxidation reactions see ref. (*153*). Although PhIO was depicted as a very efficient oxygen atom donor for manganese porphyrins and Schiff bases in catalytic oxidations, it is poorly soluble in water contrary to the three others. Water solubility of the oxygen atom donor under physiological conditions is a key point for the efficiency of the reaction of DNA damage.

The $Mn^{V}{=}O$ entity is a very powerful oxidant able to mediate two different types of oxidative reactions: transfer of an oxygen atom on a substrate (mimic of cytochrome P450, Fig. 21A) or electron abstraction (mimic of peroxidase, Fig. 21B). It is formally written $Mn^{V}{=}O$ since it corresponds to a two-electron oxidation with respect to the initial Mn^{III} complex. However,

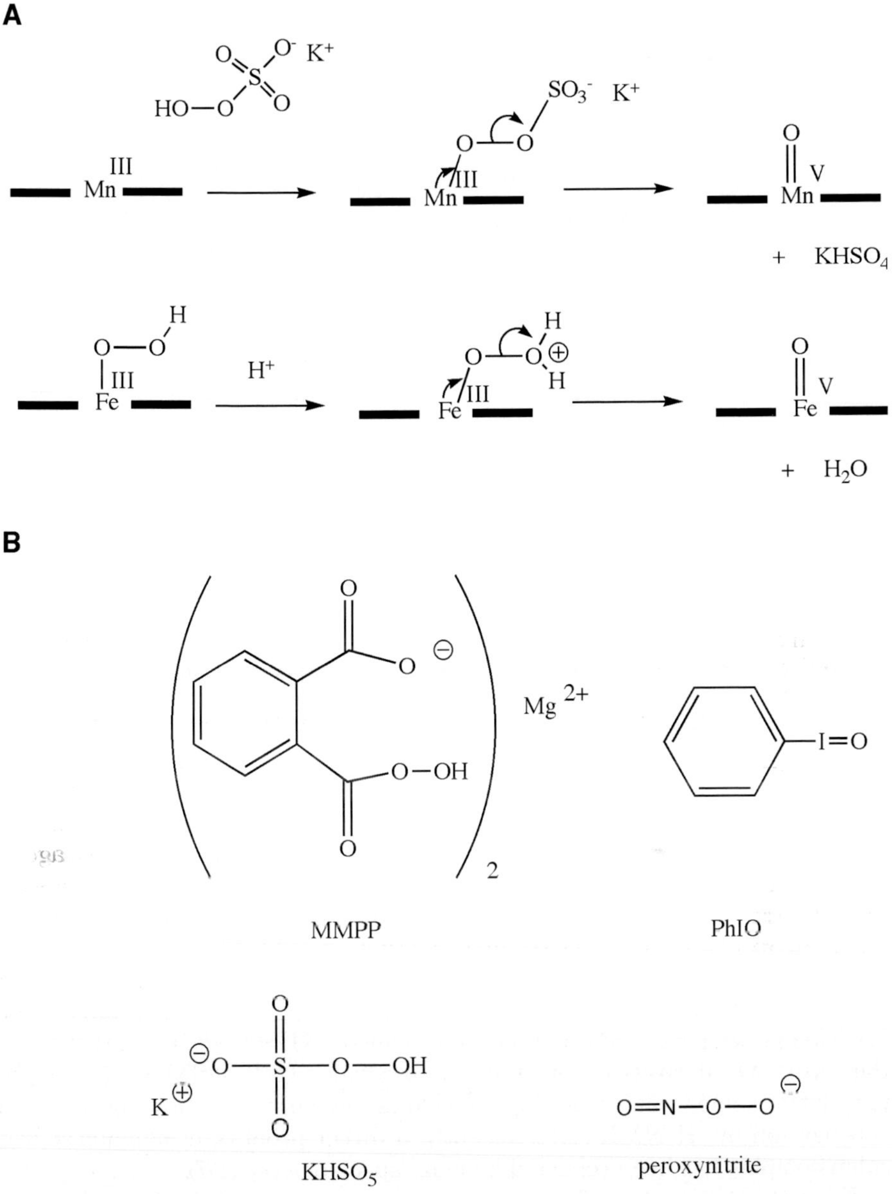

FIG. 20. (A) Activation of a Mn^{III} complex into a high-valent Mn^{V}=O complex: heterolytic cleavage of O–O bond in the presence of an "oxygen atom donor", $KHSO_5$ or during the last step of the formation of the active species of heme enzymes (cytochrome P450 and peroxidases). The porphyrin ligand is simplified. (B) Structure of some classical "oxygen atom donors" used as cofactors for oxidative DNA damage with manganese complexes: magnesium monoperphthalate (MMPP), iodosylbenzene (PhIO), potassium monopersulfate ($KHSO_5$) and peroxinitrite ($ONOO^-$).

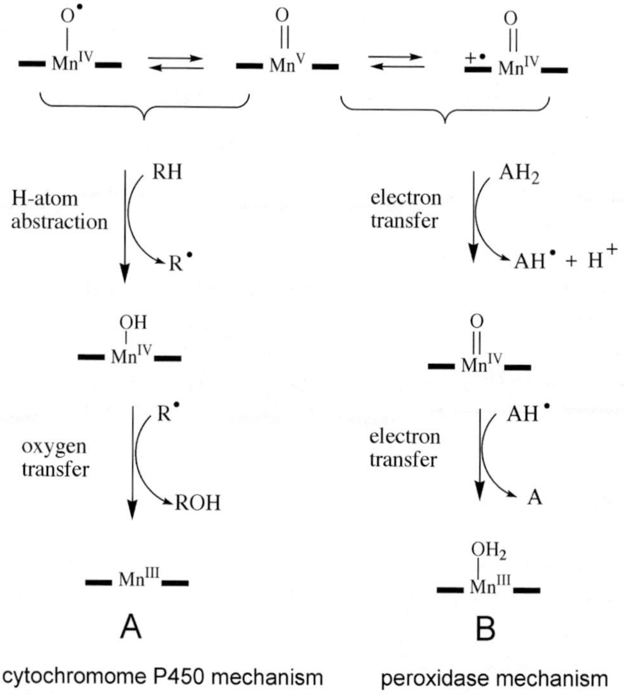

FIG. 21. Mechanism of the oxidative chemistry mediated by high-valent manganese-oxo porphyrin. A, oxygen atom transfer mechanism; B, two-electron abstraction mechanism. The porphyrin ligand is simplified. AH_2 is a two-electron donor.

it may also be considered as a Mn^{IV}-oxo complex with a radical cation located on the porphyrin ring, resembling compound I of heme peroxidases. The cytochrome P450 chemistry may be due to a Mn^{IV}-oxyl form.

The first detection of a Mn^{V}-oxo porphyrin intermediate under catalytic conditions was achieved by using rapid-mixing stopped-flow techniques with Mn-TMPyP and *meta*-chloroperbenzoic acid or $KHSO_5$ (*165*). Furthermore, the $Mn^{V}{=}O$ derivative of the *meso*-tetrakis-(2-*N*-methylpyridiniumyl)-porphyrin isomer was surprisingly stable (a few minutes) allowing its characterization by ^{1}H NMR (*166*). Recently, a direct proof of a manganese-oxo salen complex has been reported by mass spectrometry (*167*).

From the oxygen transfer mechanism shown on Fig. 21A, it can be seen that the oxygen atom incorporated into the substrate comes from the $Mn^{V}{=}O$ entity, and consequently, from the oxygen atom donor. However, in the particular case of oxidation reactions performed in aqueous media with water soluble metalloporphyrins, the catalyst was shown to transfer an oxygen atom coming either from the oxygen atom donor either from the labeled bulk water $H_2{}^{18}O$, respectively in a ratio 1/1 (*147,168–171*). Since the exchange of the oxygen atom of the metal-oxo entity with bulk water is slow, an intramolecular exchange between the oxo and a water molecule was proposed to occur *via* a mechanism referred to as the oxo-hydroxo tautomerism

(Fig. 22). It corresponds to a shift of two electrons and one proton from a hydroxo axial ligand to the oxo on the opposite side of the porphyrin ring.

The chemistry of manganese complexes (electron abstraction vs. oxygen atom transfer) on DNA will follow the same rules as those known for heme enzymes. Oxygen atom transfer is only possible when the substrate can approach the metal-oxo center (like in cytochromes P450). It will be illustrated by hydroxylation reactions of deoxyribose units. Electron transfer occurs when the substrate cannot reach the metal-oxo center inside the active site of the enzyme (like in peroxidases). This will be illustrated in the case of guanine oxidation. Thus, the mechanism of the oxidative DNA damage mediated by the manganese complexes will be highly dependent on the interaction of the metal complex with the various sequences and structures of DNA.

B. Manganese Salen or Related Complexes

The Mn^{III} complex of salen, bis-(salicylidene)ethylenediamine [or (*N,N'*-ethylenebis(salicylideneaminato)] (**28**, Fig. 23) was reported to produce DNA

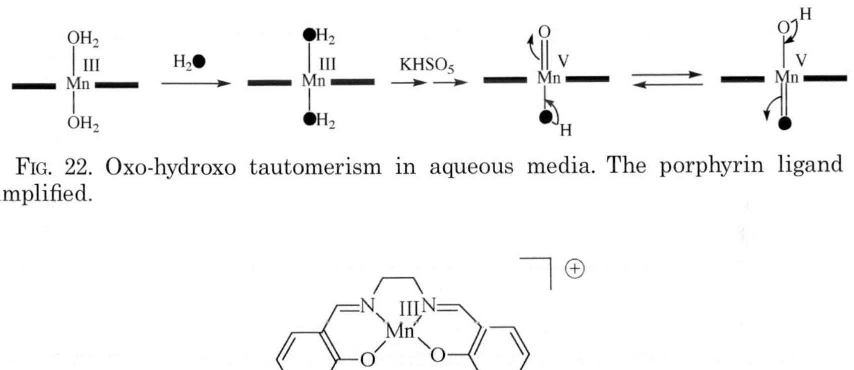

FIG. 22. Oxo-hydroxo tautomerism in aqueous media. The porphyrin ligand is simplified.

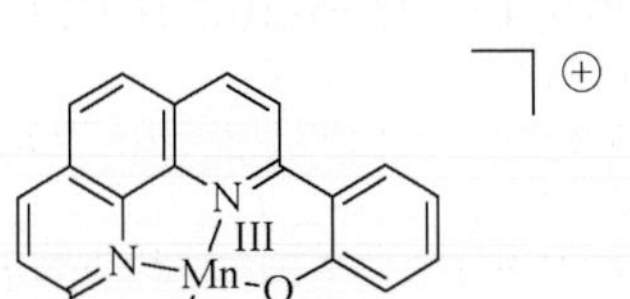

28

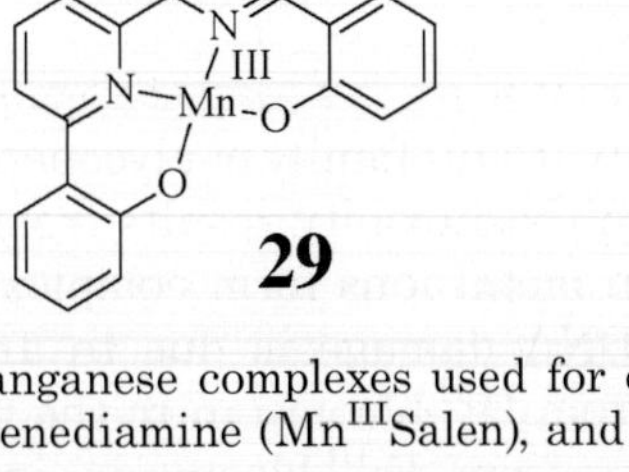

29

FIG. 23. Structure of manganese complexes used for oxidative DNA damage. **28**: Mn^{III}-bis-(salicylidene)ethylenediamine (Mn^{III}Salen), and **29**: Mn^{III}-2,9-bis(2-hydroxyphenyl)1,10-phenanthroline.

cleavage in the presence of MMPP (*172*). It was also possible to activate this complex with other oxidants like H_2O_2, potassium monopersulfate, and iodosylbenzene. The DNA cleavage pattern indicated that this complex interacted with DNA in the minor groove of AT-rich regions of DNA. Attempts to improve the primary complex by chemical modification did not give rise to significantly better DNA cleaving agents nor did it change the overall sequence preference of the compounds (*173,174*). The chemical modifications included (i) the addition of two positive charges on the ligand (*174*), (ii) variations on the bridge, (iii) ring- and imine-substitutions, (iv) chiral bridge-substitutions (*173*).

The efficiency of the Mn^{III} bis-(salicylidene)ethylenediamine complex was compared to other chemical nucleases (*175*), iron-methidiumpropyledta (*176,177*), or antitumor agents like neocarzinostatin (*178,179*), iron-bleomycin (*45,159–162*). It was also directly compared with Mn-TMPyP in the presence of MMPP under similar experimental conditions (*180*). It was clearly shown that the porphyrin derivative was a better DNA cleaving agent than the salen derivative. To achieve similar levels of double-strand scission the salen complex had to be used in higher concentration and for longer time compared to the Mn-TMPyP complex. The observed difference in activity was proposed to derive from two reasons (i) the pentacationic Mn-TMPyP had greater affinity for DNA and (ii) Mn-TMPyP was more robust oxidant. The molecular mechanism of DNA damage by the manganese salen complex was not clearly established.

Later, a new type of manganese-based agent that combined the features of both salen and phenanthroline ligand framework, the manganese complex of 2,9-bis(2-hydroxyphenyl)1,10-phenanthroline, was prepared (**29**, Fig. 23). It was tested for DNA cleavage in the presence of a reducing agent (ascorbate) or in the presence of an oxidant like potassium monopersulfate ($KHSO_5$) or MMPP (*181*). From EPR experiments, the cleavage of DNA was attributed to the production of hydroxyl radicals when the reaction was carried out in the presence of ascorbate (Fenton reaction). However, this compound was even less active than the salen derivative **28**.

For an approximate (since the experimental conditions are not identical) yet rapid comparison of the relative efficiencies of the manganese complexes toward DNA cleavage, the circular supercoiled plasmid cleavage test is convenient. About the same yield of plasmid cleavage was reached with 100 μM of **29** (1 h, 37°C) (*181*), or 10 μM of **28** (5 min, 37°C) (*174*) and finally, 250 nM of Mn-TMPyP (1 min, 20°C) (*182*).

C. Manganese Bleomycin

Bleomycin is a clinically useful family of glycopeptide antibiotic congeners with antitumoral activity. Cytotoxicity results from oxidative DNA damage. Bleomycin and transition metal ions form complexes that react with dioxygen and oxidize DNA. DNA damage is due to an activated form of iron bleomycin which forms from Fe^{II}-bleomycin in the presence of dioxygen and a source of electrons or from Fe^{III}-bleomycin in the presence of H_2O_2.

An activated bleomycin analogous to peroxidase compound I explains many characteristics of bleomycin-mediated DNA degradation reactions. Since Fe-bleomycin shared the same type of oxidative chemistry as cytochrome P450 and since manganese porphyrin complexes were shown to be good mimics of cytochrome P450 enzymes, two groups tested the DNA cleavage reaction of bleomycin with manganese instead of iron. The products of DNA oxidation by activated iron bleomycin originate from oxidation at the carbon C4′ of deoxyribose, and consist in a free base associated with a 4′-hydroxylated abasic site **23** or base propenal **13** associated with 3′-phosphoglycolate ends **11** (*45,159–162*). Hecht *et al.* (163) showed that Mn^{III}-bleomycin, in the presence of iodosylbenzene, or in the presence of molecular oxygen and ascorbate, was able to mediate epoxidation of *cis*-stilbene, cyclohexene and norbornene. The products were very similar to those observed with a manganese porphyrin under the same experimental conditions. This result was in favor of an activated species of manganese bleomycin resembling the activated manganese-oxo porphyrin. DNA degradation was observed when manganese bleomycin was tested in the presence of molecular oxygen and ascorbate or in the presence of PhIO. DNA degradation consisted of strand breaks, but no malondialdehyde was detected. The manganese bleomycin was shown to be 10-fold less active than iron bleomycin under the used experimental conditions.

DNA degradation with Mn^{II}-bleomycin in the presence of H_2O_2 was also reported by Burger *et al.* (*164*). The DNA oxidation products included free bases and **11**, as with iron bleomycin, but in different proportions (base/base propenal is 5/1 for manganese instead of 1/1 for iron bleomycin). No reaction occurred when Mn^{II}-bleomycin was tested in the presence of 2-mercaptoethanol and air. It was found that the manganese complex of bleomycin was 100-fold less active than iron bleomycin under the used experimental conditions. However, an assay of DNA damage based on the detection of base propenals **13** may not be the most appropriate in this particular case. These DNA degradation products have only been observed, with no ambiguity, with Fe-bleomycin. DNA damage by some other DNA oxidizing compounds, known to attack DNA at the C4′ carbon of deoxyribose, did not produce observable base propenals (*49,51,177,178*).

In summary, manganese bleomycin is able to mediate DNA degradation in the presence of ascorbate (but not with 2-mercaptoethanol for an unknown reason) or iodosylbenzene. It is less efficient than its iron bleomycin congener. The detailed mechanism of the DNA damage by manganese bleomycin as well as the reason for its lower efficiency were not investigated.

D. Manganese Porphyrin

D.1. Early Work: Selection of the Best Metalloporphyrin, Mn-TMPyP, and the Best Associated Oxidant

The first DNA strand scission activity reported for a metalloporphyrin was that of a series of iron cationic porphyrins in the presence of a reducing agent (dithiothreitol, DTT) and molecular oxygen (*183*). The most active compound was substituted with 4-*N*-methylpyridiniumyl on the four *meso* positions, Fe-TMPyP. Negatively charged water soluble metalloporphyrins were less efficient because charge repulsion with the negatively charged phosphate groups of DNA lowered their reactivity (*183,184*).

The use of manganese porphyrins as artificial nucleases followed rapidly (*185–187*). The comparison of the efficiency of several manganese porphyrins led to the selection of the *meso*-tetrakis(4-*N*-methylpyridiniumyl)-porphyrin (H_2-TMPyP) as the best ligand for DNA targeting (*185*). In this pioneering work, the manganese porphyrins were reacted in the presence of potassium superoxide (KO_2) and DNA damage was assayed by the analysis of strand breaks by polyacrylamide gel electrophoresis. Analysis of the cleavage sites on both strands of DNA showed that (i) the cleavage sites were located in AT rich regions of DNA and (ii) the site of cleavage on one strand was associated with a cleavage site on the other strand that was shifted in the direction of the 3′-end. This fact was immediately recognized as indicative of the binding of the manganese porphyrins in the minor groove of AT rich regions. The manganese porphyrins, due to the presence of two axial ligands on the metal, do not intercalate between DNA base pairs, but show outside binding. An X-ray structure of Mn-TMPyP demonstrates the presence of two axial water molecules (*188*).

The number of charges on the porphyrin macrocycle was important for the affinity for DNA. In a series of porphyrins having different substituents at the *meso* position, a neutral *para*-toluene moiety or a positively charged 4-*N*-methylpyridiniumyl moiety, the porphyrin derivative carrying four pyridinium substituents was more active than the porphyrin with three charges and so forth (*185*). The shape of the porphyrin was important too. Porphyrin derivatives with 2-*N*-methylpyridiniumyl groups (that are perpendicular to the plane of the porphyrin ring) are compounds with a broad axial dimension (~9 Å) compared to the compounds in the toluene/pyridinium series (~6 Å) and showed less pronounced preference for AT rich sequences. While changes in total charge and charge arrangement do not significantly influence specificity, a shape change which blocks close ligand contact with the minor groove relaxes the original AT specificity. Since these porphyrins are devoid of hydrogen bond donor and acceptor possibility, it was proposed that electrostatic and steric effects associated with the minor groove of DNA were the factors controlling their affinity and specificity. The surface of minor groove of AT-rich regions of DNA possesses a high negative potential that attracts positively charged molecules (*189,190*). The smallest Mn-TMPyP cleavage site is composed of AT base pairs and is three base pairs in length (*187*). This preferred binding site is the place of a very specific cleavage reaction that will be detailed below.

Mn- or Fe-TMPyP were also activated with ascorbate, KO_2 or PhIO (*187*). In the presence of ascorbate and molecular oxygen the iron porphyrins appeared to be more efficient whereas manganese porphyrins were better activated with PhIO (*187,191*). However, PhIO is poorly soluble in water, and consequently, the next association of Mn-TMPyP with $KHSO_5$ (*182,192,193*), or

MMPP (*194*), both oxygen atom donors soluble in water at physiological pH, proved to be the most efficient systems.

Recently, peroxynitrite, $ONOO^-$, was also reported to be able to activate Mn-TMPyP for DNA strand scission (*195*). Beside, the high reactivity of the manganese porphyrins with $ONOO^-$ led to their use as $ONOO^-$ detectors (*196*) or as $ONOO^-$ decomposition catalysts in the presence of biological antioxidants (*197*). However, it must be noted that the use of Mn-TMPyP as a quencher of peroxynitrite leads to the formation of a Mn^V=O species that may be as reactive (unless it is reduced by an added reductant) as peroxynitrite itself toward biological molecules.

Finally, the replacement of $KHSO_5$ by a more biocompatible reactant, sulfite (Na_2SO_3) associated to O_2 was described (*198*).

In summary, for oxidative DNA damage, Mn-TMPyP (Fig. 19) associated with a convenient oxygen atom donor $KHSO_5$, was selected among other porphyrin derivatives owing to the higher affinity binding constant of this metalloporphyrin with DNA and to the efficiency of the formation of the high-valent manganese-oxo species with $KHSO_5$ under physiological conditions.

D.2. Interaction of the Manganese-oxo Porphyrin with the Minor Groove of AT Rich Sequences of DNA

The binding of Mn-TMPyP in the minor groove of AT regions of DNA is not supported by structural data. Only one example of this mode of binding was observed with Ni-TMPyP (*199*), but this interaction was a consequence of crystal packing. The crystal structure of a B-type DNA self-complementary oligonucleotide (5′-CCTAGG) with Ni-TMPyP showed that the porphyrin was stacked onto the ends of the short duplex. This is logic for this square planar complex that does not have any axial ligands. A second mode of binding for the porphyrin was apparent which located the porphyrin ligand in the minor groove of an adjacent duplex. Numerous close contacts between the porphyrin and the atoms of the floor and walls of the groove were at distances < 3.5 Å. The two inner *N*-methylpyridiniumyl groups were oriented almost parallel to the groove floor. The metal at the center of the porphyrin was close to the C1′ and to the O3′ of deoxyribose.

This interaction was observed on a DNA duplex devoid of AT rich minor groove. Consequently, molecular modeling was undertaken to elucidate the specificity of the interaction between the Mn^V=O species of Mn-TMPyP and the minor groove of a 5′-TCGTCAAACCGC/5′-GCGGTTTGACGA duplex (*200*). This duplex contained the minimum minor groove interacting site for Mn-TMPyP, namely, three consecutive AT base pairs (*187,201*). The manganese-oxo porphyrin was docked within the minor groove of B-DNA with only minor structural changes (Fig. 24). The porphyrin was deeply buried in the minor groove and the metal-oxo entity was positioned to reach C–H bonds of deoxyriboses. The simulations showed that the oxygen atom of the metal-oxo was at a similar average distance from the hydrogen atoms of C4′–H and C5′–H bonds of deoxyribose. This tight interaction between the

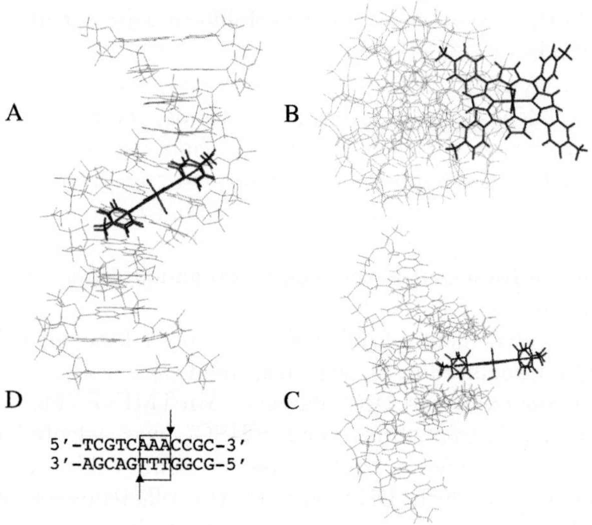

FIG. 24. Modelisation of the interaction of a high-valent manganese-oxo porphyrin (with a *trans* hydroxo ligand) (red) with the minor groove of the three consecutive AT base pairs of 5′-TCGTCAAACCGC/5′-GCGGTTTGACGA duplex (blue). A, side view; B, top view; C, side view, rotation by 90° compared to A; D, sequence of the duplex with the cleavage site indicated by arrows (*200*).

manganese porphyrin and the minor groove of three AT base pairs sequence is in accordance with the higher affinity of Mn-TMPyP for AT regions of DNA compared to GC regions, affinity constant of 12×10^4 and $0.2 \times 10^4\ M^{-1}$, respectively (*202*). This binding constant for the so-called strong site (three consecutive AT base pairs) was even found higher $\sim 10^7\ M^{-1}$ (*203*).

Furthermore, it can be seen on the model that there is enough space for an oxygen atom donor ($KHSO_5$ or another ligand) to coordinate at the axial position of the metal (Fig. 24C). Since the manganese-oxo may be located on either side of the porphyrin macrocycle (Fig. 22) the oxidation of either strand of DNA is possible without moving the complex from its site of interaction (Fig. 24A, 24C).

A detailed study of the chemical reactivity of Mn-TMPyP has shown conclusively that Mn-TMPyP is able to cleave the two strands of DNA, from the minor groove, at that minimum strong binding site consisting of three consecutive AT base pairs (*201,204*). This reaction is described in the next paragraph.

D.3. Cleavage of DNA by Mn-TMPyP/$KHSO_5$ at the Three Consecutive AT Base Pairs Site, Hydroxylation of C5′

The cleavage of DNA by Mn-TMPyP associated to $KHSO_5$, at the three consecutive AT base pairs site, is due to a mechanism of oxygen atom

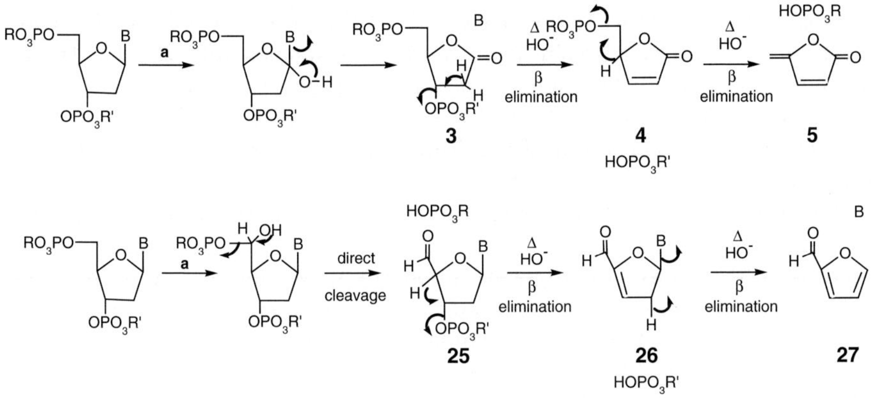

FIG. 25. Mechanism of C1′ and C5′ direct hydroxylation of deoxyribose mediated by the Mn^{V}=O species of Mn-TMPyP in the presence of $KHSO_5$. **a** stands for Mn-TMPyP + $KHSO_5$ oxidation. Δ refers to a heating step at 90°C at pH = 8 for 15 min. To be compared with Figs 4 and 8.

transfer from the Mn^{V}=O porphyrin to DNA. This reaction is reminiscent to an enzymatic reaction where the catalyst would be the small partner and would interact with a large substrate. The exquisite positioning of the metal-oxo in the vicinity of its target substrate leads to a very precise hydroxylation of carbon C5′ of deoxyribose of the nucleotide unit that is located immediately 3′ to the last base pair of the three AT base pairs sequence (Fig. 24D, Fig. 25) (*201,205,206*). This mechanism of DNA oxidation consists of direct strand breaks either on one strand or on both strands depending on the experimental conditions. The 3′-shift observed in the cleavage site on the complementary strands is at present the only unambiguous signature of the porphyrin location in the minor groove.

When DNA damage was assayed under low oxidation conditions (500 nM Mn-TMPyP and 750 nM $KHSO_5$) on long restriction DNA fragments, the C5′ chemistry was the only one that could be detected (*198*). The very strong affinity of the manganese porphyrin for the AT sequences present on the restriction fragment of DNA directs the mechanism of DNA degradation under these experimental conditions. However, under more drastic oxidation conditions other mechanisms of DNA damage can be observed.

D.4. Hydroxylation of C1′ by Mn-TMPyP/$KHSO_5$

Oxygen atom transfer was also observed on C1′ position of the deoxyribose units with high molecular weight DNA (Fig. 25). This second site of hydroxylation was evidenced by the release of the corresponding sugar oxidation product, 5-methylene-2-furanone (5-MF, **5**) (*204,207*). This sugar oxidation product is readily detected by HPLC after a heating step at 90°C and at pH = 8. Labeling experiments showed that the mechanism of C1′

hydroxylation by Mn-TMPyP activated in the presence of $KHSO_5$ was a direct oxygen atom transfer from a $Mn^V{=}O$ species into the C1′–H bond of deoxyribose (*171*). The same type of labeling experiments could not be conclusive in the case of C5′ hydroxylation due to the fast exchange with solvent of the oxygen atom of the 5′-aldehyde function formed in that case (Fig. 25).

The ratio of C5′/C1′ hydroxylation depended on the sequence of DNA. It can be measured by the ratio of the two products of sugar oxidation furfural/5-MF (**27**/**5**). On single-stranded DNA, the percentage of 5-MF production (C1′ oxidation) was about 90 % of the total sugar oxidation products (sum of **27** and **5**). On double-stranded DNA with only GC base pairs the percentage of **5** was 95 % whereas on a DNA substrate with only AT base pairs the percentage of **5** decreased to 30–40 % with a concomitant percentage of **27** (C5′ oxidation) at 60–70 % (*204,208*). Under the used experimental conditions, the reactivity of the metalloporphyrin towards these different DNA substrates was around 10 % of degradation of nucleosides according to the concentration of the DNA substrate expressed in bases. Thus, when the metalloporphyrin was not in the presence of AT rich sequences, offering the special minor groove interaction site described in the preceding paragraph, the oxidation of C1′ of deoxyribose units could be observed.

The best way to detect C1′ oxidation is the HPLC assay for its corresponding sugar degradation product 5-MF **5**. For comparison, C1′ oxidation may not be detected when DNA degradation is analyzed by polyacrylamide gel electrophoresis. Firstly, this mechanism of DNA damage does not lead to a direct strand break, but strand cleavage will occur only after a heating step at pH = 8 and 90°C. Secondly, the heat induced breakage of DNA may not be visible if this C1′ chemistry happens to be completely random along the DNA sequence. It is only detectable if its occurrence at some sugar units reaches a threshold value for band detection. This is not the case for Mn-TMPyP/$KHSO_5$ since C1′ oxidation was not observed on electrophoresis gels.

The C1′–H1′ bond is located so deeply in the minor groove of DNA that it seems difficult to envision a direct attack of the metal-oxo on that bond. It is tentatively proposed that the C1′ chemistry is observed after the DNA has been destabilized by first events of oxidative degradation. However, the oxidation of the tertiary carbon C1′ is easier than that of the secondary carbon C5′. This explains why C1′ oxidation is the major process on single-stranded DNA where the two C–H bonds are equally accessible to the oxidizing agent. What chemistry could induce double-stranded DNA to melt enough so that the C–H bond at C1′ becomes accessible to the porphyrin? One hypothesis would be that some C5′ chemistry starts the degradation of DNA and that this partially melted DNA is then susceptible to undergo attack at C1′. Another possibility for melting DNA would be oxidative base damage and will be detailed in the next paragraph.

The C4′ oxidation was never detected with the Mn-TMPyP/$KHSO_5$ system on double-stranded DNA. The classical base propenals **13** (*208*) or the related

3′-phosphoglycolate ending fragments **11** (*45,49,51,159–162,177,178*), were not observed. The product of a hydroxylation of the C4′–H bond, namely, a 4′-hydroxylated abasic site **23**, is also known from literature and would be expected if the Mn^{V}=O species would be able to reach the C4′–H bond of deoxyribose. This 4′-hydroxylated abasic site does not give rise to any detectable oxidized residue of deoxyribose, but could have been detected as an oxidized DNA fragment with short oligonucleotides and this was not the case (*209*). It was concluded that the Mn^{V}=O species of Mn-TMPyP did not react with the C4′–H bond of deoxyriboses of double-stranded DNA, but reacted only with C5′–H and C1′–H.

D.5. Guanine Oxidation by Mn-TMPyP/$KHSO_5$

The covalent attachment of the cationic manganese porphyrin onto an oligonucleotide allowed us to discover that the Mn^{V}=O species was able to mediate guanine oxidation by an electron transfer mechanism. An oligonucleotide carrying the manganese porphyrin attached at the 5′-end was hybridized with a complementary single-stranded DNA in order to perform sequence specific DNA damage (Fig. 26A) (*210–213*). The covalently bound manganese porphyrin was not free to interact with DNA in the same manner as Mn-TMPyP and the porphyrin was forced to react on a target sequence that would not be usually selected by the free reagent. We observed, in that case, a cleavage mechanism that was only located at G residues and was consequently due to base oxidation (Fig. 26B) (*210*). Thus, when the manganese porphyrin is maintained in the vicinity of a GC rich sequence, the highly reactive manganese-oxo entity reacts by an outer sphere electron transfer mechanism and abstracts electrons from the most oxidizable base of DNA, guanine (*214*).

The redox potential of the guanine base of guanosine was measured at 1.29 V vs. NHE and is significantly lower (0.5 V) than that of the nearest nucleoside, adenosine (*214*). However, depending on the DNA sequence, the guanine bases may be more sensitive to oxidation. The oxidation potential of the 5′-G of a 5′-GG sequence in double-stranded DNA is lower than that of an isolated G by 0.5 V (*215*). Even though guanine is the most sensitive base toward oxidation by electron transfer its oxidation needs strong oxidants like type I photosensitizers (oxidizing excited state of the drug), ionizing radiations, or some oxidizing reagents endowed with high redox potentials (*216–233*) including some metal complexes (*234–240*). For some review articles, see *55*, *241–243*.

The one-electron oxidation of guanine by electron transfer is at the origin of the well-documented possibility of charge transfer reactions within the highly organized double helix of DNA (*244–246*). Additionally, oxidation of guanine by electron transfer leads to the formation of modified base residues on DNA referred to as lesions. This oxidative damage is usually sensitive to an alkaline treatment, which induces the cleavage of the glycosidic bond of a modified nucleoside and gives rise to a strand scission

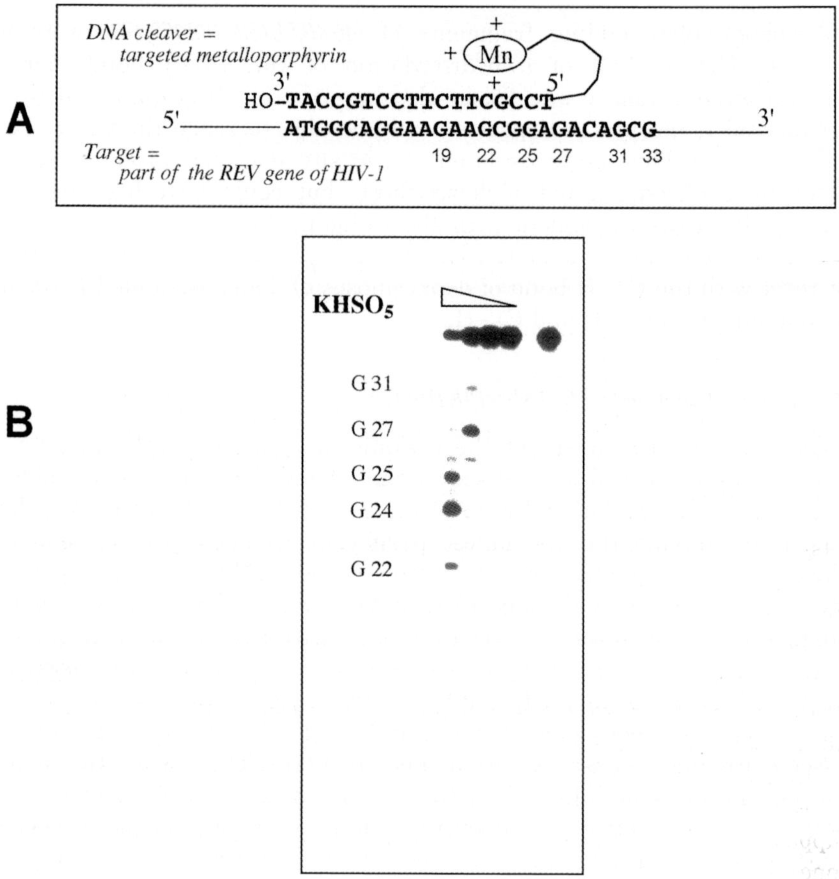

FIG. 26. Guanine oxidation by a cationic manganese porphyrin covalently linked to an oligonucleotide in the presence of $KHSO_5$. A shows the hybridization of the conjugated molecule with its complementary strand. B is the electrophoresis gel with the cleavage fragments at G residues upon piperidine treatment. ^{32}P-labeled 35-mer single-stranded DNA target was 10 nM, Mn-porphyrin conjugate (10 nM), calf thymus DNA (0.2 mM bp), 100 mM NaCl, phosphate buffer pH = 7, reaction lasted 1 h at 37°C. $KHSO_5$ concentration decreased from 1, 0.1, 0.01, to 0.001 mM (left to right) and the last lane on the right was a control without $KHSO_5$ (*211*).

at the damaged residue. Some exceptions exist, for example a 8-oxoguanine residue is not an alkali-labile site (*247*). This property of DNA cleavage upon alkaline treatment allows the detection and the localization of guanine damage, but does not give any indication on the nature of the chemical modifications of oxidized guanine residues (Fig. 26B).

The Mn^{V}=O entity of manganese porphyrin is a powerful chemical oxidant capable of abstracting electrons from guanine. However, the products of the reaction clearly result from a two-electron oxidation mechanism.

D.6. Oxidation of 2′-deoxyguanosine by Mn-TMPyP/KHSO$_5$

The two-electron oxidation mechanism of guanine by Mn-TMPyP/KHSO$_5$ was evidenced with a model compound, 2′-deoxyguanosine (dG). Incubation of dG with Mn-TMPyP/KHSO$_5$ led to the formation of one product of reaction in nearly quantitative yield, 2-amino-5-[(2-deoxy-β-D-*erythro*-pentafuranosyl)amino]-4*H*-imidazol-4-one (imidazolone nucleoside or dIz for short) (90 %) (*248*). The reaction was very rapid (1 min). It was analyzed by HPLC coupled to electrospray ionization mass spectrometry (LC/ESI-MS). The product was identified based on its retention time compared to a standard, prepared from previously published procedure (*249–251*), and its in-line ESI-MS spectrum. This product of guanine oxidation by electron transfer had been previously characterized with a photosensitizer as oxidative reagent, benzophenone (*249,250*). The mechanism of its formation involved an intermediate guanine radical cation, $G^{+\bullet}$, that deprotonated immediately ($pK_a = 3.9$ (*252*)) under the experimental conditions (*249*) and gave rise to a neutral guanine radical, $(G–H)^\bullet$, that is trapped by a molecule of dioxygen. Indeed, one oxygen atom from O_2 was shown to be incorporated in the product. The mechanism of the formation of dIz was due to the trapping of the intermediate guanine radical at the former C5 of guanine as shown on Fig. 27.

On the contrary, in the case of the oxidation of dG by Mn-TMPyP/KHSO$_5$, the formation of dIz was not dependent upon the presence of molecular oxygen. Furthermore, LC/ESI-MS analysis of the reaction showed that when the reaction was performed in the presence of labeled molecular oxygen ($^{18}O_2$) dIz was not labeled. The fact that no intermediate radical species was trapped by molecular oxygen indicated that the radical intermediate due to a one–electron oxidation of guanine, $(G–H)^\bullet$, was rapidly further oxidized, in this reaction medium, into a non-radical cationic species, $(G–H)^+$. This second electron abstraction is faster than the trapping by molecular oxygen. The non-radical cationic species was to be trapped by nucleophiles. However, when the reaction was performed in labeled water $H_2{}^{18}O$, dIz did

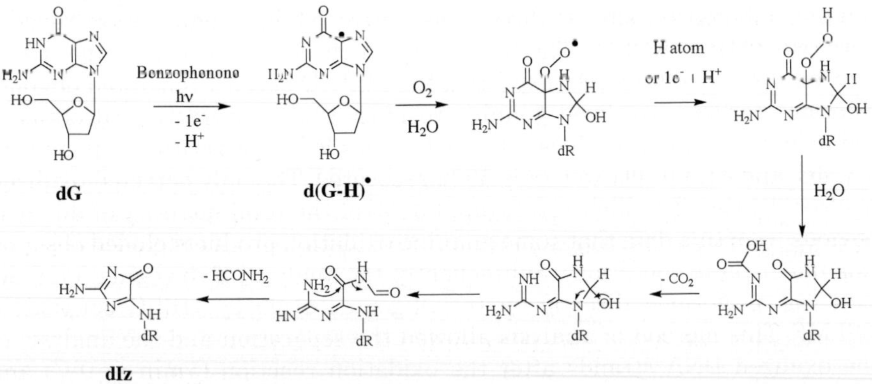

FIG. 27. Oxidation of 2′-deoxyguanosine by photoactivated benzophenone, a one-electron mechanism. dR stands for deoxyribose (*249*).

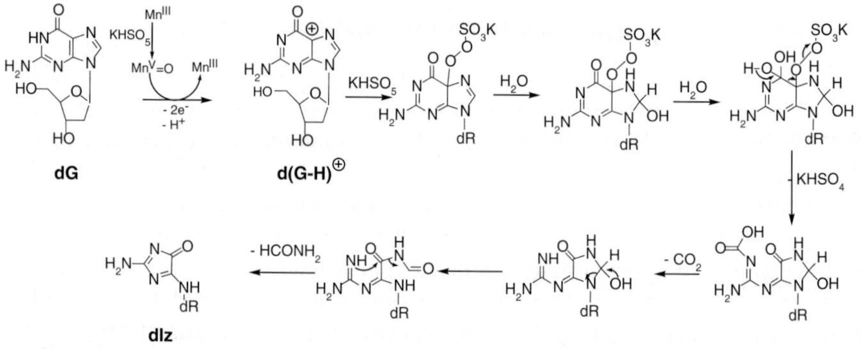

FIG. 28. Oxidation of dG by Mn-TMPyP/$KHSO_5$, a two-electron mechanism. dR stands for deoxyribose (*248*).

not incorporate an atom of ^{18}O from labeled water. Consequently, the only rational origin of the oxygen atom at the former C5 of guanine in dIz, was from the peroxide $KHSO_5$. Peroxides are better nucleophiles than water molecules. These labeling studies showed unambiguously that, with Mn-TMPyP/$KHSO_5$, the oxidation of guanine was a direct two-electron oxidation. The mechanism of the formation of dIz through the two-electron oxidation of dG is shown in Fig. 28. After the abstraction of a first electron, a guanine radical cation is formed. This radical cation deprotonates immediately and gives rise to a guanine neutral radical that is then rapidly oxidized by the abstraction of a second electron. This leads to an intermediate non-radical cationic derivative of guanine, $(G–H)^+$, which is trapped by the most electrophilic entity in the reaction medium, namely $KHSO_5$. After several steps, the imidazolone derivative formed after the release of formamide which could also be detected.

D.7. Oxidation of Guanine in Double-Stranded Oligonucleotides with Mn-TMPyP/KHSO$_5$

Short double-stranded oligonucleotide duplexes, devoid of minor groove binding interaction site of three consecutive AT base pairs, were reacted with Mn-TMPyP/$KHSO_5$. The sequence was chosen to favor guanine oxidation vs. sugar oxidation. As expected, the Mn^V=O species generated oligonucleotides carrying oxidized guanine residues (*253*). Enzymatic hydrolysis of the oxidized oligonucleotides into nucleosides led to quantitative amount of dA, dC, and dT, but dG was only 50 % recovered. The only detected oxidized nucleoside was dIz which represented 30 % of the total quantity of dG. It is obvious from this data that some guanine oxidation products eluded observation under these conditions. Subsequently, the analysis of the oxidation reaction was performed directly on the oligonucleotide by HPLC coupled to ESI-MS. This method of analysis allowed the separation and the analysis of the oxidized DNA strands after the oxidation reaction (5 min at 0 °C) and without further work-up. From the enzymatic hydrolysis experiment has appeared that the mass of the damaged strands of DNA corresponded to

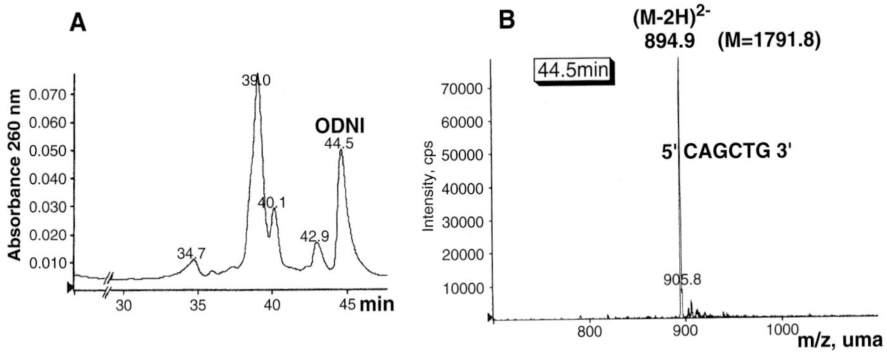

FIG. 29. LC/ESI-MS analysis of the oxidation of a short double-stranded oligonucleotide by Mn-TMPyP/$KHSO_5$. ODN I stands for the self complementary oligonucleotide 5′-CAGCTG. The oxidation was carried out in 50 mM Tris/HCl buffer pH = 7, NaCl 100 mM with the following concentrations of reactants: duplex ODN I, Mn-TMPyP, $KHSO_5$, 200 μM, 200 μM, 2 mM, respectively. The reaction lasted 5 min at 0°C. A: HPLC trace; B: in-line ESI-MS spectrum of ODN I (retention time, 44.5 min).

damage only at the guanine bases. The direct analytic HPLC/ESI-MS procedure had the advantage to allow the detection of all the products of guanine oxidation since they were separated as modified DNA strands and to allow the characterization of unstable oxidation products since the analysis was rapid. An example of a chromatogram of the reaction is shown on Fig. 29A. The determination of the molecular mass of the separated peaks was possible with high accuracy (0.02 %) (Fig. 29B). The labeling of the products showed that the oxidation of the guanine bases, within duplex DNA, followed the same two-electron oxidation mechanism as in the case of the mononucleoside model. The reaction was not dependent upon the presence of molecular oxygen. However, the non-radical guanine cation intermediate $(G–H)^+$, reacted, in that case, with a water molecule (incorporation of a ^{18}O atom from $H_2{}^{18}O$) instead of $KHSO_5$. The negative charges of DNA precluded the approach of $KHSO_5$ *(254)*.

Within double stranded oligonucleotides, the imidazolone derivative was not the major product. The HPLC peak detected at 260 nm (Fig. 29A) and containing this lesion (in-line ESI-MS data) eluted at a retention time of 40.1 min. This product incorporated one oxygen atom from labeled water and the mechanism of its formation could be attributed to a nucleophilic attack of a water molecule at the former C5 of guanine *(253)*.

The main HPLC peak at 39 min corresponded to oligonucleotide strands carrying a modified residue corresponding to an increase of 4 Dalton compared to the starting oligonucleotide **31**. It incorporated two ^{18}O atoms from $H_2{}^{18}O$. This major product of guanine oxidation by Mn-TMPyP/$KHSO_5$ on double-stranded oligonucleotides was identified as a dehydroguanidinohydantoin residue (Fig. 30) (see below). This product of guanine oxidation **31** was also reported in the literature with other oxidation systems in the case of guanine or 8-oxoguanine oxidation *(255,256)*.

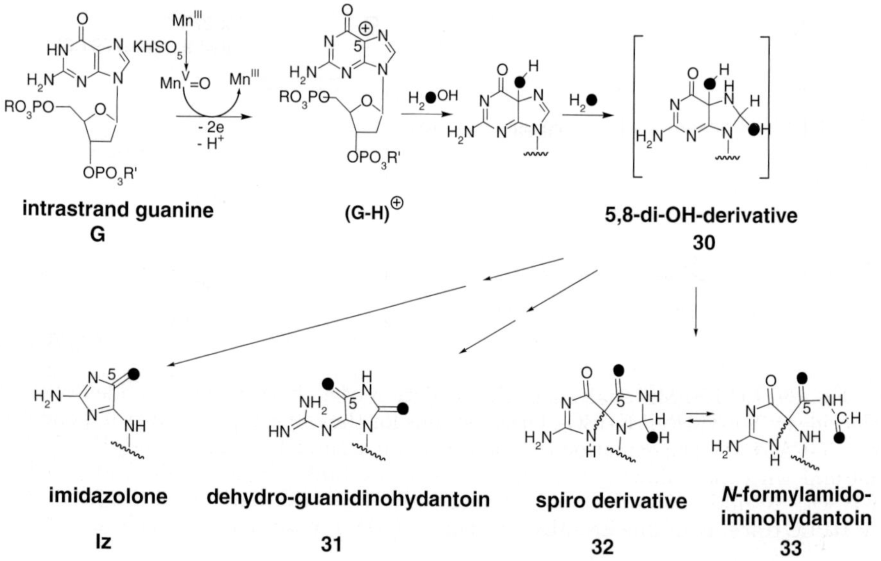

FIG. 30. Products of guanine oxidation in double-stranded oligonucleotides generated by Mn-TMPyP/$KHSO_5$. Incorporation of labeled oxygen atoms in the products of the reaction in a labeling experiment in $H_2{}^{18}O$ is illustrated. Numbering of carbon atoms corresponds to that of guanine.

A minor product was also present in the main HPLC peak at 39 min, corresponding to an increase in mass of 34 Dalton according to its molecular mass compared to the starting oligonucleotide. This product incorporated two ^{18}O atoms from $H_2{}^{18}O$, but one of them exchanged with the aqueous eluent during the chromatography. This lesion (**32**,**33**) has never been observed previously with other reagents. The proposed *N*-formylamido-iminohydantoin structure **33** due to the opening of a spiro derivative **32** (Fig. 30) is deduced from labeling data and from literature on a related product found from guanine oxidation (*257*) or 8-oxoguanine oxidation, namely, a spiroiminodihydantoin derivative (*258–260*).

Labeling experiments under different reaction conditions showed that the dehydro-guanidinohydantoin derivative was not a precursor of imidazolone and that these three products arose independently from a common unstable precursor, the 5,8-diOH-derivative (**30**), as shown in Fig. 30 (*254*). The *N*-formylamido-iminohydantoin derivative **33** would form by a rearrangement of **30**. The mechanism of the formation of the imidazolone and **31**, based on the trapping of the $(G–H)^+$ intermediate by a molecule of water leading to the intermediate **30** is proposed in Fig. 31. From the 5,8-di-OH-derivative (**30**), the attack of a third molecule of water at the former C6 of guanine would give rise to the guanidinohydantoin intermediate **34**. Cyclization by attack of the guanidinium amine onto the former C5 of guanine, opening of the imidazole ring, and the release of formamide would lead to the imidazolone derivative (Iz) after a second oxidation reaction due to the strong oxidizing conditions. On the other hand, in double-stranded DNA, the hydrogen

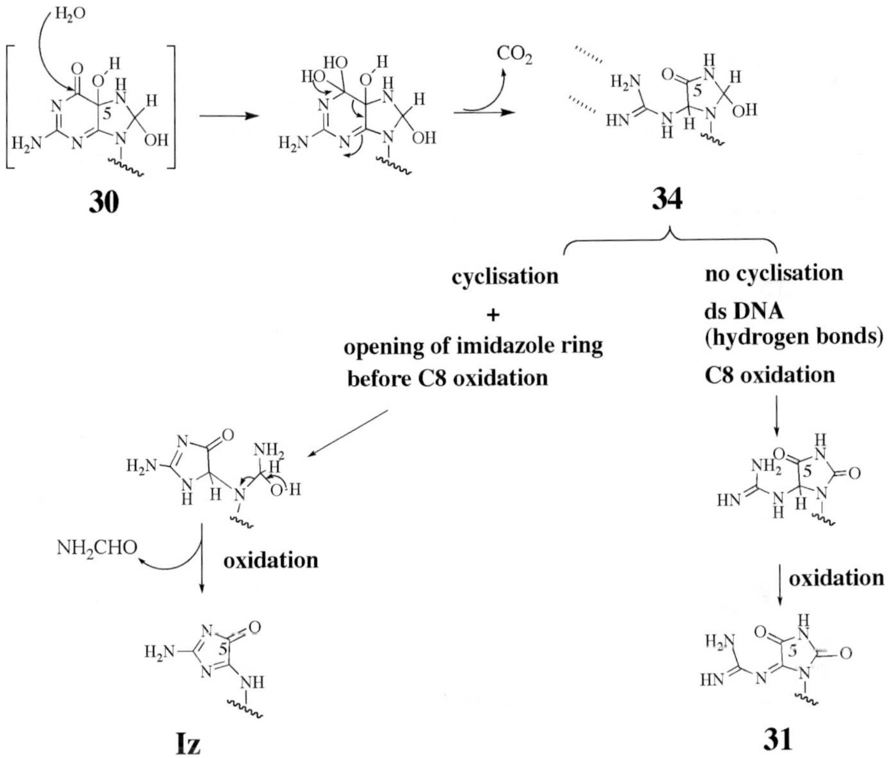

FIG. 31. Proposed mechanism for the formation of dehydro-guanidinohydantoin and imidazolone from intermediate **30** accounting for the ^{18}O labeling data. Numbering of carbon atoms corresponds to that of guanine.

bonding of the guanidinium moiety with the complementary cytosine would preclude the cyclisation and favor the oxidation of the former C8 of guanine. The dehydro-guanidinohydantoin **31** would form after a second oxidation event under the strong oxidizing conditions. This mechanism is compatible with the major formation of **31** within oligonucleotides compared to the case of the nucleoside oxidation. The second oxidation of these two compounds was possible in our case because most of the oxidation reaction occurred at the terminal guanine on the short duplex substrate. On longer DNA strands, **34** may be observed.

The isolation, mass and NMR characterization of the lesions was performed with the dinucleotide model compound d(GpT) since the chromatographic separation of the products of guanine oxidation was followed easily by UV-VIS spectrophotometry (thymine base was not modified by the oxidation) and the very polar oxidized guanine residues could be retained on the reverse phase HPLC column by the dinucleotide structure of the substrate. The oxidation of the guanine moiety of a dinucleoside monophosphate under appropriate conditions afforded the imidazolone (Iz) and the dehydro-guanidinohydantoin derivative **35** in amounts compatible with NMR analysis (*261*). However, **35** is not stable and was first stabilized in two ways for the

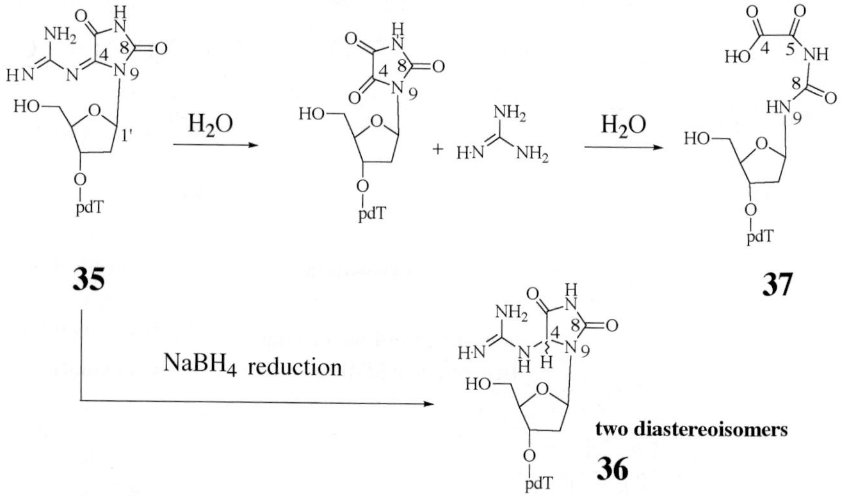

FIG. 32. Characterization of the dehydro-guanidinohydantoin derivative by its hydrolysis into an oxaluric acid derivative or its reduction into a mixture of two diastereoisomers of a guanidinohydantoin derivative on a dinucleotide. Numbering of carbon atoms corresponds to that of guanine.

NMR analysis. Its imine function is susceptible to hydrolysis with a half-life of 5 h at 37°C (*255*) or 8 h at 0°C (*262*). Stabilization was obtained by its reduction to a guanidinohydantoin derivative **36** (*262*), or its hydrolysis to an oxaluric acid derivative **37** (*263*) (Fig. 32). The structure of these two products confirmed the proposed structure for the dehydro-guanindinohydantoin **35** (Fig. 32). The guanidinohydantoin lesion **34**, was characterized as a guanine oxidation product for other DNA oxidizing reagents (*228,256*). However, in the case of the strong oxidant Mn-TMPyP/$KHSO_5$ the compound is reoxidized into the dehydro-guanidinohydantoin **31** (*262*).

In summary, Mn-TMPyP/$KHSO_5$ was able to mediate guanine oxidation in double-stranded oligonucleotides by a two electrons abstraction mechanism. A non-radical guanine cation intermediate, $(G–H)^+$, reacted with the nucleophiles present in the reaction medium like $KHSO_5$ or H_2O depending on the experimental conditions and on the substrate. The main guanine lesions that resulted included imidazolone, dehydro-guanidinohydantoin and a proposed *N*-formylamido-iminohydantoin derivative. Imidazolone was the major product when the trapping at the former C5 of guanine was due to $KHSO_5$, dehydro-guanidinohydantoin was the major product on double-stranded DNA, and the *N*-formylamido-iminohydantoin derivative was only observed on double-stranded DNA.

D.8. Oxidation of Quadruplex DNA by Mn-TMPyP/$KHSO_5$

The tetracationic non-metalated porphyrin, *meso*-tetrakis(4-*N*-methylpyridiniumyl)-porphyrin, (H_2-TMPyP) was reported to bind to quadruplex DNA and to be able to degrade it by photochemical processes (*264*). The sequence

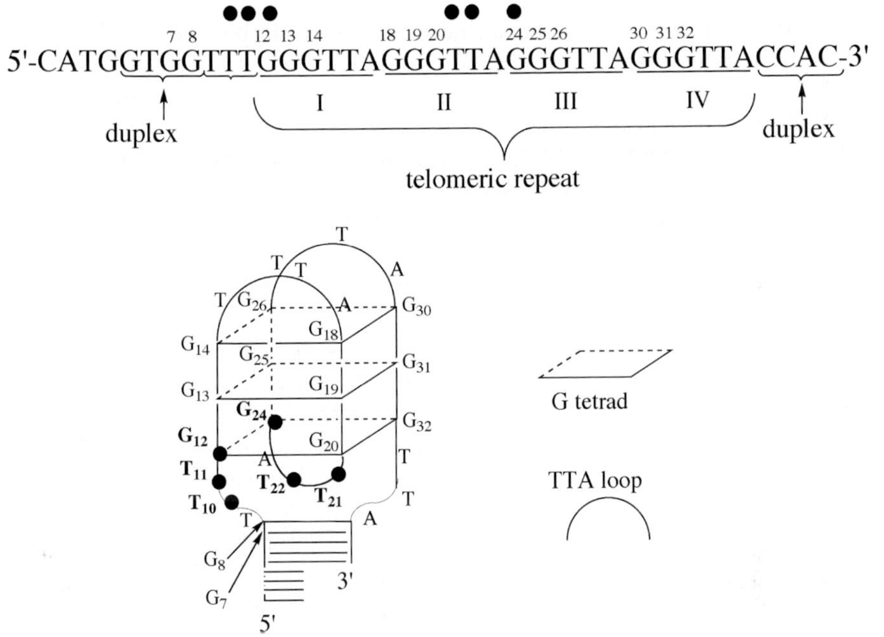

FIG. 33. Human telomeric DNA model. A, sequence of the oligonucleotide; B, proposed folded structure for an intrastrand quadruplex in the presence of K^+ ions (KCl 200 mM). The oxidized residues are highlighted (*265*).

and structure of the DNA target is shown on Fig. 33. Under irradiation the non-metalated porphyrin generates singlet dioxygen that can mediate oxidative DNA damage at the vicinity of the porphyrin location (*55,242,243*). According to the results of this photochemical footprinting, the flat porphyrin was stacked externally with the last guanine tetrad of an antiparallel quadruplex DNA.

The Mn-TMPyP/$KHSO_5$ system was tested onto the same quadruplex DNA mimicking the human telomeric sequence (*265*). The reaction was monitored by PAGE and by HPLC. It was found that the Mn-TMPyP associated with $KHSO_5$ was able to degrade the human telomeric sequence at guanine residues of the last guanine tetrads (G_{12}, G_{24} residues) and at deoxyribose units on the single-stranded loops (T_{10}, T_{11}, T_{21}, T_{22} residues). The sites of DNA oxidation revealed the positioning of the manganese porphyrin. It was located at the junction between the quadruplex and the duplex region of the target (Fig. 33). The intensity of damage was an indication of the binding affinity of the manganese porphyrin with this particular DNA target. It was found that the cleavage efficiency at the telomeric DNA sequence was of the same order of magnitude as the one reported for a three consecutive AT base pairs site on double-stranded DNA. This binding site for the manganese porphyrin is thus considered as being a strong one.

Both mechanisms of oxidation of the manganese-oxo porphyrin, electron transfer and oxygen atom transfer were observed on the same DNA target.

The product of deoxyribose oxidation at C1′, **5**, could be observed by HPLC. However, it was noted that a minor amount of C4′ hydroxylation may be observed as well at some residues in the loops. The oxidation of the C4′-H bond of some deoxyribose units in the single-stranded loops was due to an hydroxylation (formation of 4′-hydroxylated abasic site, **23**), no 3′-phosphoglycolate were detected. The C4′ oxidation mechanism was evidenced, on polyacrylamide gels, by the protection of $NaBH_4$ toward piperidine sensitive cleavage at some T residues.

In conclusion, the strong oxidative system, Mn-TMPyP/$KHSO_5$, that generates in solution a manganese-oxo species, is able to perform oxidative damage of DNA. The notable efficiency of this system is due to (i) a high affinity of the reagent for DNA, (ii) a high efficiency in the generation of the $Mn^V{=}O$ species with $KHSO_5$, (iii) the robustness of the complex toward bleaching. The mechanism of DNA degradation was shown to involve C–H bond direct hydroxylation at the deoxyribose residues (oxygen atom transfer) and electron abstraction at guanine bases. With this powerful oxidant, free radical chemistry was not observed. Trapping of intermediate radicals on the guanine base or at the sugars by molecular oxygen did not occur, contrarily to what was observed with other DNA oxidative agents.

IV. Conclusion

The copper complex Cu(1,10-phenanthroline)$_2$ and the manganese porphyrin Mn-TMPyP showed high reactivity in oxidative DNA damage. They allowed better knowledge on the resulting lesions on DNA. The mechanism of oxidative DNA damage depends on the nature, the strength and the location on DNA of the activated metal complex. The way the oxidizing reagent interacts with DNA governs the site of attack, deoxyribose versus bases. In both cases, one may distinguish two types of oxidative mechanisms, a "1-electron oxidation" chemistry and a "2-electron oxidation" chemistry. The "1-electron oxidation" mechanism prevails when the life span of a radical intermediate on DNA is long enough to allow it to be trapped by molecular oxygen whereas the "2-electron oxidation" mechanism will occur when the first radical on DNA is further oxidized by a second one-electron oxidation on DNA. The generation of a radical can originate from hydrogen atom abstraction from the deoxyriboses (homolytic cleavage of a C–H bond) that creates a carbon-centered radical. On aromatic systems, hydrogen atom abstraction is not possible due to the high energy of aromatic C–H bond compared to aliphatic C–H. Alternatively, the addition of hydroxyl radicals on the double bonds of the bases (it can occur at any base) leads also to a radical at one nucleic base. Finally, one electron abstraction (electron transfer mechanism) is guanine specific and leads to a guanine radical cation $G^{+\bullet}$, that may deprotonate into a neutral radical $(G{-}H)^\bullet$. The trapping of the radical cation $G^{+\bullet}$ by H_2O was proposed, but this event leads also to

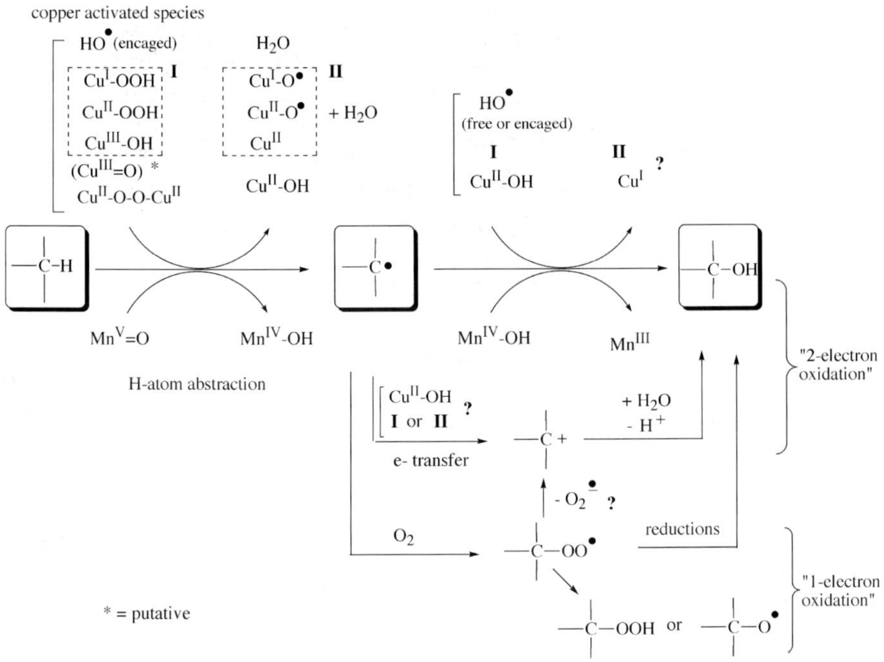

FIG. 34. DNA oxidation at deoxyribose units by a "1-electron oxidation" vs. a "2-electron oxidation" mechanism.

another radical intermediate (GOH•). The two mechanisms are illustrated in Figs 34 and 35, for deoxyribose and guanine, respectively.

The formation of a radical at deoxyribose moieties (one-electron oxidation of DNA) can be performed by HO• radicals or by activated species of oxidized metal complexes. Depending on the nature of the oxidizing species the fate of the deoxyribose radical will differ.

If the oxidant does not mediate rapid further oxidation of the radical, the latter can be trapped by molecular oxygen ("1-electron oxidation" chemistry). The products resulting from this pathway will incorporate an oxygen atom derived from molecular oxygen. At some particular sites (C1′ carbon), the stabilization of a positive charge at the carbon center makes the formation of a carbon-centered cation possible by the release of superoxide anion. In that case, after a one-electron oxidation of a deoxyribose unit, molecular oxygen plays the role of a second oxidant and this pathway becomes similar to a "2-electron oxidation" mechanism.

However, the oxidized metal complex may be able to perform a second oxidative event at the deoxyribose radical ("2-electron oxidation" mechanism). This oxidative event may be an electron transfer or an oxygen rebound mechanism (direct hydroxylation by Mn^{IV}–OH species). Alternatively, hydroxyl radical can combine with a carbon-centered radical. Oxygen rebound, reaction of hydroxyl radicals, or the attack of a molecule of water at a carbon cation, give rise to an alcohol at the oxidized carbon of the deoxyribose.

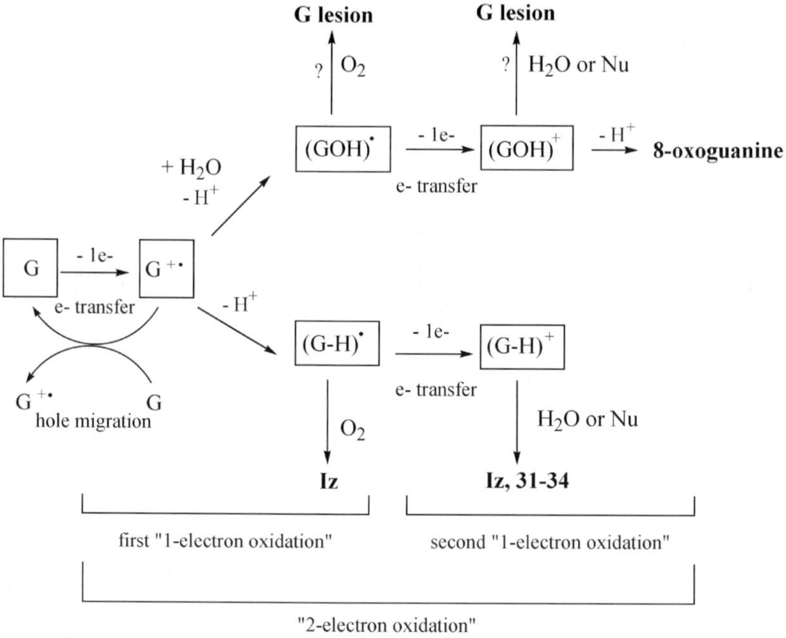

FIG. 35. DNA oxidation at guanine bases by a "1-electron oxidation" vs. a "2-electron oxidation" mechanism. Nu stands for nucleophilic species.

In the case of oxidative damage at the bases of DNA, the addition of hydroxyl radicals on the double bonds of the bases, that can occur at any base, leads to typical products of "1-electron oxidation" mechanism. A radical intermediate can be generated also by electron transfer, but this mechanism is guanine specific (Fig. 35). As in the mechanism described for deoxyribose, the one-electron oxidation of guanine leads to a radical species. After deprotonation of the initial guanine radical cation, $G^{+\bullet}$, the neutral radical $(G–H)^{\bullet}$ formed may be trapped by molecular oxygen, leading to imidazolone. Alternatively, the trapping of $G^{+\bullet}$ by H_2O was proposed to account for the formation of 8-oxoguanine. This event leads also to a radical intermediate, $(GOH)^{\bullet}$, that must then be oxidized again to explain the formation of 8-oxoguanine. The second oxidation event of $(G–H)^{\bullet}$ leads to a non-radical cation $(G–H)^{+}$ that can be trapped by nucleophilic reagents (water molecules, peroxides, amine groups) ("2-electron oxidation" mechanism).

The manganese complexes activated in the presence of appropriate oxygen atom donors are clearly capable of "2-electron oxidation" processes. Direct hydroxylation of deoxyribose (oxygen rebound mechanism) or two electrons abstraction from guanine were reported. The formation of high-valent metal-oxo species was possible for Mn-TMPyP by the means of dissymmetric peroxides with good leaving groups. This chemistry of activation precludes the formation of hydroxyl radicals (free or encaged) and allows the formation of oxidative species that are two redox equivalents above the initial state of the complex. These activated complexes are powerful oxidants not only in terms of electron transfer, but also in oxygen transfer reactions. No radical chemistry is observed, the trapping of intermediate radical species by O_2 on the deoxyribose units or on the bases does not occur.

In the case of copper complexes, the chemistry is less well understood. High valent copper-oxo species are not likely to be formed, consequently copper-hydroperoxo or copper-hydroxo species are usually proposed as active species in DNA oxidation. These species are thus more susceptible to homolytic cleavage of the peroxide or the metal-hydroxo bond and consequently, to "1-electron oxidation" mechanism. However, the labeling of the product of deoxyribose oxidation at C1′ by Cu(1,10-phenanthroline)$_2$ clearly demonstrated that these complexes can mediate a "2-electron oxidation" mechanism of DNA damage since the oxygen atom incorporated in DNA originates from H_2O.

The mechanisms of oxidative DNA damage reported *in vitro* with these two chemical nucleases illustrate the diversity of the processes that one may consider for the study of *in vivo* DNA damage.

Abbreviations

8-oxoguanine, 8-oxo-7,8-dihydroguanine; DTT, dithiothreitol; H_4edta, ethylenediamine tetra-acetic acid; GC/MS, gas chromatography coupled to mass spectrometry; LC/ESI-MS, liquid chromatography coupled to electrospray ionization mass spectrometry; MMPP, magnesium monoperphthalate; PAGE, polyacrylamide gel electrophoresis; phen, 1,10-phenanthroline; PhIO, iodosylbenzene; Fe-TMPyP, Fe^{III}-bis(aqua)*meso*-tetrakis(4-*N*-methylpyridiniumyl)-porphyrin; Mn-TMPyP, Mn^{III} bis(aqua)*meso*-tetrakis(4-*N*-methylpyridiniumyl)-porphyrin; TMPyP, *meso*-tetrakis(4-*N*-methylpyridiniumyl)-porphyrinato dianion; H_2-TMPyP, *meso*-tetrakis(4-*N*-methylpyridiniumyl)-porphyrin.

Acknowledgments

The authors thank Dr. Bernard Meunier for his continuing support and guidance. Prof. Jean Bernadou and Dr. Philippe Arnaud, Laboratoire de Chimie de Coordination, Toulouse, are acknowledged for fruitful discussions and for providing the modelisation views of Fig. 24 (PA).

References

1. Sigman, D. S.; Graham, D. R.; D'Aurora, V.; Stern, A. M. *J. Biol. Chem.* **1979**, *254*, 12269.
2. Sigman, D. S. *Acc. Chem. Res.* **1986**, *19*, 180.
3. Sigman, D. S.; Mazunder, A.; Perrin, D. M. *Chem. Rev.* **1993**, *93*, 2295.
4. Pratviel, G.; Bernadou, J.; Meunier, B. *Angew. Chem. Int. Ed.* **1995**, *34*, 746.
5. Bree, A. P.; Murphy, J. A. *Free Rad. Biol. Med.* **1995**, *18*, 1033.
6. Sigman, D. "*DNA and RNA Cleavers and Chemotherapy of Cancer and Viral Diseases*"; Kluwer Academic Publishers: Dordrecht/Boston/London, **1996**, p. 119.

7. Pogozelski, W. K.; Tullius, T. D. *Chem. Rev.* **1998,** *98*, 1089.
8. Chen, C. B.; Milne, L.; Landgraf, R.; Perrin, D. M.; Sigman, D. S. *ChemBioChem.* **2001,** *2*, 735.
9. James, B. R.; Williams, R. J. P.; *J. Chem. Soc.* **1961**, 2007.
10. Sillen, L. G.; Martell, A. E. "*Stability Constants of Metal-ion Complexes*"; The Chemical Society London Publication: Oxford, **1971**.
11. Dobson, J. F.; Green, B. E.; Healy, P. C.; Kennard, C. H. L.; Pakawatchai, C.; White, A. H. *J. Chem. Soc. Dalton Trans.* **1984,** *37*, 649.
12. Nakai, H.; Deguchi, Y. *Bull. Chem. Soc. Jpn.* **1975,** *48*, 2557.
13. Veal, J. M.; Rill, R. L. *Biochemistry* **1991,** *30*, 1132.
14. Veal, J. M.; Merchant, K.; Rill, R. L. *Nucleic Acids Res.* **1991,** *19*, 3383.
15. Marshall, L. E.; Graham, D. R.; Reich, K. A.; Sigman, D. S. *Biochemistry* **1981,** *20*, 244.
16. Johnson, G. R. A.; Nazhat, N. B. *J. Am. Chem. Soc.* **1987,** *109*, 1990.
17. Williams, L. D.; Thivierge, J.; Goldberg, I. H. *Nucleic Acids Res.* **1988,** *16*, 11607.
18. Dizdaroglu, M.; Aruoma, O. I.; Halliwell, B. *Biochemistry* **1990,** *29*, 8447.
19. Thederahn, T. B.; Kuwabara, M. D.; Larsen, T. A.; Sigman, D. S. *J. Am. Chem. Soc.* **1989,** *111*, 4941.
20. Pope, L. M.; Reich, K. A.; Graham, D. R.; Sigman, D. S. *J. Biol. Chem.* **1982,** *257*, 12121.
21. Kuwabara, M.; Yoon, C.; Goyne, T.; Thederahn, T.; Sigman, D. S. *Biochemistry* **1986,** *25*, 7401.
22. Zelenko, O.; Gallagher, J.; Xu, Y.; Sigman, D. S. *Inorg. Chem.* **1998,** *37*, 2198.
23. Oyoshi, T.; Sugiyama, H. *J. Am. Chem. Soc.* **2000,** *122*, 6313.
24. Reich, K. A.; Marshall, L. E.; Graham, D. R.; Sigman, D. S. *J. Am. Chem. Soc.* **1981,** *103*, 3582.
25. Goldstein, S.; Czapski, G. *J. Am. Chem. Soc.* **1986,** *108*, 2244.
26. Veal, J. M.; Rill, R. L. *Biochemistry* **1988,** *27*, 1822.
27. Veal, J. M.; Rill, R. L. *Biochemistry* **1989,** *28*, 3243.
28. Stocker, J. C. *Theor. Biol.* **1989,** *37*, 107.
29. Yoon, C.; Kuwabara, M. D.; Law, R.; Wall, R.; Sigman, D. S. *J. Biol. Chem.* **1988,** *263*, 8458.
30. Spassky, A.; Sigman, D. S. *Biochemistry* **1985,** *24*, 8050.
31. Sigman, D. S.; Spassky, A.; Rimsky, S.; Buc, H. *Biopolymers* **1985,** *24*, 183.
32. Pope, L. E.; Sigman, D. S. *Proc. Natl. Acad. Sci. USA* **1984,** *81*, 3.
33. Murakawa, G. J.; Chen, C. H.; Kuwabara, M. D.; Nierlich, D. P.; Sigman, D. S. *Nucleic Acids Res.* **1989,** *17*, 5361.
34. Goyne, T. E.; Sigman, D. S. *J. Am. Chem. Soc.* **1987,** *109*, 2846.
35. Meijler, M. M.; Zelenko, O.; Sigman, D. S. *J. Am. Chem. Soc.* **1997,** *119*, 1135.
36. Zheng, Y.; Sheppard, T. L. *Chem. Res. Toxicol.* **2004,** *17*, 197.
37. Chen, T.; Greenberg, M. M. *J. Am. Chem. Soc.* **1998,** *120*, 3815.
38. Bales, B. C.; Pitié, M.; Meunier, B.; Greenberg, M. C. *J. Am. Chem. Soc.* **2002,** *124*, 9062.
39. Roupioz, Y.; Lhomme, J.; Kotera, M. *J. Am. Chem. Soc.* **2002,** *124*, 9129.
40. Gonzalez-Alvarez, M.; Alzuet, G.; Borras, J.; Pitié, M.; Meunier, B. *J. Biol. Inorg. Chem.* **2003,** *8*, 644.
41. Patwardhan, A.; Cowan, J. A.; *Chem. Commun.* **2001**, 1490.
42. John, D. C. A.; Douglas, K. T. *Biochem. J.* **1993,** *289*, 463.
43. Pitié, M.; Burrows, C. J.; Meunier, B. *Nucleic Acids Res.* **2000,** *28*, 4856.
44. Cheng, C. C.; Goll, J. G.; Neyhart, G. A.; Welch, T. W.; Singh, P.; Thorp, H. H. *J. Am. Chem. Soc.* **1995,** *117*, 2970.
45. Burger, R. M. *Chem. Rev.* **1998,** *98*, 1153.
46. Clausen, C. A.; Long, E. C. *Chem. Rev.* **1999,** *99*, 2797.
47. Giese, B.; Beyrich-Graf, X.; Erdmann, P.; Giraud, L.; Imwinkelried, P.; Müller, S. N.; Schwitter, U. *J. Am. Chem. Soc.* **1995,** *117*, 6146.
48. Giese, B.; Beyrich-Graf, X.; Erdmann, P.; Petretta, M.; Schwitter, U. *Chem. Biol.* **1995,** *2*, 367.
49. Rashid, R.; Langfinger, D.; Wagner, R.; Schuchmann, H. P.; von Sonntag, C. *Int. J. Radiat. Biol.* **1999,** *75*, 101.
50. Dussy, A.; Meggers, E.; Giese, B. *J. Am. Chem. Soc.* **1998,** *120*, 7399.

51. Giloni, L.; Takeshita, M.; Johnson, F.; Iden, C.; Grollman, A. P. *J. Biol. Chem.* **1981**, *256*, 8608.
52. Miaskiewicz, K.; Osman, R. *J. Am. Chem. Soc.* **1994**, *116*, 232.
53. Sy, D.; Savoye, C.; Begusova, M.; Michalik, V.; Charlier, M.; Spotheim-Maurizot, M. *Int. J. Radiat. Biol.* **1997**, *72*, 147.
54. Balasubramanian, B.; Pogozelski, W. K.; Tullius, T. D. *Proc. Natl. Acad. Sci. USA* **1998**, *95*, 9738.
55. Burrows, C. J.; Muller, J. G. *Chem. Rev.* **1998**, *98*, 1109.
56. Frelon, S.; Douki, T.; Favier, A.; Cadet, J. *Chem. Res. Toxicol.* **2003**, *16*, 191.
57. Yamamoto, K.; Kawanishi, S. *J. Biol. Chem.* **1989**, *264*, 15435.
58. Yamamoto, K.; Kawanishi, S. *J. Biol. Chem.* **1991**, *266*, 1509.
59. Oikawa, S.; Kawanishi, S. *Biochemistry* **1996**, *35*, 4584.
60. Spencer, J. P. E.; Jenner, A.; Aruoma, O. I.; Evans, P. J.; Kaur, H.; Dexter, D. T.; Jenner, P.; Lees, A. J.; Marsden, D. C.; Halliwell, B. *FEBS Lett.* **1994**, *353*, 246.
61. Bal, W.; Lukszo, J.; Kasprzak, K. S. *Chem. Res. Toxicol.* **1997**, *10*, 915.
62. Muller, J. G.; Burrows, C. J. *Inorg. Chim. Acta* **1998**, *275, 276*, 314.
63. Milne, L.; Perrin, D. M.; Sigman, D. S. *Methods* **2001**, *23*, 160.
64. Pearson, L.; Chen, C. H.; Gaynor, R. P.; Sigman, D. S. *Nucleic Acids Res.* **1994**, *22*, 2255.
65. Sun, J. S.; François, J. C.; Lavery, R.; Saison-Behmoras, T.; Montenay-Garestier, T.; Thuong, N. T.; Hélène, C. *Biochemistry* **1988**, *27*, 6039.
66. François, J. C.; Saison-Behmoras, T.; Barbier, C.; Chassignol, M.; Thuong, N. T.; Hélène, C. *Proc. Natl. Acad. Sci. USA* **1989**, *86*, 9702.
67. Chen, C. H.; Mazumder, A.; Constant, J. F.; Sigman, D. S. *Bioconjugate Chem.* **1993**, *4*, 69.
68. Shimizu, M.; Inoue, H.; Ohtsuka, E. *Biochemistry* **1994**, *33*, 606.
69. François, J. C.; Hélène, C. *Biochemistry* **1995**, *34*, 65.
70. Bergstrom, D. E.; Gerry, N. P. *J. Am. Chem. Soc.* **1994**, *116*, 12067.
71. Perrin, D. M.; Mazumder, A.; Sadeghi, F.; Sigman, D. S. *Biochemistry* **1994**, *33*, 3848.
72. Pfau, J.; Arvidson, D. N.; Youderian, P.; Pearson, L. L.; Sigman, D. S. *Biochemistry* **1994**, *33*, 11391.
73. Penderrast, P. S.; Ebright, Y. W.; Ebright, R. H. *Science* **1994**, *265*, 959.
74. Bruice, T. W.; Wise, J. G.; Rosser, D. S. E.; Sigman, D. S. *J. Am. Chem. Soc.* **1991**, *113*, 5446.
75. Pan, C. Q.; Johnson, R. D. *Biochemistry* **1996**, *35*, 4326.
76. Xiao, G.; Cole, D. L.; Gunsalus, R. P.; Sigman, D. S.; Chen, C. H. B. *Protein Science* **2002**, *11*, 2427.
77. Zaid, A.; Sun, J. S.; Nguyen, C. H.; Bisagni, E.; Garestier, T.; Grierson, D. S.; Zain, R. *Chem. Bio. Chem.* **2004**, *5*, 1550.
78. François, J. C.; Saison-Behmoaras, T.; Chassignol, M.; Thuong, N. T.; Hélène, C. *J. Biol. Chem.* **1989**, *264*, 5891.
79. Hirai, M.; Shinozuka, K.; Sawai, H.; Ogawa, S. *Chem. Lett.* **1992**, 2023.
80. Federlin, P.; Kern, J. M.; Rastegar, A.; Dietrich-Buchecker, C.; Marnot, P. A.; Sauvage, J. P. *New J. Chem.* **1990**, *14*, 9.
81. Pitié, M.; Donnadieu, B.; Meunier, B. *Inorg. Chem.* **1998**, *37*, 3486.
82. Pitié, M.; Sudres, B.; Meunier, B. *Chem. Commun.* **1998**, 2597.
83. Pitié, M.; Meunier, B. *Bioconjugate Chem.* **1998**, *9*, 604.
84. Pitié, M.; Boldron, C.; Gornitzka, H.; Hemmert, C.; Donnadieu, B.; Meunier, B. *Eur. J. Inorg. Chem.* **2003**, 528.
85. Ross, S. A.; Pitié, M.; Meunier, B. *Eur. J. Inorg. Chem.* **1999**, 557.
86. Boldron, C.; Ross, S. A.; Pitié, M.; Meunier, B. *Bioconjugate Chem.* **2002**, *13*, 1013.
87. Pitié, M.; Van Horn, J. D.; Brion, D.; Burrows, C. J.; Meunier, B. *Bioconjugate Chem.* **2000**, *11*, 892.
88. Dervan, P. B. *Science* **1986**, *232*, 464.
89. Sreedhara, A.; Cowan, J. A. *J. Biol. Inorg. Chem.* **2001**, *6*, 337.
90. Eppley, H. J.; Lato, S. M.; Ellington, A. D.; Zaleski, J. M. *Chem. Comm.* **1999**, 2405.
91. Dhar, S.; Senapati, D.; Das, P. K.; Chattopadhyay, P.; Nethaji, M.; Chakravarty, A. R. *J. Am. Chem. Soc.* **2003**, *125*, 12118.

92. Sagripanti, J. L. "*Metals Ions in Biological Systems*"; Ed. Sigel, H.; Marcel Dekker. Inc./ New York: Basel. Hong-Kong, **2002**, p. 196.
93. Baudoin, O.; Teulade-Fichou, M. P.; Vigneron, J. P.; Lehn, J. M. *Chem. Comm.* **1998**, 2349.
94. Humphreys, K. J.; Karlin, K. D.; Rokita, S. E. *J. Am. Chem. Soc.* **2002**, *124*, 6009.
95. Humphreys, K. J.; Karlin, K. D.; Rokita, S. E. *J. Am. Chem. Soc.* **2002**, *124*, 8055.
96. Hemmert, C.; Pitié, M.; Gornitzka, H.; Soulet, S.; Meunier, B. *J. Biol. Inorg. Chem.* **2001**, *6*, 14.
97. Gonzallez-Alvarez, M.; Alzuet, G.; Borras, J.; Macias, B.; Castineiras, A. *Inorg. Chem.* **2003**, *42*, 2992.
98. Labudova, O.; Labuda, J.; Fenikova, L.; Kollarova, M.; Durackova, Z. *J. Inorg. Biochem.* **1996**, *61*, 227.
99. Routier, S.; Bernier, J. L.; Catteau, J. P.; Bailly, C. *Med. Chem. Lett.* **1997**, *7*, 1729.
100. Lamour, E.; Routier, S.; Bernier, J. L.; Catteau, J. P.; Bailly, C.; Vezin, H. *J. Am. Chem. Soc.* **1999**, *121*, 1862.
101. Dhar, S.; Charkravarty, A. R. *Inorg. Chem.* **2003**, *42*, 2483.
102. Naik, A. D.; Reddy, P. A. N.; Nethaji, M.; Chakravarty, A. R. *Inorg. Chim. Acta* **2003**, *349*, 149.
103. Kawanishi, S.; Hiraku, Y.; Murata, M.; Oikawa, S. *Free Rad. Biol. Med.* **2002**, *32*, 822.
104. Flowers, L.; Ohnishi, S. T.; Penning, T. M. *Biochemistry* **1997**, *36*, 8640.
105. Mack, D. P.; Iverson, B. L.; Dervan, P. B. *J. Am. Chem. Soc.* **1988**, *110*, 7572.
106. Bailly, C.; Sun, J. S.; Colson, P.; Houssier, C.; Hélène, C.; Waring, M. J.; Hénichart, J. P. *Bioconjugate Chem.* **1992**, *3*, 100.
107. Shullenberger, D. F.; Eason, P. D.; Long, E. C. *J. Am. Chem. Soc.* **1993**, *115*, 11038.
108. Oikawa, S.; Kurasaki, M.; Kojima, Y.; Kawanishi, S. *Biochemistry* **1995**, *34*, 8763.
109. Chand, D. K.; Schneider, H. J.; Bencini, A.; Bianchi, A.; Giorgi, C.; Ciattini, S.; Valtancoli, B. *Chem. Eur. J.* **2000**, *6*, 4001.
110. Detmer, C. A.III; Pamatong, F. V.; Bocarsly, J. R. *Inorg. Chem.* **1997**, *36*, 3676.
111. Borah, S.; Melvin, M. S.; Lindquist, N.; Manderville, R. A. *J. Am. Chem. Soc.* **1998**, *120*, 4557.
112. Melvin, M. S.; Wooton, K. E.; Rich, C. C.; Saluta, G. R.; Kucera, G. L.; Lindquist, N.; Manderville, R. A. *J. Inorg. Biochem.* **2001**, *87*, 129.
113. Asad, S. F.; Singh, S.; Ahmad, A.; Hadi, S. M. *Toxicol. Lett.* **2002**, *131*, 181.
114. Furstner, A. *Angew. Chem. Int. Ed.* **2003**, *42*, 3582.
115. Jezowska-Bojczuk, M.; Lesniak, W.; Szczepanik, W.; Gatner, K.; Jezierski, A.; Smoluch, M.; Bal, W. *J. Inorg. Biochem.* **2001**, *84*, 189.
116. Lesniak, W.; Harris, W. R.; Kravitz, J. Y.; Schacht, J.; Pecoraro, V. L. *Inorg. Chem.* **2003**, *42*, 1420.
117. Mizutani, H.; Oikawa, S.; Hiraku, Y.; Murata, M.; Kojima, M.; Kawanishi, S. *Cancer Sci.* **2003**, *94*, 686.
118. Ehrenfeld, G. M.; Shipley, J. B.; Heimbrook, D. C.; Sugiyama, H.; Long, E. C.; van Boom, J. H.; van der Marel, G. A.; Oppenheimer, N. J.; Hecht, S. M. *Biochemistry* **1987**, *26*, 931.
119. Singh, U. S.; Scannel, R. T.; An, H.; Carter, B. J.; Hecht, S. M. *J. Am. Chem. Soc.* **1995**, *117*, 12691.
120. Deng, J. Z.; Newman, D. J.; Hecht, S. M. *J. Nat. Prod.* **2000**, *63*, 1269.
121. Starck, S. R.; Deng, J. Z.; Hecht, S. M. *Biochemistry* **2000**, *39*, 2413.
122. Iyengar, B. S.; Takahashi, T.; Remers, W. A.; Bradner, W. T. J. *Med. Chem.* **1986**, *29*, 144.
123. Ohnishi, S.; Kawanishi, S. *Biochem. Biophys. Res. Comm.* **2002**, *290*, 778.
124. Seiko, K.; Murata, M., Oikawa, S.; Hiraku, Y.; Hirakawa, K.; Kawanishi, S. *Chem. Res. Toxicol.* **2003**, *16*, 1470.
125. Sigel, H. Ed., *"Metal Ions in Biological Systems"*; vol. 12; Marcel Dekker, Inc.: New York, Basel, **1981**.
126. Sigel, H. Ed., "*Metal Ions in Biological Systems*"; vol. 13; Marcel Dekker, Inc.: New York, Basel, **1981**.
127. Masarwa, M.; Cohen, H.; Meyerstein, D.; Hickman, D. L.; Bakac, A.; Espenson, J. H. *J. Am. Chem. Soc.* **1988**, *110*, 4293.

128. Detmer, C. A.III; Pamatong, F. V.; Bocarsly, J. R. *Inorg. Chem.* **1996**, *35*, 6292.
129. Drouin, R.; Rodriguez, H.; Gao, S. W.; Gebreyes, Z.; O'Connor, T. R.; Holmquist, G. P.; Akman, S. A. *Free Radic. Biol. Med.* **1996**, *21*, 261.
130. Yamamoto, K.; Kawanishi, S. *J. Biol. Chem.* **1989**, *264*, 15435.
131. Kawanishi, S.; Hiraku, Y.; Murata, M.; Oikawa, S. *Free Radic. Biol. Med.* **2002**, *32*, 822.
132. Schweigert, N.; Acero, J. L.; von Guten, U.; Canonica, S.; Zehnder, J. B.; Eggen, R. I. L. *Environ. Mol. Mutagen.* **2000**, *36*, 5.
133. Flowers, L.; Ohnishi, T.; Penning, T. M. *Biochemistry* **1997**, *36*, 8640.
134. Bossu, F. P.; Chellappa, K. L.; Margerum, D. W. *J. Am. Chem. Soc.* **1977**, *99*, 2195.
135. Meunier, B. "*Biomimetic Oxidations Catalyzed by Transition Metal Complexes Imperial*"; College Press: London, **1998**.
136. Amine, A.; Atmani, Z.; El Hallaoui, A.; Giorgi, M.; Pierrot, M.; Réglier, M. *Bioorg. Med. Chem. Lett.* **2002**, *12*, 57.
137. Mayer, J. M. *Comments Inorg. Chem.* **1988**, *8*, 125.
138. Lewis, E. A.; Tolman, W. B. *Chem. Rev.* **2004**, *104*, 1047.
139. Prigge, S. T.; Mains, R. E.; Eipper, B. A.; Amzel, L. M. *Cell. Mol. Life Sci.* **2000**, *57*, 1236.
140. Klinman, J. P. *Chem. Rev.* **1996**, *96*, 2541.
141. Blackburn, N. J.; Rhames, F. C.; Ralle, M.; Jaron, S. *J. Biol. Inorg. Chem.* **2000**, *5*, 341.
142. Kitajima, N.; Koda, T.; Iwata, Y.; Moro-oka, Y. *J. Am. Chem. Soc.* **1990**, *112*, 8833.
143. Osako, T.; Nagato, S.; Tachi, Y.; Kitagawa, T.; Itoh, S. *Angew. Chem. Int. Ed.* **2002**, *41*, 4325.
144. Theophanides, T.; Anastassopoulou, J. *Critical Rev. Oncol. Hematol.* **2002**, *42*, 57.
145. Hatcher, L. Q., Karlin, K. D. *J. Biol. Inorg. Chem.* **2004**, *9*, 669.
146. Rorabacher, D. B. *Chem. Rev.* **2004**, *104*, 651.
147. Meunier, B.; de Visser, S. P.; Shaik, S. *Chem. Rev.* **2004**, *104*, 3947.
148. Everse, J.; Everse, K. E.; Grisham, M. B. Eds. In: "*Peroxidases in Chemistry and Biology*"; vol. I and II, CRC Press: Boca Raton, **1991**.
149. Groves, T. J.; Han, Y. Z. In: "*Cytochrome P450. Structure, Mechanism and Biochemistry*"; 2nd edn.; Ortiz de Montellano, P. R.; Plenum Press: New York, **1995**, pp. 3–48.
150. Mansuy, D. *Coord. Chem. Rev.* **1993**, *125*, 129.
151. Dolphin, D.; Traylor, T. G.; Xie, L. Y. *Acc. Chem. Res.* **1997**, *30*, 251.
152. McLain, J. L., Lee, J.; Groves, J. T. In: "*Biomimetic Oxidations Catalyzed by Transition Metal Complexes*"; Ed. Meunier, B. Imperial College Press: London, **2000**, pp. 91–169.
153. Meunier, B.; Robert, A.; Pratviel, G.; Bernadou, J. In: "*The Porphyrin Handbook*"; Eds. Kadish, K.M.; Smith, K.M.; Guilard, R. vol. 4; Academic Press: San Diego, **2000**, pp. 119–187.
154. Srinivasan, K.; Michaud, P.; Kochi, J. K. *J. Am. Chem. Soc.* **1986**, *108*, 2309.
155. Jacobsen, E. N.; Zhang, W.; Muci, A. R.; Ecker, J. R.; Deng, L. *J. Am. Chem. Soc.* **1991**, *113*, 7063.
156. Palucki, M.; Finney, N. S.; Pospisil, P. J.; Güler, M. L.; Ishida, T.; Jacobsen, E. N. *J. Am. Chem. Soc.* **1998**, *120*, 948.
157. Jacobsen, E. N. In: "*Comprehensive Organometallic Chemistry II*"; Eds. Wilkinson, G.; Stone, F. G. A.; Abel, E. W.; Hegedus, L. S.; vol. 12; Pergamon Press: New York, **1995**.
158. Katsuki, T. *Coord. Chem. Rev.* **1995**, *140*, 189.
159. Stubbe, J.; Kozarich, J. W. *Chem. Rev.* **1987**, *87*, 1107.
160. Hecht, S. M. *Acc. Chem. Res.* **1986**, *19*, 383.
161. Chen, J.; Stubbe, J. *Nature Rev. Cancer* **2005**, *5*, 102.
162. Burger, R. M. *Struct. Bonding* **2000**, *97*, 289.
163. Ehrenfeld, G. M.; Murugesan, N.; Hecht, S. M. *Inorg. Chem.* **1984**, *23*, 1496.
164. Burger, R. M.; Freedman, J. H.; Horwitz, S. B.; Peisach, J. *Inorg. Chem.* **1984**, *23*, 2215.
165. Groves, T. J.; Lee, J.; Marla, S. S. *J. Am. Chem. Soc.* **1997**, *119*, 6269.
166. Jin, N.; Groves, T. J. *J. Am. Chem. Soc.* **1999**, *121*, 2923.
167. Feichtinger, D.; Plattner, D. A. *Angew. Chem. Int. Ed. Engl.* **1997**, *36*, 1718.
168. Bernadou, J.; Fabiano, A.-S.; Robert, A.; Meunier, B. *J. Am. Chem. Soc.* **1994**, *116*, 9375.
169. Bernadou, J.; Meunier, B.; *Chem. Commun.* **1998**, 2167.
170. Meunier, B.; Bernadou, J. *Struct. Bonding* **2000**, *97*, 1.

171. Pitié, M.; Bernadou, J.; Meunier, B. *J. Am. Chem. Soc.* **1995**, *117*, 2935.
172. Gravert, D. J.; Griffin, J. H. *J. Org. Chem.* **1993**, *58*, 820.
173. Gravert, D. J.; Griffin, J. H. *Inorg. Chem.* **1996**, *35*, 4837.
174. Mandal, S. S.; Varshney, U.; Bhattacharya, S. *Bioconjugate Chem.* **1997**, *8*, 798.
175. Griffin, J. H. *Bioorg. Med. Chem. Lett.* **1995**, *5*, 73.
176. Hertzberg, R. P.; Dervan, P. B. *J. Am. Chem. Soc.* **1982**, *104*, 313.
177. Hertzberg, R. P.; Dervan, P. B. *Biochemistry* **1984**, *23*, 3934.
178. Goldberg, I. H. *Acc. Chem. Res.* **1991**, *24*, 191.
179. Xi, Z.; Goldberg, I. H. In: "*Comprehensive Natural Products Chemistry*"; Eds. Barton, D. H. R.; Nakanishi, K. vol. 7; Elsevier Science: Oxford, **1999**, pp. 553–592.
180. Gravert, D. J.; Griffin, J. H. *Bioorg. Med. Chem. Lett.* **1996**, *6*, 889.
181. Routier, S.; Joanny, V.; Zaparucha, A.; Vezin, H.; Catteau, J.-P.; Bernier, J.-L.; Bailly, C. *J. Chem. Soc. Perkin Trans. 2* **1998**, 863.
182. Bernadou, J.; Pratviel, G.; Bennis, F.; Girardet, M.; Meunier, B. *Biochemistry* **1989**, *28*, 7268.
183. Fiel, R. J.; Beerman, T. A.; Mark, E. H.; Datta-Gupta, N. *Biochem. Biophys. Res. Commun.* **1982**, *107*, 1067.
184. Banfi, S.; Cassani, E.; Caruso, E.; Cazzaro, M. *Bioorg. Med. Chem.* **2003**, *11*, 3595.
185. Bromley, S. D.; Ward, B. W.; Dabrowiak, J. C. *Nucleic Acids Res.* **1986**, *14*, 9133.
186. Ward, B.; Skorobogaty, A.; Dabrowiak, J. C. *Biochemistry* **1986**, *25*, 7827.
187. Ward, B.; Skorobogaty, A.; Dabrowiak, J. C. *Biochemistry* **1986**, *25*, 6875.
188. Prince, S.; Körber, F.; Cooke, P. R.; Lindsay Smith, J. R.; Mazid, M. A. *Acta Crystallogr. Sect. C* **1993**, *49*, 1158.
189. Lavery, R.; Pullman, B. *J. Biomol. Struct. Dynamics* **1985**, *2*, 1021.
190. Weiner, P. K.; Langeridge, R.; Blaney, J. M.; Schaefer, P. A.; Kollman, P. A. *Proc. Natl. Acad. Sci. USA* **1982**, *79*, 3754.
191. Rodriguez, M.; Kodadek, T.; Torres, M.; Bard, A. J. *Bioconjugate Chem.* **1990**, *1*, 123.
192. Fouquet, E.; Pratviel, G.; Bernadou, J.; Meunier, B. *J. Chem. Soc. Chem. Commun.* **1987**, 1169.
193. Ward, B.; Rehfuss, R.; Dabrowiak, J. C. *J. Biomol. Struct. Dynamics* **1987**, *4*, 685.
194. Pratviel, G.; Bernadou, J.; Ricci, M.; Meunier, B. *Biochem. Biophys. Res. Commun.* **1989**, *160*, 1212.
195. Groves, J. T.; Marla, S. S. *J. Am. Chem. Soc.* **1995**, *117*, 9578.
196. Marla, S. S.; Lee, J.; Groves, T. J. *Proc. Natl. Acad. Sci. USA* **1997**, *94*, 14243.
197. Lee, J.; Hunt, J. A.; Groves, T. J. *Bioorg. Med. Chem. Lett.* **1997**, *7*, 2913.
198. Wietzerbin, K.; Muller, J. G.; Jameton, R. A.; Pratviel, G.; Bernadou, J.; Meunier, B.; Burrows, C. *Inorg. Chem.* **1999**, *38*, 4123.
199. Bennet, M.; Krah, A.; Wien, F.; Garman, E.; Mckenna, R.; Sanderson, M.; Neidle, S. *Proc. Natl. Acad. Sci. USA* **2000**, *97*, 9476.
200. Arnaud, P.; Zakrzewska, K.; Meunier, B. *J. Comput. Chem.* **2003**, *24*, 797.
201. Pitié, M.; Pratviel, G.; Bernadou, J.; Meunier, B. *Proc. Natl. Acad. Sci. USA* **1992**, *89*, 3967.
202. Ding, L.; Bernadou, J.; Meunier, B. *Bioconjugate Chem.* **1991**, *2*, 201.
203. Dabrowiak, J. C.; Ward, B.; Goodisman, J. *Biochemistry* **1989**, *28*, 3314.
204. Pratviel, G.; Pitié, M.; Bernadou, J.; Meunier, B. *Angew. Chem. Int. Ed. Engl.* **1991**, *30*, 702.
205. Pratviel, G.; Duarte, V.; Bernadou, J.; Meunier, B. *J. Am. Chem. Soc.* **1993**, *115*, 7939
206. Pratviel, G.; Pitié, M.; Périgaud, C.; Gosselin, G.; Bernadou, J.; Meunier, B. *J. Chem. Soc. Chem. Commun.* **1993**, 149.
207. Bernadou, J.; Lauretta, B.; Pratviel, G.; Meunier, B. *C. R. Acad. Sci.* **1989**, *309 III*, 409.
208. Pratviel, G.; Pitié, M.; Bernadou, J.; Meunier, B. *Nucleic Acids Res.* **1991**, *19*, 6283.
209. Angeloff, A.; Dubey, I.; Pratviel, G.; Bernadou, J.; Meunier, B. *Chem. Res. Toxicol.* **2001**, *14*, 1413.
210. Mestre, B.; Pratviel, G.; Meunier, B. *Bioconjugate Chem.* **1995**, *6*, 466.
211. Mestre, B.; Jakobs, A.; Pratviel, G.; Meunier, B. *Biochemistry* **1996**, *35*, 9140.

212. Mestre, B.; Pitié, M.; Loup, C.; Claparols, C.; Pratviel, G.; Meunier, B. *Nucleic Acids Res.* **1997**, *25*, 1022.
213. Dubey, I.; Pratviel, G.; Meunier, B. *J. Chem. Soc. Perkin Trans. 1* **2000**, 3088.
214. Steenken, S.; Jovanovic, S. V. *J. Am. Chem. Soc.* **1997**, *119*, 617.
215. Saito, I.; Takayama, M.; Sugiyama, H.; Nakatani, K. *J. Am. Chem. Soc.* **1995**, *117*, 6406.
216. Adam, W.; Grimm, G. N.; Marquardt, S.; Saha-Möller, C. R. *J. Am. Chem. Soc.* **1999**, *121*, 1179.
217. Wan, C.; Fiebig, T.; Schiemann, O.; Barton, J. K.; Zewail, A. H. *Proc. Natl. Acad. Sci. USA* **2000**, *97*, 14052.
218. Schiemann, O.; Turro, N. J.; Barton, J. K. *J. Phys. Chem. B* **2000**, *104*, 7214.
219. Knör, G. *Inorg. Chem. Commun.* **2001**, *4*, 160.
220. Luo, W.; Muller, J. G.; Burrows, C. J. *Org. Lett.* **2001**, *3*, 2801.
221. Travascio, P.; Witting, P. K.; Mauk, A. G.; Sen, D. *J. Am. Chem. Soc.* **2001**, *123*, 1337.
222. Suzuki, T.; Masuda, M.; Friesen, M. D.; Fenet, B.; Ohshima, H. *Nucleic Acids Res.* **2002**, *30*, 2555.
223. Suzuki, T.; Friesen, M. D.; Ohshima, H. *Chem. Res. Toxicol.* **2003**, *16*, 382.
224. Joffe, A.; Mock, S.; Yun, B. H.; Kolbanovskiy, A.; Geacintov, N. E.; Shafirovich, V. *Chem. Res. Toxicol.* **2003**, *16*, 966.
225. Joffe, A.; Geacintov, N. E.; Shafirovich, V. *Chem. Res. Toxicol.* **2003**, *16*, 1528.
226. Kobayashi, K.; Tagawa, S. *J. Am. Chem. Soc.* **2003**, *125*, 10213.
227. Ravanat, J.-L.; Saint-Pierre, C.; Cadet, J. *J. Am. Chem. Soc.* **2003**, *125*, 2030.
228. Ye, Y.; Muller, J. G.; Luo, W.; Maynes, C. L.; Shallop, A. J.; Jones, R. A.; Burrows, C. J. *J. Am. Chem. Soc.* **2003**, *125*, 13926.
229. Williams, T. T.; Dohno, C.; Stemp, E. D. A.; Barton, J. K. *J. Am. Chem. Soc.* **2004**, *126*, 8148.
230. Liu, C.-S.; Hernandez, R.; Schuster, G. B. *J. Am. Chem. Soc.* **2004**, *126*, 2877.
231. Niles, J. C.; Wishnok, J. S.; Tannenbaum, S. R. *Chem. Res. Toxicol.* **2004**, *17*, 1501.
232. Niles, J. C.; Wishnok, J. S.; Tannenbaum, S. R. *Chem. Res. Toxicol.* **2004**, *17*, 1510.
233. McCallum, J. E. B.; Kuniyoshi, C. Y.; Foote, C. S. *J. Am. Chem. Soc.* **2004**, *126*, 16777.
234. Stemmler, A. J.; Burrows, C. J. *J. Am. Chem. Soc.* **1999**, *121*, 6956.
235. Farrer, B. T.; Thorp, H. H. *Inorg. Chem.* **2000**, *39*, 44.
236. Sugden, K. D.; Campo, C. K.; Martin, B. D. *Chem. Res. Toxicol.* **2001**, *14*, 1315.
237. Choi, S.; Delaney, S.; Orbai, L.; Padgett, E. J.; Hakemian, A. S. *Inorg. Chem.* **2001**, *40*, 5481.
238. Joudah, L.; Moghaddas, S.; Bose, R.N. *Chem. Commun.* **2002**, 1742.
239. Choi, S.; Cooley, R. B.; Hakemian, A. S.; Larrabee, Y. C.; Bunt, R. C.; Maupas, S. D.; Muller, J. G.; Burrows, C. J. *J. Am. Chem. Soc.* **2004**, *126*, 591.
240. Li, L.; Karlin, K. D.; Rokita, S. E. *J. Am. Chem. Soc.* **2005**, *127*, 520.
241. Steenken, S. *Chem. Rev.* **1989**, *89*, 503.
242. Cadet, J.; Berger, M.; Douki, T.; Ravanat, J.-L. *Rev. Physiol. Biochem. Pharmacol.* **1997**, *131*, 1.
243. Armitage, B. *Chem. Rev.* **1998**, *98*, 1171.
244. Nunez, M. E.; Hall, D. B.; Barton, J. K. *Chem. Biol.* **1999**, *6*, 85.
245. Schuster, G. B. *Acc. Chem. Res.* **2000**, *33*, 253.
246. Giese, B. *Annu. Rev. Biochem.* **2002**, *71*, 51.
247. Muller, J. G.; Duarte, V.; Hickerson, R. P.; Burrows, C. J. *Nucleic Acids Res.* **1998**, *26*, 2247.
248. Vialas, C.; Pratviel, G.; Claparols, C.; Meunier, B. *J. Am. Chem. Soc.* **1998**, *120*, 11548.
249. Cadet, J.; Berger, M.; Buchko, G. W.; Joshi, P. C.; Raoul, S.; Ravanat, J.-L. *J. Am. Chem. Soc.* **1994**, *116*, 7403.
250. Raoul, S.; Berger, M.; Buchko, G. W.; Joshi, P. C.; Morin, B.; Weinfeld, M.; Cadet, J. *J. Chem. Soc. Perkin Trans. 2* **1996**, 371.
251. Vialas, C.; Pratviel, G.; Meyer, A.; Rayner, B.; Meunier, B. *J. Chem. Soc. Perkin Trans. 1* **1999**, 1201.

252. Candeias, L. P.; Steenken, S. *J. Am. Chem. Soc.* **1989**, *111*, 1094.
253. Vialas, C.; Claparols, C.; Pratviel, G.; Meunier, B. *J. Am. Chem. Soc.* **2000**, *122*, 2157.
254. Lapi, A.; Pratviel, G.; Meunier, B. *Met.-Based Drugs* **2001**, *8*, 47.
255. Duarte, V.; Gasparutto, D.; Yamaguchi, L. F.; Ravanat, J.-L.; Martinez, G. R.; Medeiros, M. H. G.; Di Mascio, P.; Cadet, J. *J. Am. Chem. Soc.* **2000**, *122*, 12622.
256. Luo, W.; Muller, J. G.; Rachlin, E. M.; Burrows, C. J. *Chem. Res. Toxicol.* **2001**, *14*, 927.
257. Adam, W.; Arnold, M. A.; Grüne, M.; Nau, W. M.; Pischel, U.; Saha-Möller, C. R. *Org. Lett.* **2002**, *4*, 537.
258. Luo, W.; Muller, J. G.; Rachlin, E. M.; Burrows, C. J. *Org. Lett.* **2000**, *3*, 613.
259. Niles, J. C.; Wishnok, J. S.; Tannenbaum, S. R. *Org. Lett.* **2001**, *3*, 963.
260. Suzuki, T.; Masuda, M.; Friesen, M. D.; Ohshima, H. *Chem. Res. Toxicol.* **2001**, *14*, 1163.
261. Chworos, A.; Seguy, C.; Pratviel, G.; Meunier, B. *Chem. Res. Toxicol.* **2002**, *15*, 1643.
262. Chworos, A.; Coppel, Y.; Dubey, I.; Pratviel, G.; Meunier, B. *J. Am. Chem. Soc.* **2001**, *123*, 5867.
263. Seguy, C.; Pratviel, G.; Meunier, B. *Chem. Commun.* **2001**, 2116.
264. Wheelhouse, R. T.; Sun, D.; Han, H.; Han, F. X.; Hurley, L. H. *J. Am. Chem. Soc.* **1998**, *120*, 3261.
265. Vialas, C.; Pratviel, G.; Meunier, B. *Biochemistry* **2000**, *39*, 9514.

LIGAND INFLUENCES IN COPPER-DIOXYGEN COMPLEX-FORMATION AND SUBSTRATE OXIDATIONS

LANYING Q. HATCHER and KENNETH D. KARLIN*

The Johns Hopkins University, Department of Chemistry, 3400 N. Charles Street, Baltimore, MD 21218, USA

I. Introduction

The reaction of Cu(I) with molecular oxygen is of considerable interest for understanding the inherent reactivity tendencies of copper ion in biological enzymes as well as in synthetic systems. The low cost and availability, and the broad range of oxidative potential makes copper reagents or catalysts potentially very attractive. Comprehensive knowledge of Cu^I/O_2 reactivity will lay the foundation for designing better, perhaps even "green" oxidation catalysts. With this in mind, we seek to understand the kinetics and thermodynamics of the copper(I) mediated interactions and reduction of dioxygen. It is of interest to elucidate different possible copper-dioxygen compound structures, and to understand their bonding interactions and associated spectroscopic signatures. Finally, we strive to generate efficient oxidative

ADVANCES IN INORGANIC CHEMISTRY
VOLUME 58 ISSN 0898-8838 / DOI: 10.1016/S0898-8838(05)58004-8

compounds, not only for practical applications, but also for mechanistic interrogation, to gain a fundamental understanding of the behavior of discrete copper-dioxygen intermediates. In this review, which is not intended to be all inclusive, we describe our progress in pursuit of such goals, highlighting much of our most recent work. For comprehensive recent reviews, the reader is referred to the excellent articles by Stack (*1*), Tolman (*2*), and their co-workers, which include nearly all of the recent advances in the field of copper-dioxygen chemistry through mid-2003.

This review contains two main parts with the first section emphasizing the importance of the ligand coordinating to the copper center. We will first describe how ligand modifications can often result in drastically different dioxygen reactivity. The rates and thermodynamics of O_2-binding are influenced by the structural, electronic, or geometric properties of the supporting ligand. In addition, the characteristics of the Cu_nO_2 adduct changes in terms of structure, and the degree of reduction of O_2 (i.e., superoxide $(O_2)^-$, peroxide $(O_2)^{2-}$, or oxide $(O^{2-})_2$), depending on the ligand utilized. We will examine several examples to highlight the ligand aspects such as chelate ring size, electronics, and nitrogen donor type, and their consequences on the reactivity of the copper(I) complex towards O_2.

The second part of the article will discuss ligand effects on the subsequent reactivity of the dioxygen adduct produced. Intramolecular and intermolecular oxidations by copper-dioxygen compounds are well documented. We first discuss general reactivity patterns of Cu_2O_2 compounds with a review of our early findings, and then describe our more recent efforts focusing on detailed mechanistic investigations of substrate oxidations.

A. Biological Inspiration

Oxidative transformations carried out by the reaction of copper(I) and O_2 in Nature are widespread in terms of the variety of substrates and in copper active-site structures. The archetype for dicopper-dioxygen activation has been the dioxygen carrier for arthropods and mollusks, hemocyanin (Hc), which has a dinuclear copper(I) active site where the copper centers are *ca.* 4.5 Å apart. Dioxygen binding results in the two-electron reduced peroxide in a side-on bound dicopper(II) structure (Fig. 1). Crystallographic information is (also) established for the reduced, as well as the oxidized "met" forms (*3–5*). Based on their spectroscopic similarities, a similar active-site structure is proposed for the monooxygenase tyrosinase (Tyr); X-ray structures are available for catechol oxidase (CO) depicting a similar active-site structure to Hc and Tyr (*6*). However, unlike the reversible binding reactivity of hemocyanin, tyrosinase catalyzes the aromatic hydroxylation of tyrosine to 3,4-dihydroxyphenylalanine (L-DOPA), and can further perform catechol oxidase activity which oxidizes the catechol to the quinone (Fig. 2) (*7*).

Also extensively studied are dicopper monooxygenases involved in the regulation of neurotransmitters. Dopamine β-hydroxylase (DβH) catalyzes the stereospecific benzylic hydroxylation of dopamine for the biosynthesis of the hormone norepinephrine, while peptidylglycine α-hydroxylating

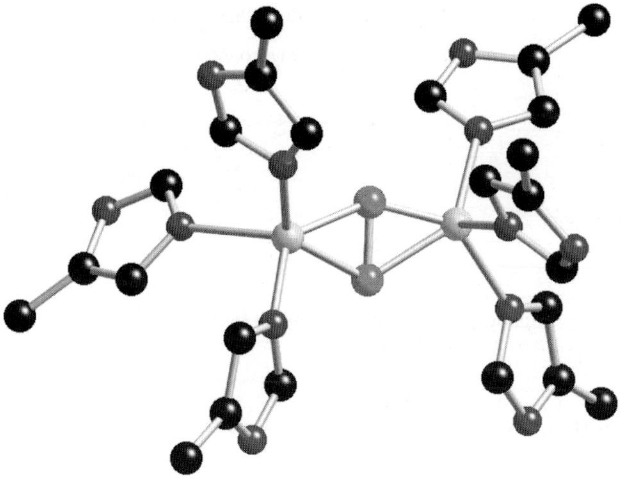

FIG. 1. Crystal structure of the oxyhemocyanin Cu_2-O_2 active site (PDB code: 1OXY).

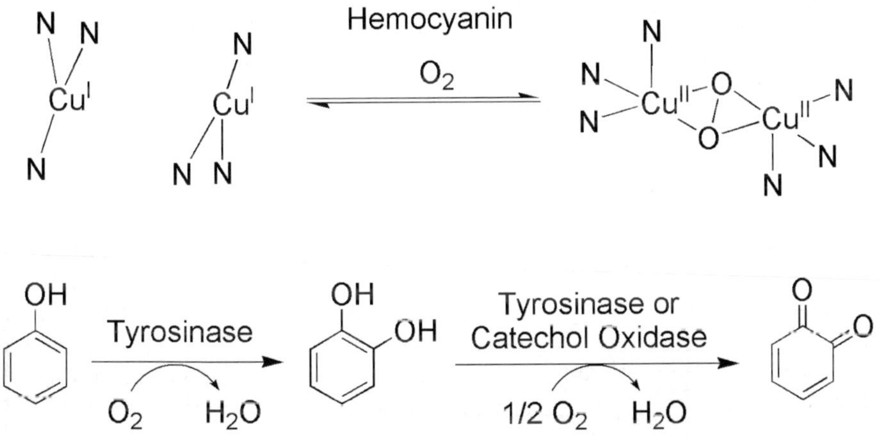

FIG. 2. Reversible binding and oxidative function of dicopper proteins.

monooxygenase (PHM) catalyzes the first step in peptide amidation leading to active biopeptides, following hydroxylation of the α-carbon of a terminal glycine residue (Fig. 3) (*8*).

The two enzymes are similar in function and exhibit considerable sequence homologies (*9*). In contrast to Tyr and CO, the copper ions in DβH and PHM activate O_2 *via* an uncoupled mechanism; the crystal structure of PHM reveals that the distance between the copper ions is 11 Å (Fig. 3) (*10*). One copper(I) site (Cu_M, also known as Cu_B) binds O_2, forming

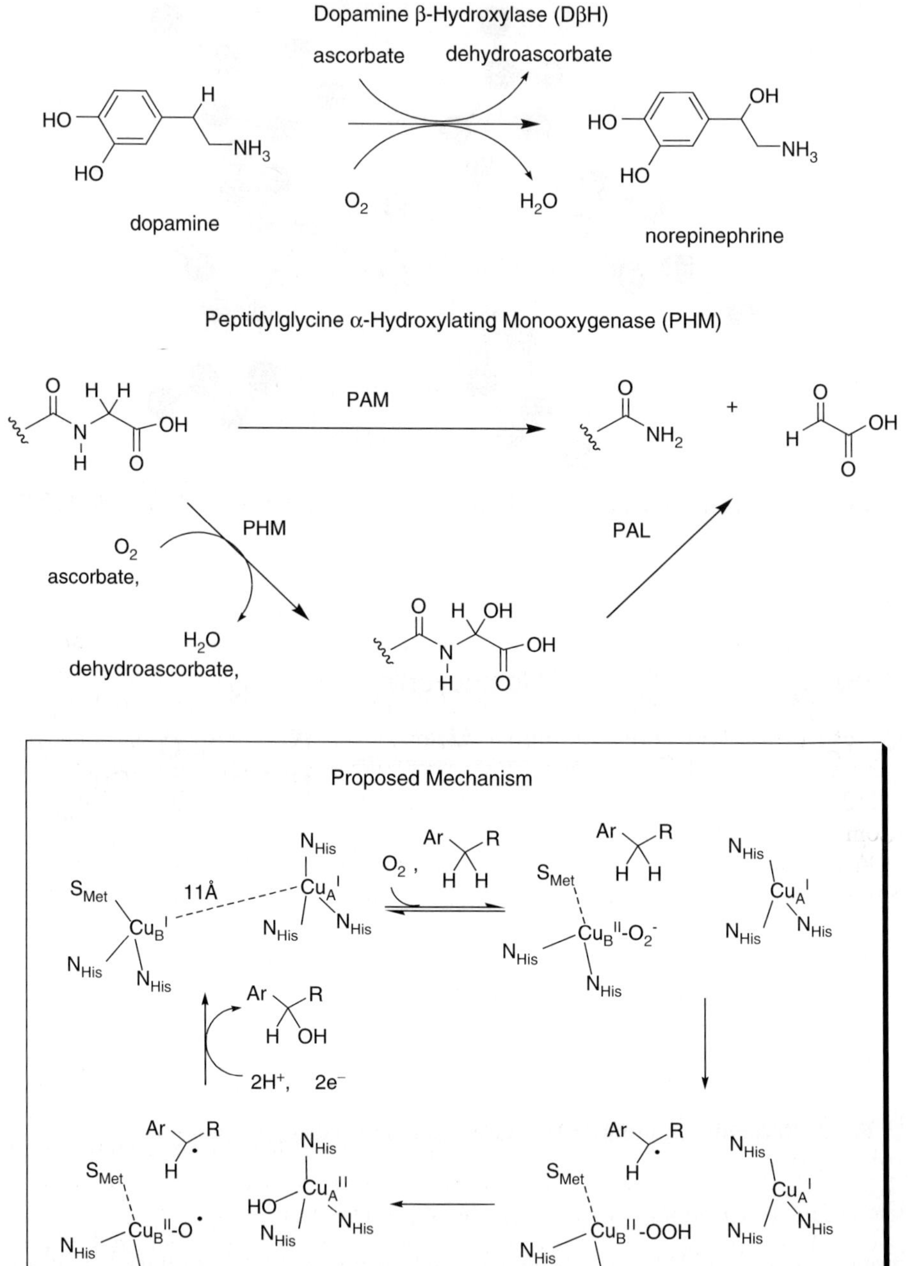

FIG. 3. Reactivity and proposed mechanism of dopamine β-hydroxylase and peptidylglycine α-hydroxylating monooxygenase.

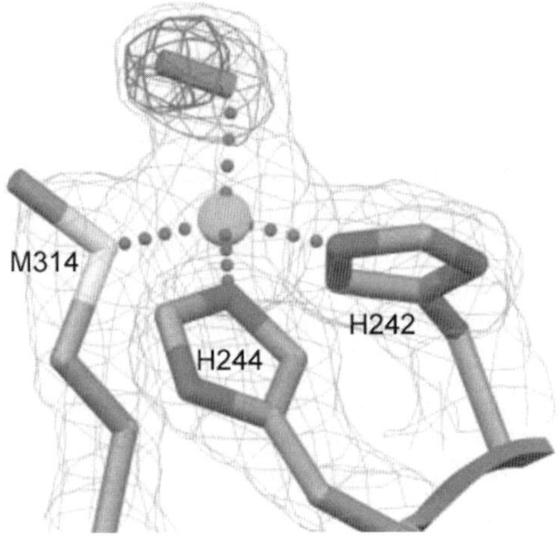

FIG. 4. PHM Cu_M active site with bound O_2^- ligand which might be relevant to catalysis.

what many researchers now believe to be a copper(II)-superoxide moiety as the active-site substrate hydrogen atom abstractor (*11–14*). The second copper(I) ion (Cu_Z, or Cu_A) serves as a reducing agent providing a second electron (Fig. 3). A recent dioxygen and substrate analogue bound crystal structure of PHM shows a copper-superoxide moiety (Fig. 4) in close proximity to the substrate (not shown); if the superoxide were rotated by $\sim 110^\circ$, the terminal O atom would be only 2.2 Å from the *pro-S* hydrogen atom of the glycine substrate (*15*).

Another enzyme of interest is the monocopper-containing galactose oxidase (GO) for its unique coordination environment and reactivity (*8,16,17*). Secreted by fungi, GO mediates the oxidation of D-galactose and other primary alcohols while reducing O_2 to dihydrogen peroxide Eq. (1). The active site of the enzyme contains a pentacoordinate copper ion; there is a weakly coordinating axial position tyrosine and the equatorial ligands are comprised of two histidine residues, a water, or acetate molecule, and a post-translationally modified tyrosine residue containing a cross-linked cysteine. The active form of the enzyme is a Cu^{II} ion coupled with a tyrosyl radical, thus a Cu^{II}-($^\bullet$OAr) molecule effects the two-electron oxidation including substrate hydrogen-atom abstraction. Following this and production of the reduced active site (i.e., Cu^{I}-(HOAr), the enzyme is recycled by O_2 which gets reduced to H_2O_2 while restoring the Cu^{II}-tyrosyl radical active form.

$$RCH_2OH + O_2 \longrightarrow RCHO + H_2O_2 \qquad (1)$$

B. SYNTHETIC Cu_n:O_2 STRUCTURAL/SPECTROSCOPIC TYPES

In the past two decades, our understanding of the fundamentals of copper-dioxygen reactivity has flourished through coordination chemistry efforts and in particular with the employment of low-temperature spectroscopy to interrogate the properties of complexes generally stable only under these conditions. The biomimetic approach has been overwhelmingly successful in discovering discrete copper-dioxygen complex types (*18*). These studies have revealed trends in copper ion ligation and associated redox properties, and their consequences on O_2 reactivity including thermodynamic, kinetic, and bonding properties of derived copper-dioxygen adducts (*19*).

Inorganic model studies have shown that several different O_2 binding modes and Cu/O_2 reaction stoichiometries exist in synthetic complexes derived from copper(I)/O_2 reactions (Fig. 5). Dioxygen-adduct structures of 1 to 4 copper per O_2 are known. These species each have distinctive spectroscopic properties as well as reactivity tendencies. The spectroscopy and general properties of the most important or well understood types are summarized below.

B.1. Mononuclear Copper-Dioxygen Compounds

In the equimolar Cu^I:O_2 stoichiometry, the initial reaction of one copper(I) ion with one mole-equivalent of O_2 usually (but see below) leads to the one-electron reduction of dioxygen to form a copper(II)-superoxide complex where the superoxo ion can bind either an end-on $Cu^{II}(\eta^1\text{-}O_2^-)$, or side-on $Cu^{II}(\eta^2\text{-}O_2^-)$ geometry (Fig. 5). The structural properties and reactivity attributes of mononuclear copper-superoxo complexes are not so well established, due to instability, in particular because these have a propensity to be reduced further to dicopper(II) peroxide by reaction with another molecule of Cu(I). With only a few exceptions (*20–22*), Cu-superoxo complexes are not isolated but are only transiently observed on route to dicopper peroxo formation and they are implicated by kinetic or mechanistic data for the formation of dicopper(II) peroxo complexes (*22–25*). Putatively generated but not structurally observable, end-on superoxo complexes have a LMCT feature at $\sim$400 nm and resonance Raman (rR) active (O–O) bond stretching vibration at $\sim 1100\,\text{cm}^{-1}$ similar to other metal superoxides (*26*). A side-on superoxo complex was stabilized (and structurally characterized) using a sterically demanding pyrazoyl borate ligand to prevent superoxide reduction from a second Cu(I) molecule. This complex has an O–O stretching frequency of $1112\,\text{cm}^{-1}$ and an O–O distance of 1.22 Å (*20,27*). In a related compound with a sterically demanding β-diketiminate ligand, spectral/structural features and DFT calculations suggest a further reduced dioxygen moiety, with a $Cu^{III}\text{-}(O_2)^{2-}$ peroxide formulation (Fig. 5) (*28,29*). The only known reactivity of mononuclear Cu^{III}-peroxo compounds or a mononuclear Cu^{II}-peroxo compound (the latter suggested to form transiently by deprotonation of a Cu^{II}-hydroperoxo complex) is the generation of dimeric Cu_2O_2 species by the reaction with a second monomeric copper complex (*29,30*). Finally, the other monomeric copper-O_2 structural type is a copper(II)-hydroperoxide, for which there are several examples in the literature (*31–35*) with one being crystallographically characterized after its stabilization by a ligand environment with H-bonding groups (*36*). However, of the several

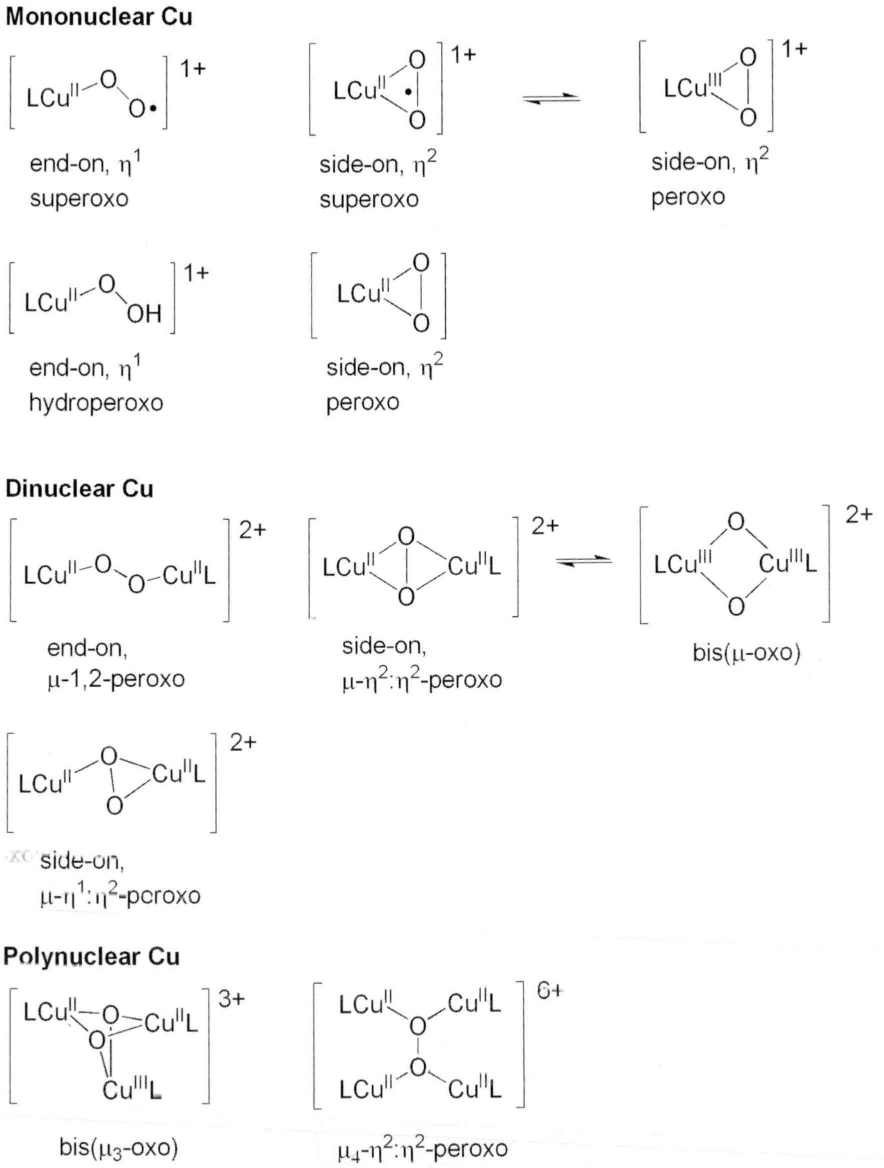

FIG. 5. Copper-O_2 structural motifs; spectroscopic types and/or crystallographically characterized structures (see discussion).

mononuclear copper(II)-hydroperoxo compounds in the literature, none derive from copper(I)-dioxygen reactivity, but by Cu^{II} and H_2O_2 chemistry (*1*).

Limited substrate reactivity is known for synthetic Cu^{II}-O_2(H) compounds. The hydrogen atom abstraction reactivity at the "mononuclear" active-sites of DβH and PHM are intriguing and warrant further considerations of oxidative chemistry by monocopper synthetic analogues.

B.2. Dinuclear Copper-Dioxygen Compounds

By contrast to the study of mononuclear Cu^{II}-O_2(H) species, the literature is now quite rich in systems containing dinuclear Cu_2-O_2 moieties. There are four known structural types of dicopper-dioxygen compounds, three of which are well studied and characterized by X-ray diffraction (Fig. 5). The first type to be crystallographically characterized was the Cu^{II}_2(μ-1,2-O_2^{2-}) "end-on" peroxo complex of the ligand TMPA (tris(2-pyridyl)methylamine) (see Fig. 8, below) formed by the low-temperature reaction of the Cu^I complex with O_2 in propionitrile. The oxygenation reaction is a reversible process whereby O_2 is released with application of vacuum with slight heating, or by bubbling with carbon monoxide. In the solid state, the copper(II) centers are 4.359 Å apart in a distorted trigonal bipyramidal geometry where the *trans*-peroxide atoms serve as axial ligands and the intraperoxide distance is 1.432 Å (Fig. 6) (*37,38*). The compound has also been probed spectroscopically in great detail (*39*). In propionitrile solvent, three peroxo to copper(II) charge transfer transitions were assigned in the UV-VIS spectra: two intense bands at 520 nm ($\varepsilon \sim$ 11,300 $M^{-1}cm^{-1}$) and 615 nm ($\varepsilon \sim$ 5,800 $M^{-1}cm^{-1}$) for the π^*_σ and π^*_v to Cu-d transitions, and a LMCT at ~435 nm ($\varepsilon \sim$ 1,700 $M^{-1}cm^{-1}$) assigned as a spin forbidden triplet π^*_v to Cu-d transition. Resonance Raman (rR) spectroscopy revealed a peroxide O–O stretching frequency of 832 cm^{-1} which shifts to 788 cm^{-1} when $^{18}O_2$ is used as the dioxygen source; the Cu–O vibration at 561 cm^{-1} shifts to 535 cm^{-1} with labeled $^{18}O_2$ (*39*). There are now ample examples of dicopper(II) end-on peroxo complexes, including a second crystallographically characterized structure (*22*), all having typical LMCT bands in the range of 520–540 nm and a shoulder centered around 600 nm, and exhibit $^{18}O_2$ sensitive O–O and Cu–O vibrations at around 812–847 cm^{-1} and 530–560 cm^{-1}, respectively (*1*).

Soon after the structural characterization of end-on peroxo complexes, another breakthrough in copper/dioxygen studies came with the X-ray

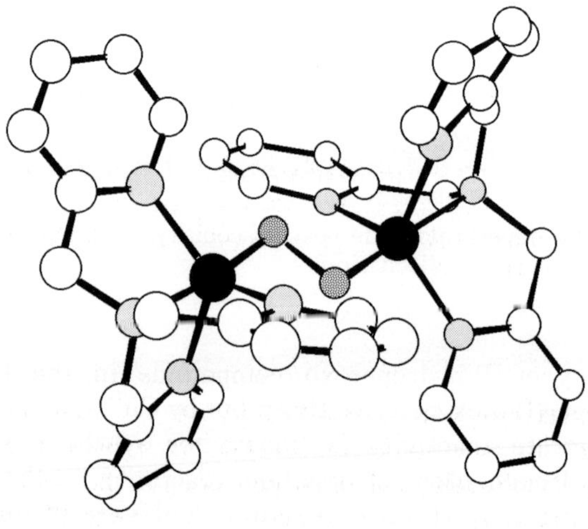

FIG. 6. X-ray crystal structure of $[\{Cu^{II}(TMPA)\}_2(O_2^{2-})]^{2+}$.

structure of a μ-η^2:η^2 "side-on" peroxo dicopper(II) complex by the late Nobumasa Kitajima, Kiyoshi Fujisawa, and co-workers (Fig. 5) (*40*). With the ligand tris(3,5-isopropyl-pyrazoyl)-borate (Tp^{iPr}) (see Fig. 8, below), the reaction of either Cu^I with O_2 or with H_2O_2 and a dicopper(II) precursor led to a complex with spectroscopic features strikingly similar to those of oxy-hemocyanin, with optical features at 349 nm ($\varepsilon = 21{,}000\ M^{-1}cm^{-1}$) and 551 nm ($\varepsilon = 790\ M^{-1}cm^{-1}$), and ν(O–O) = 741 cm^{-1}. The structure of the $Cu^{II}_2(\mu$-η^2:η^2-$O_2)$ complex portrays a shortened Cu...Cu distance relative to the end-on peroxo compound with a Cu...Cu separation of ~3.560 Å and a peroxidic O–O bond distance of 1.412 Å. The structure not only shed new light in copper-dioxygen chemistry and bonding, but also correctly predicted the structure of the active site of oxyhemocyanin, based on the close match in spectroscopic properties and Cu...Cu distance, known at that time from EXAFS spectroscopy.

Another distinct copper-dioxygen structural type is one in which dioxygen is fully reduced to oxide (O^{2-}) by two copper(I) complexes formally generating a $Cu^{III}_2(\mu$-$O)_2$ compound. Tolman and co-workers (*41*) discovered the novel structure with copper(I) complexes of substituted triazacyclononane ligands, R_3-TACN (see Fig. 8, below), and the $[\{Cu^{III}(Bn_3\text{-TACN})\}_2(O)_2]^{2+}$ complex was crystallographically characterized. The peroxide bond cleavage is evident in the crystal structures of such moieties illustrating O...O distances of 2.3 Å and extraordinarily short Cu...Cu distances of 2.8 Å. The Cu^{III} formulation is confirmed by K-edge X-ray absorption spectroscopy (XAS) which shows the 1s-3p transition at 8981 eV, 2 eV higher in energy than for the known Cu^{II} ion (*42a*). This unique core also has distinct optical and vibrational features in comparison to its $Cu^{II}_2(O_2)^{2-}$ isomer; bis(μ-oxo) dicopper(III) cores are characterized by their intense LMCT bands at ~300 nm and ~400 nm and the symmetric core "breathing" vibration around 600 cm^{-1} (*42b,43*).

The bis(μ-oxo) core is often involved in a fast equilibrium with the isoelectronic dicopper(II) side-on peroxo species (see Section III) (*44*). Calculations and experimental observations portray the two species being thermodynamically comparable with a low barrier for interconversion (*45–47*). The relative stability of the high-valent copper(III) arises from strengthened Cu–O bonds present when the peroxide O–O bonding interactions are lost (*42*). The bis(μ-oxo) dicopper(III) d^8 compounds are best stabilized by strong σ-donating alkyl nitrogen donors preferably in a square planar environment. Several examples in the literature exist where both Cu_2O_2 cores are in a facile equilibrium and the supporting ligand plays a large role in biasing one species vs. the other (*1*).

B.3. Polynuclear Copper-Dioxygen Compounds

Copper-dioxygen compounds or interactions of higher nuclearity exist in the multicopper oxidases ($Cu{:}O_2 = 3{:}1$) such as in ascorbate oxidase (AO), laccase, and ceruloplasmin, all of which couple the four electron reduction of O_2 to water with the oxidation of a substrate (*7*). A number of AO protein X ray studies are available, including a derivative with an end-on

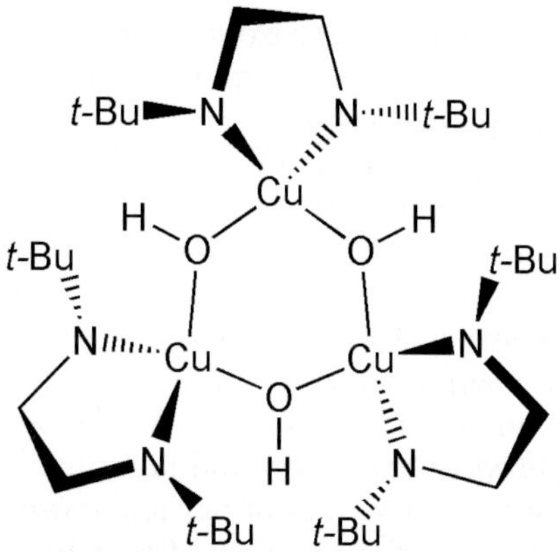

FIG. 7. A crystallographically characterized trimeric $[Cu_3(\mu_3\text{-}OH)_2]^{3+}$ complex.

(hydro)peroxo moiety present (*48*). Recently, structures of laccase have been crystallographically characterized revealing a diatomic (dioxygen or peroxide) moiety near to the trinuclear center (*49,50*). The modeling of these tricopper cores is not as vast as the Cu_2O_2 cores which are of greater relevance to Hc, CO, and Tyr. The successful generation of trinuclear copper-dioxygen compounds relies on the careful interplay of sterics and $[Cu^I]/[O_2]$ relative concentrations. Compounds must have non-sterically demanding ligands to accommodate the trinuclear cluster and the initial concentration of copper(I) must be significantly higher than that of dissolved dioxygen. A trinuclear copper-dioxygen adduct was successfully prepared and structurally characterized by Stack and co-workers utilizing an *N,N*-dimethyl-cyclohexyl-1,2-diamine ligand. The four electron reduction of dioxygen is achieved by three Cu(I) centers forming a $[Cu_3(\mu_3\text{-}O)_2]^{3+}$ core having two Cu^{II} ions and one Cu^{III} ion (*51*). Just recently, the same group also crystallographically characterized a $[Cu_3(\mu_3\text{-}OH)_3]^{3+}$ compound (Fig. 7) as a structural model for the proposed native intermediate in multicopper oxidases (*52*).

There are a few cases where O_2-adducts of discrete trinuclear compounds have been detected (*53–55*), but many new efforts in this direction are needed.

Two structural types of tetranuclear copper-dioxygen compounds exist where the peroxide is bridged by four Cu(II) ions in either a *cis* (*56,57*) or *trans* (*58*) environment (Fig. 5). However, the mechanism of formation of these species from Cu^I/O_2 chemistry is not well understood considering that reduction to peroxide only requires two electron equivalents yet the compound provides 4 electrons from Cu(I).

II. Tetradentate Ligand Influences

Since biological copper ion is often coordinated by the imidazole group of histidine residues, typical copper complexes designed to model enzyme

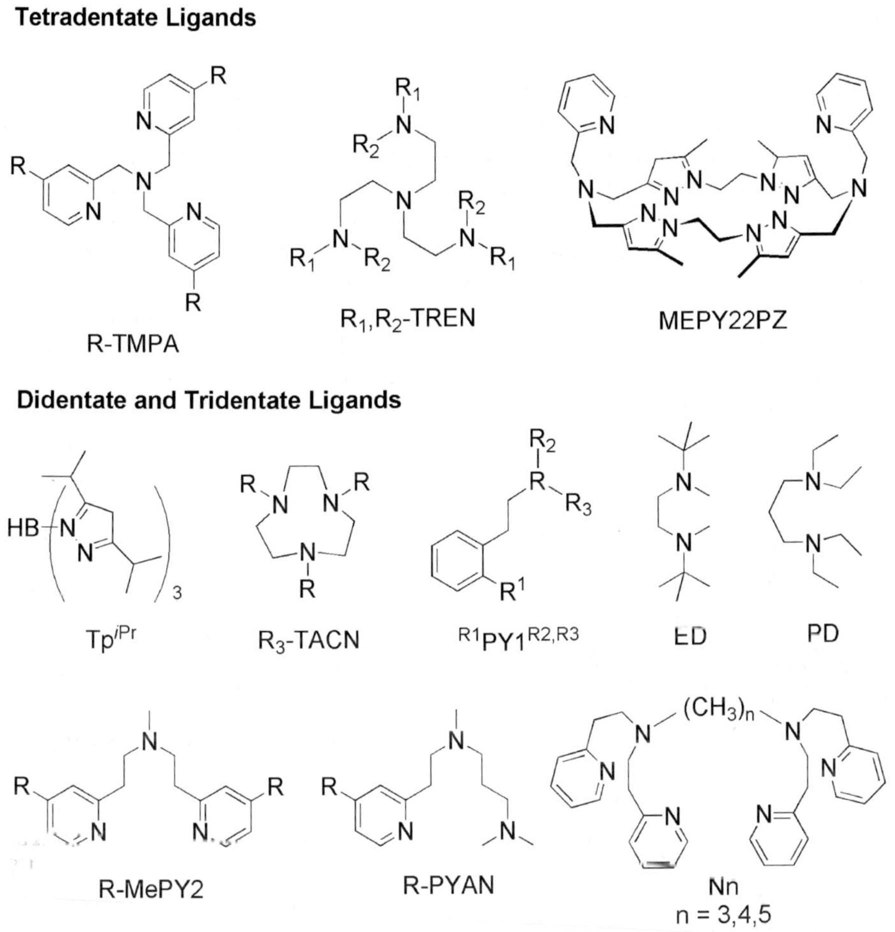

FIG. 8. Representative example of amine ligands used for Cu_n-O_2 reactivity.

active sites contain nitrogen donating ligands. Pyrazole, imidazole, pyridines, and secondary and tertiary alkyl amines are among the variety of nitrogen donors in the ligands used for copper(I) and its reactivity with O_2. It was apparent from early on that the successful assembly of dicopper-dioxygen intermediates strongly depended on ligand design. The ligands depicted in Fig. 8 are only a representative group of the multitude of ligand varieties seen in the literature that support copper-dioxygen reactivity. In this article, we hope to show how minor ligand modifications oftentimes can have profound implications for copper-dioxygen reactivity – or lack thereof, and certain ligand properties can lead to the prediction of the copper-dioxygen structure following a (ligand)Cu^I/O_2 reaction. Herein we highlight various aspects of ligand features in copper-dioxygen reactivity and discuss the chemistry of these compounds towards substrates.

A. Chelate Ring Size Effects

The chelate ring size of a given ligand proves to play a significant factor in the O_2-reactivity of copper(I) complexes. This effect is manifested in a series of tripodal tetradentate pyridylamine ligands. In an early study we uncovered dramatic effects on dioxygen reactivity by a trivial ligand modification, with the sequential addition of a second methylene unit between the central amine and the pyridine arms, forming the ligands: PMEA, PMAP, and TEPA (Fig. 9) (*59*). These effectively impose two changes in the copper complexes: (1) The redox potential of the copper(I) and copper(II) (*60*) complexes increase as the ligand increases in arm length. This effect is understood as arising from the relative stability (i.e., formation constants) of the Cu(II) ions for a 5-compared to a 6-membered chelate ring (*61,62*). (2) The increased chelate ring size leads to structural changes of corresponding copper(II) complexes; $[Cu^{II}(TMPA)Cl]^+$ is a trigonal bipyramidal structure with axial chloride, while $[Cu^{II}(TEPA)Cl]^+$ possesses a square pyramidal structure, with chloride in the equatorial position. These structures are also retained in the solution state (*60*).

The ligand changes effect large variations in their dioxygen reactivity: The reaction of $[Cu(TMPA)(MeCN)]^+$ with O_2 in propionitrile at low temperatures initially generates an unstable copper(II)-superoxo compound (λ_{max} 410 nm, $\varepsilon \sim 4{,}000\ M^{-1}cm^{-1}$, 747 nm $\varepsilon \sim 1{,}000\ M^{-1}cm^{-1}$) and then forms a more stable dicopper(II) end-on peroxo compound (λ_{max} 525 nm,

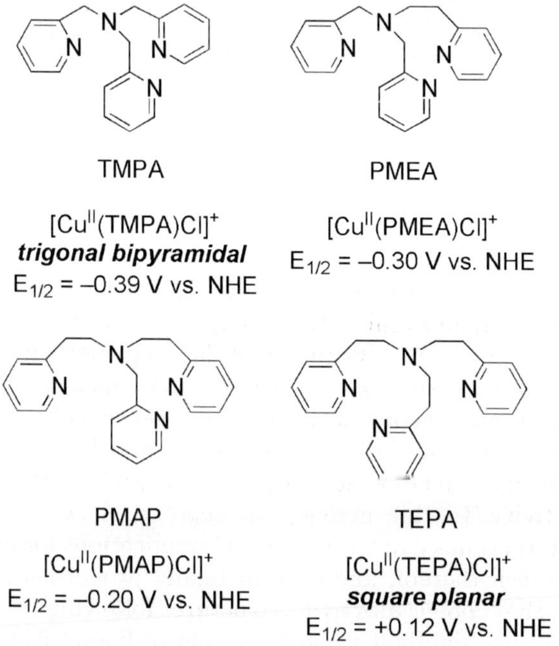

Fig. 9. Ligand modifications of increasing chelate ring size and resulting redox potentials.

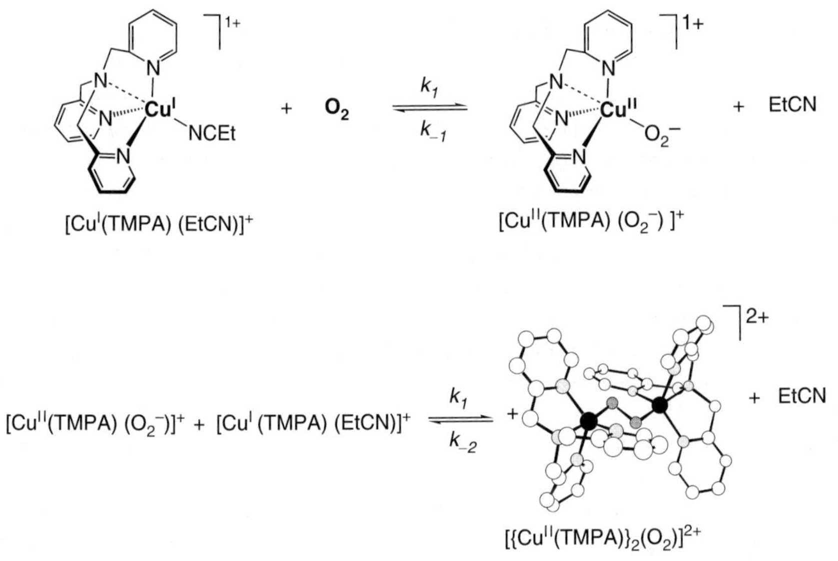

SCHEME 1.

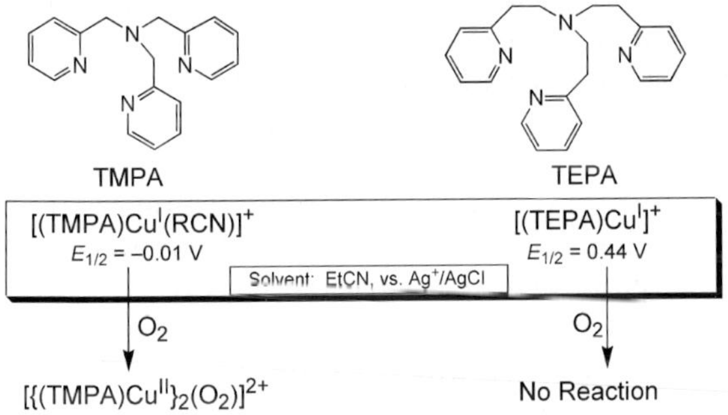

FIG. 10. Differential O_2 reactivity with the ligands TMPA vs. TEPA.

$\varepsilon \sim 11{,}500\ M^{-1}cm^{-1}$, λ_{max} 590 nm, $\varepsilon \sim 7{,}600\ M^{-1}cm^{-1}$) (Scheme 1) (*37,38*). The lifetime of the O_2-adduct of $[Cu(PMEA)]^+$ is diminished by comparison; a short lived end-on peroxo intermediate (λ_{max} 536 nm, $\varepsilon \sim 5000\ M^{-1}cm^{-1}$) can only be observed with stopped flow UV-VIS spectroscopy. No dioxygen-intermediate was observed for $[Cu(PMAP)]^+$ on the millisecond timescale, and $[Cu(TEPA)]^+$ is completely unreactive towards dioxygen. Although the effects are not completely understood, the shut down of O_2-reactivity demonstrates the effects of redox potential and ligand chelate ring size of derived copper(I) complexes (Fig. 10).

B. ELECTRONIC EFFECTS IN TETRADENTATE LIGAND SYSTEMS

More recently, we modified the TMPA ligand in terms of its electronic donating ability to the copper center and studied the effects on dioxygen reactivity. The ligands, R-TMPA (R = Me, *t*Bu-, MeO, NMe_2) (Fig. 8) containing electron donating substituents on the *para*-position of the pyridine donors, were synthesized. A ligand effect was first noted in the coordination environment of the copper(I) center: The Cu^I complexes of H-TMPA, Me-TMPA, *t*Bu-TMPA, and MeO-TMPA, are all isolated as 5-coordinate copper(I) compounds with the 5th ligand being acetonitrile (from the Cu^I starting material $[Cu^I(MeCN)_4]^+$). In contrast, the complex with the most electron rich ligand, NMe_2-TMPA, is stable as a 4-coordinate complex as $[Cu(NMe_2\text{-TMPA})]^+$ is readily isolated even when prepared in acetonitrile as solvent. The somewhat varying electronic properties of the $[Cu(R\text{-TMPA})]^+$ complexes are also manifest by their redox properties. As the donor ability of the ligand increases, (i.e., as the pyridyl substituent was made more electron donating), the Cu^{II}/Cu^I redox potentials ($E_{1/2}$ from cyclic voltammetry) decrease from –0.40 V to –0.46 V, to –0.49 V, to –0.70 V vs. Fc/Fc^+ in MeCN, for R = H, *t*Bu, MeO, and NMe_2, respectively. Thus, as expected, the most electron donating ligand $[Cu(NMe_2\text{-TMPA})]^+$ is the easiest to oxidize from the Cu^I to the Cu^{II} redox state. The complexes also differed in the relative amount of π-backbonding to a π-acceptor ligand such as carbon monoxide. The C–O stretching frequencies of the carbonyl adducts $[Cu(R\text{-TMPA})(CO)]^+$ ranged from 2092, 2089, 2087, to 2079 cm^{-1} for R = H, *t*Bu, MeO, and NMe_2, respectively; $[Cu(NMe_2\text{-TMPA})(CO)]^+$ exhibits the most copper(I) backbonding, possessing the weakest C–O bond and lowest stretching frequency.

The R-substituent effected changes described above for the $[Cu(R\text{-TMPA})]^+$ compounds are accompanied by changes in their dioxygen reactivity. Oxygenation reactions were carried out in propionitrile (EtCN) from −90°C to 10°C and monitored with a UV-VIS spectrometer connected with a stopped-flow mixing unit (*63*). On the millisecond timescale, similar kinetic processes as described in Scheme 1 were observed for $[Cu(R\text{-TMPA})]^+$, R = Cl, H, MeO, and NMe_2. At low temperature (< 70°C), oxygenation occurs with initial reversible binding of O_2 forming a mononuclear superoxocopper(II) species which rapidly reacts with a second Cu(I) molecule to form the end-on peroxo complex. The rate constants for superoxo formation (k_1, Scheme 1) range from $1.18 \times 10^4\ M^{-1}s^{-1}$ to $3.11 \times 10^4\ M^{-1}s^{-1}$ at 183 K. The small variation in k_1 with respect to the R-substituent, together with relatively high activation enthalpies and positive activation entropies suggest that propionitrile dissociation plays a large role in the initial O_2-binding process. The rate constant for the back reaction, k_{-1}, shows a modest ligand effect; k_{-1} decreases by an order of magnitude as the R group becomes more electron donating (e.g., R = H vs. R = MeO). This corresponds to a larger superoxo binding constant ($K_1 = k_1/k_{-1}$) for complexes with electron donating substituents (Scheme 1). For the dicopper(II) end-on peroxide $(O_2)^{2-}$ formation, the k_2 rate constants (at 183 K) varied between $1.34 \times 10^{-4}\ M^{-1}s^{-1}$ and $4.10 \times 10^4\ M^{-1}s^{-1}$ with k_2 increasing as the complex is made more electron rich. The equilibrium constant for end-on peroxo formation (K_2 at 183 K) for the complex with R = MeO is roughly one order of magnitude greater than for the complex with R = H.

A dramatic ligand effect was observed for the oxygenation reaction of $[Cu(NMe_2\text{-TMPA})]^+$; formation of the superoxo compound occurs within the mixing timescale, and peroxo formation is delayed for two reasons: 1) the equilibrium K_1 is considerably larger than for R = MeO and H, and 2) the rapid rate (according to k_1) depletes Cu^I as most of it is bound as Cu^{II}-$(O_2)^-$ hence a low concentration of Cu^I is left to reduce copper-superoxide to form the dicopper(II) end-on peroxo complex. To summarize, the copper-dioxygen adducts (Cu^{II}-superoxide and Cu_2^{II}-peroxide) of the more electron rich ligands showed higher thermodynamic stability, while their kinetics of formation was only slightly enhanced due to the competition with EtCN (i.e., the solvent) as a copper(I) ligand.

The latter point is illustrated when a switch in reaction solvent is carried out. In oxygenation experiments run in the much weaker coordinating solvent, tetrahydrofuran (THF), the rate of oxygenation of $[Cu(H\text{-TMPA})]^+$ increased dramatically; complete superoxo formation occurred within the stopped-flow mixing time, ~1 ms. Temperature-dependent studies afforded thermodynamic parameters showing considerably enhanced thermodynamic stability of the superoxo and end-on peroxo complexes: For $[Cu^{II}(H\text{-TMPA})(O_2)^-]^+$ formation, $\Delta H = -41$ kJ/mol in THF vs. –29.8 kJ/mol in EtCN, and for $[\{Cu^{II}(H\text{-TMPA})\}_2(O_2)^{2-}]^{2+}$ formation, $\Delta H = -94$ kJ/mol in THF vs. –77 kJ/mol in EtCN.

The ligand substituents also result in changes in the bonding interactions of the end-on peroxo complexes $[\{Cu(R\text{-TMPA})\}_2(\mu\text{-}1,2\text{-}O_2^{2-})]^{2+}$. The 4-R-pyridyl substituents (R = H, OMe, and NMe_2) effect changes in the stretching frequencies of the Cu–O and O–O bonds (*64*). All compounds exhibited $^{18}O_2$ sensitive rR features assigned as the ν(O–O) and ν(Cu–O) stretching vibrations, and each decreased with increased ligand donor ability. The parent complex, $[\{Cu(H\text{-TMPA})\}_2(\mu\text{-}1,2\text{-}O_2)]^{2+}$, exhibits an intraperoxide stretching frequency at 827 cm^{-1}, which is higher in energy than that for the complexes with the OMe-TMPA and NMe_2-TMPA ligand; ν(O–O) = 822 cm^{-1} and 812 cm^{-1}, respectively. The same trend was observed for ν(Cu–O) which decreased from 561 cm^{-1} to 557 cm^{-1}, to 551 cm^{-1}, for $[\{Cu(R\text{-TMPA})\}_2(\mu\text{-}1,2\text{-}O_2)]^{2+}$ where R = H, OMe, and NMe_2, respectively. The decrease in O–O and Cu–O bond stretching frequencies is ascribed to the increased σ-donation from the ligand to the Cu^{II} d orbital, which decreases the amount of σ-donation received from the peroxide π* orbital. Hence, the Cu–O bond decreases, and the added electron density on the antibonding π* orbital of the peroxide causes the O–O bond to decrease as well (*64*).

C. Tetradentate vs. Tridentate Ligand Influences in Dioxygen Reactivity

Early copper(I)/dioxygen reactivity studies were done with the "classic" tripodal pyridyl-amine ligand TMPA (Fig. 8) (see above). The low-temperature reaction of $[Cu(TMPA)(RCN)]PF_6$ (R = Me or Et) with O_2 was found to produce a dicopper-dioxygen adduct where the peroxide is bound in a *trans*(μ-1,2-O_2) "end-on" geometry (Fig. 5). Since the discovery of TMPA there have been several more recent examples of end-on peroxo dicopper(II) complexes with different tetradentate amine donor ligands (Fig. 8). The variation in ligands types is quite large: end-on peroxo dicopper(II) complexes have been generated using mononucleating and dinucleating ligands containing tripodal alkyl/pyridyl, and macrocyclic amine groups (*1*). Regardless of *N*-donor ligand type, at specified reaction conditions, most copper(I) complexes of tetradentate ligands form end-on peroxo complexes upon reaction with O_2 (after initial $Cu^{II}(O_2)^-$ formation) (*26,65–70*).

For example, a rationally designed macrocyclic ligand containing two tetradentate copper(I) centers (MePY22PZ, Fig. 8) reacts with O_2 and forms an end-on peroxo compound with exceptional room temperature stability in both aprotic and protic solvents (*66*). Also, derivatives of the TREN ligand (Fig. 8) have been used for Cu^I/O_2 chemistry: At −90°C in acetone the copper(I) complex of the hexamethyl substituted TREN ligand (Fig. 8, R_1 and $R_2 = Me$) produces an end-on peroxo dicopper complex after initial monomeric copper(II) superoxo formation as indicated by their characteristic spectroscopic features (*69*). TREN derivatives whereby $R_1 = H$ or Me, and $R_2 = Bn$ (Fig. 8) also led to end-on superoxo and peroxo formation, with the $[\{Cu(H,Bn\text{-}TREN)\}_2O(\mu\text{-}1,2\text{-}O_2)]^{2+}$ complex being the second end-on peroxo dicopper(II) complex characterized by X-ray crystallography (*22*).

D. Unusual Structures with Tetradentate Ligands

There are some interesting exceptions to the roughly general observations that copper(I) complexes of tetradentate amine ligands generate end-on peroxo compounds. With a sterically hindered tetradentate ligand, Me_2-TMPA, (with methyl groups on the 6-position of two pyridine arms) oxygenation of $[Cu(Me_2\text{-}TMPA)]^+$ at low temperature gives a bis(μ-oxo) compound, as characterized by $\nu(Cu_2\text{–}O_2) = 590$ and $\lambda_{max} = 378$ nm ($\varepsilon = 22{,}000\ M^{-1}cm^{-1}$) and 494 nm ($\varepsilon = 300\ M^{-1}cm^{-1}$). A crystal structure shows that the tetradentate ligands bind to the copper(III) ions adopting a distorted octahedral environment with elongated axial amine ligands; thus, the ligand effectively serves as a didentate ligand (Fig. 11) (*71,72*).

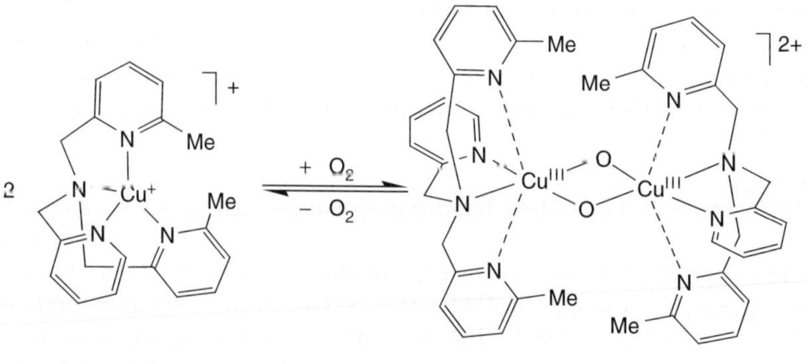

Fig. 11. A crystallographically characterized bis(μ-oxo)dicopper(III) complex with tetradentate Me_2-TMPA ligand.

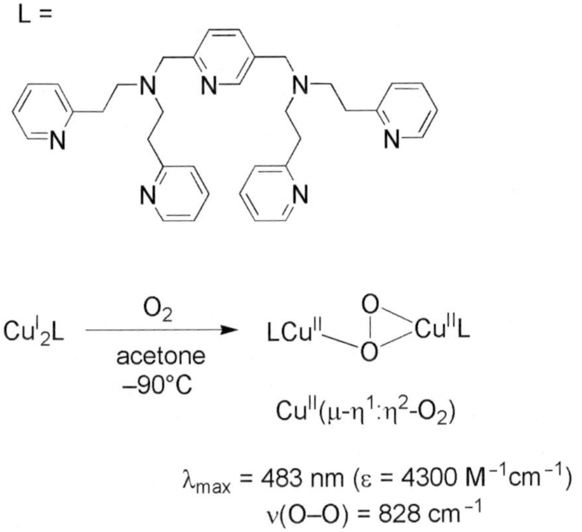

FIG. 12. Tetradentate/tridentate dinucleating ligand and Cu^I/O_2 reactivity.

The tendencies of tridentate amine Cu(I) to bind O_2 in an η^2-fashion (side-on), and tetradentate amine Cu to bind in an end-on η^1-fashion is found in a recent example from Itoh and co-workers with an unsymmetrical dinucleating ligand (Fig. 12) (*73*). The low temperature (−90°C) oxygenation reaction of the dicopper(I) complex results in a peroxide with a ν(O–O) stretching frequency of 828 cm^{-1} similar to other end-on peroxo dicopper(II) complexes, but lacking the UV-VIS spectral features of typical end-on peroxo complexes. Instead, the compound exhibits one feature at 483 nm ($\varepsilon = 4{,}300\ M^{-1}cm^{-1}$) which also differs from the characteristic LMCT peaks observed for side-on peroxo complexes (see Section I.B.2). The reactivity tendency of this novel dicopper-peroxo compound encompasses those of both end-on and side-on peroxo complexes. The reaction with acid results in the release of dihydrogen peroxide exemplifying its nucleophilic, or basic character. Yet, the compound performs oxygen atom transfer to phenyl methyl sulfide and triphenyl phosphane and couples phenols, thus also displaying an electrophilic nature (see Section IV.A). While structural details are not yet available for this compound, the unique UV-VIS spectra and remarkable combination of electrophilic and nucleophilic behavior towards external substrates suggests this novel compound to have a $\mu(\eta^1{:}\eta^2\text{-}O_2)$ binding geometry as dictated by the preferential binding of the tridentate-Cu^I and tetradentate-Cu^I complex.

III. Tridentate Ligand Influences: Side-on Peroxo vs. Bis(μ-oxo) Equilibrium

The isomeric $Cu^{II}_2(O_2)^{2-}$ "side-on peroxo" and $Cu^{III}_2(O^{2-})_2$ "bis(μ-oxo)" cores can be thermodynamically comparable and hence are often found in a facile

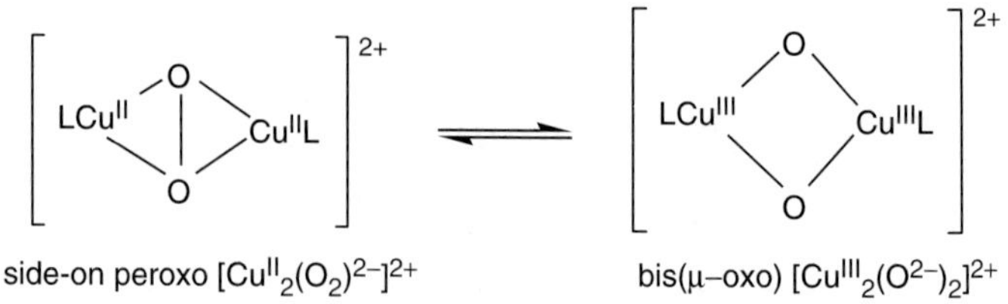

SCHEME 2.

equilibrium (Scheme 2) (*74,75*) which is highly dependent on the supporting ligand, and to a lesser extent, upon reaction conditions. The rate of interconversion between each species is notably faster than the rate of O_2 uptake (the latter having rate constants upwards of $10^4\,M^{-1}s^{-1}$ at −80°C) giving evidence for an exceptionally small barrier for core interconversion (*76*). The factors that bias the position of the equilibrium are complex and as yet to be fully understood. Herein, we present noteworthy ligand features and environmental factors (solvent/counterion) that are known to better stabilize each respective dicopper-dioxygen core.

A. Environmental Factors

A.1. Counterion

Stack and co-workers reported that an *N*-substituted ethylene diamine dicopper(III) bis(μ-oxo) compound counterbalanced with a weakly coordinating SbF_6 anion fully isomerizes to the side-on peroxo compound upon titration with a more coordinating anion, $CH_3SO_3^-$ (*77*). This effect is attributed to the side-on peroxo isomer being capable of coordinating an axial ligand due to its more accessible (i.e., larger) dicopper core. Coordinating anions may also promote side-on peroxo $Cu_2^{II}(O_2)^{2-}$ formation more so than a $Cu_2^{III}(O^{2-})_2$ complex because Cu^{III} d^8 prefers square planar geometry while Cu^{II} d^9 complexes are often observed as pentacoordinate complexes in either square pyramidal or trigonal bipyramidal geometry. However, an exception to this idea is a case where a bis(μ-oxo) dicopper(III) complex supported by a propylene diamine ligand exhibited pentacoordinate ligation with the fifth ligand being an axially coordinated counterion (triflate or perchlorate), as determined from EXAFS spectroscopic data (*78*) and in the X-ray crystal structure of an asymmetrically ligated dicopper(III) bis(μ)oxo complex containing the propylene diamine ligand and a diketiminate-based ligand (*29*). It is likely that the accessibility of the counteranion to coordinate to the metal center is dependent, in part, to the coordinated ligand (*78*). Hence, pentacoordination for $Cu_2^{III}(O^{2-})_2$ cores is observed in certain cases where ligand sterics/geometry permit it (i.e., R-TACN). Nevertheless, coordinating counterions have been shown to play a role in the side-on peroxo/bis(μ-oxo) equilibrium (Scheme 2).

A.2. Temperature

With substituted triazacyclononane ligands, Tolman and co-workers were the first to observe a temperature dependence on the equilibrium between the $Cu_2^{II}(O_2)^{2-}$ and $Cu_2^{III}(O^{2-})_2$ isomers. The reaction of $[Cu^I(iPr_3\text{-TACN})]^+$ (L = R_3-TACN where R = *i*Pr, Fig. 8) and O_2 in THF led to the formation of $[\{Cu^{II}(iPr_3\text{-TACN})\}_2(O_2)]^{2+}$ and $[\{Cu^{III}(iPr_3\text{-TACN})\}_2(O)_2]^{2+}$ in equilibrium where the relative amount of the bis(μ-oxo) isomer increased with decreasing temperature. Peak fitting to resolve the spectral bands at 355 and ~420 nm corresponding to the individual $[\{Cu^{II}(iPr_3\text{-TACN})\}_2(O_2)]^{2+}$ and $[\{Cu^{III}(iPr_3\text{-TACN})\}_2(O)_2]^{2+}$ species allowed for the estimation of K_{eq} at various temperatures. The thermodynamic parameters for the reaction $[\{Cu^{II}(iPr_3\text{-TACN})\}_2(O_2)]^{2+} \rightleftharpoons [\{Cu^{III}(iPr_3\text{-TACN})\}_2(O)_2]^{2+}$ were found to be $\Delta H^\circ = -3.8$ kJ/mol and $\Delta S^\circ = -25$ J/mol·K confirming the small energetic difference between the two isomers (*74*). Further experimental evidence for the thermodynamic equivalence between the two isomers came from Stack and co-workers with Cu_2O_2 compounds with an ethylene diamine ligand (ED, Fig. 8) estimating $\Delta H^\circ \approx -2.5$ kJ/mol and $\Delta S^\circ \approx -8.4$ J/mol·K (*75*).

By a different method our group has also shown that the bis(μ-oxo) isomer is enthalpically stabilized but entropically destabilized. The temperature-dependent kinetic analysis of the oxygenation reaction of [Cu(PYAN)(MeCN)]$^+$ (L = R-PYAN where R = H, Fig. 8) shows changing apparent epsilon values at 360 and 400 nm due to the changing relative amounts of $Cu_2^{II}(O_2)$ and $Cu_2^{III}(O)_2$ species in solution (solvent = THF or acetone). The calculated spectra are then considered an effective concentration of the relative amounts of species in solution (Fig. 13). A least squares fitting of the temperature-dependent data was employed to calculate the actual relative concentrations (and corresponding spectra) of the pure side-on peroxo and bis(μ-oxo) in solution (Fig. 13). For the reaction $[\{Cu^{II}(PYAN)\}_2(O_2)]^{2+} \rightleftharpoons [\{Cu^{III}(PYAN)\}_2(O)_2]^{2+}$ we determined thermodynamic parameters for K_{eq}: in THF: $\Delta H^\circ = -15.7$ (0.4) kJ/mol and $\Delta S^\circ = -83$ (*2*) J/mol·K, and in acetone: $\Delta H^\circ = -15.8$ (1.2) kJ/mol and $\Delta S^\circ = -76$ (*6*) J/mol·K. Similar results were found for the equilibrium of an analogous complex with a derivatized OMe-PYAN ligand (Fig. 8) (*79*). Low temperatures (enthalpically) favor the bis(μ-oxo) dicopper(III) core.

A.3. Solvent

The solvent employed has also been shown to affect the equilibrium in several ligand systems. The peroxo/oxo dicopper compounds supported by Tolman's substituted triazacyclononane ligand (iPr_3-TACN, Fig. 8) also exhibited a peculiar solvent effect whereby precipitation/aggregation of the bis(μ-oxo) isomer occurred in THF, while both species in equilibrium were soluble in CH_2Cl_2 and acetone. The findings were attributed to the overall greater effective polarity of CH_2Cl_2 and acetone (*74*). On the other hand, Stack and co-workers describe a solvent effect which correlates with the dielectric constant of the solvent. Oxygenating a copper(I) complex of an ethyl-substituted propylene diamine ligand (PD, Fig. 8) in different

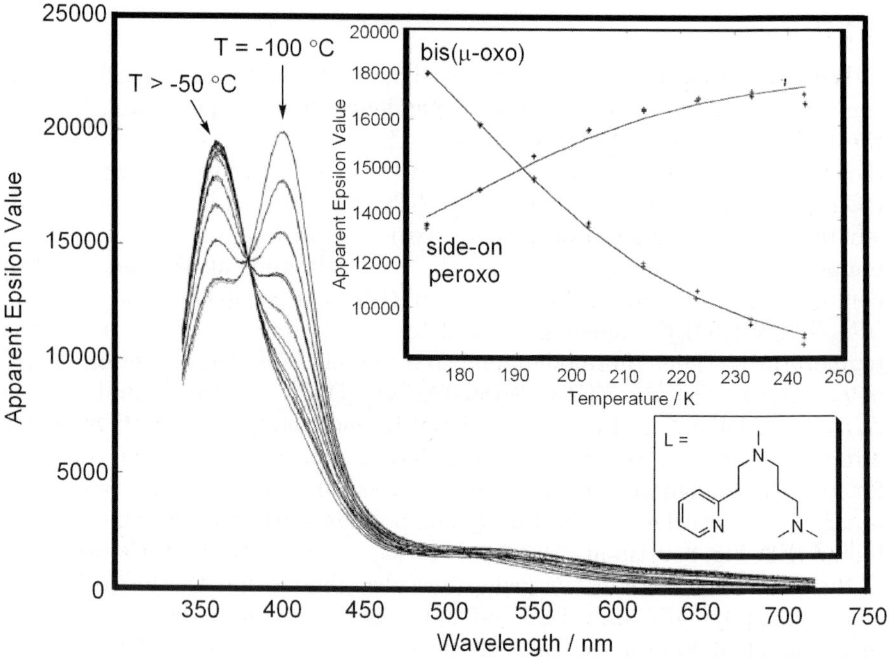

FIG. 13. Temperature dependence (–100 to –30°C) of the spectra involving the $[\{Cu^{II}(L)\}_2(O_2)]^{2+}$ ($\lambda_{max} = 360$ nm) and $[\{Cu^{III}(L)\}_2(O)_2]^{2+}$ ($\lambda_{max} = 400$ nm) equilibrium. Inset, Apparent epsilon values at 360 and 400 nm vs. temperature. See text.

solvents significantly changes the position in the $[Cu^{II}_2(O_2)]^{2+}$:$[Cu^{III}_2(O)_2]^{2+}$ equilibrium. In THF, a roughly equal mixture of each isomer exists, but in a solvent with a smaller dielectric constant, 2MeTHF, the equilibrium lies towards the side-on peroxo species. Conversely in a solvent of high dielectric constant, acetone, only the bis(μ-oxo) isomer is obtained (*77*).

Recently, in our group we encountered spectroscopic evidence for solvent interacting primarily with the side-on peroxo core (*76*). The tridentate alkyl/pyridyl ligand, PYAN (Fig. 8, R = H) with its copper(I) complex undergoes solvent dependency unlike Stack's previously reported effect on dielectric constant (see above). Following the relative absorption intensities at 360 and 400 nm, the $[\{Cu^{II}(PYAN)\}_2O_2]^{2+}$ adducts are a mixture of dicopper side-on peroxo and bis(μ-oxo) compounds, and the relative amounts of bis(μ-oxo) generated increases within the solvent series: CH_2Cl_2 < diethyl ether, < acetone, < tetrahydrofuran (THF). In other words, an increasing amount of $Cu^{III}_2(O)_2$ forms as the coordinating ability of the solvent increases (Fig. 14). A correlating trend is also observed in the rR spectra of the side-on peroxo species in the varying solvents: The satellite peak in the Cu...Cu vibration ($\nu \sim 240\,cm^{-1}$), assigned as the Cu–N_{eq} stretch, shifts to higher energy as the solvent coordinating ability increases (Fig. 14, inset). The results are interpreted to mean that the solvent interaction with a metal center of the side-on peroxo complex pushes the Cu^{II} ion more into the ligand N_2O_2 plane, allowing greater in-plane orbital overlap that results

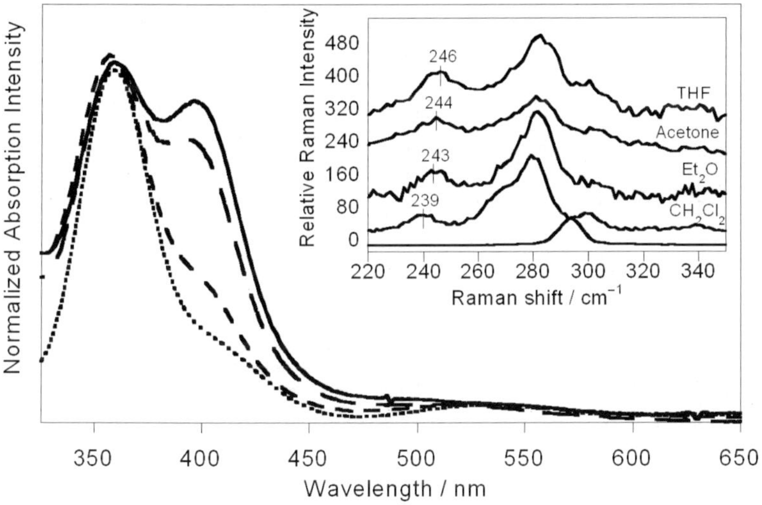

FIG. 14. Absorption spectra of $[\{Cu(PYAN)\}_2O_2]^{2+}$ in CH_2Cl_2 (⋯⋯), diethyl ether (– – –), acetone (— —), and THF (——). Inset: rR spectra of $[\{Cu^{II}(PYAN)\}_2(O_2)]^{2+}$ in different solvents (the bottom spectrum is that of CH_2Cl_2 which was subtracted from the CH_2Cl_2 trace).

in a strengthened Cu–N_{eq} bond. The greater orbital overlap can lead to more backbonding to the σ* antibonding orbital favoring O–O bond cleavage.

B. Specific Tridentate Ligand Influences on the Peroxo/bis(μ-oxo) Equilibrium

By far, the largest denominators undermining the preference between a side-on peroxo or bis(μ-oxo) dicopper structure is the exact nature of the supporting ligand, where even minor modifications can lead to a dramatic change in the position of the equilibrium. From information gathered from the large array of ligands employed for copper-dioxygen reactivity studies, one can generalize several key ligand features that result in a bias toward one isomer over the other. The nitrogen donor group becomes important; whether alkyl or pyridyl, secondary or tertiary, each has substantial implications for the resulting copper-dioxygen structure. General ligand properties such as sterics, denticity, and electronics also contribute extensively to the geometric preference between a side-on peroxo dicopper(II) species and a bis(μ-oxo) dicopper(III) compound (Scheme 2).

B.1. Ligand Denticity in Tridentate Ligand Systems

Often, bis(μ-oxo) copper compounds are stabilized by didentate ligands generating square planar $Cu_2^{III}(O)_2$ complexes favorable for d^8 copper(III) ion (*1*). "Removal" of a third nitrogen donor often can change the dioxygen adduct produced from a side-on peroxo isomer to a bis(μ-oxo) compound. For example, the copper(I) complex of the tridentate pyridyl amine ligand,

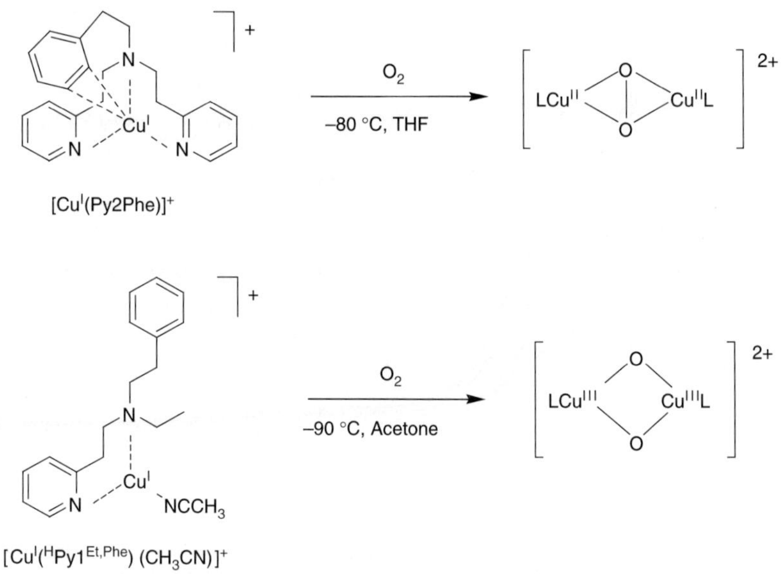

FIG. 15. Tridentate vs. didentate ligand influences on O_2 adduct formation.

$[Cu^I(Py2Phe)]^+$, (Fig. 15, top) forms a side-on peroxo dicopper complexes upon low-temperature reaction with O_2 in different solvents (*80,81*). By contrast, the closely related copper(I) analogue $[Cu^I(^HPy1^{Et,Phe})(CH_3CN)]^+$ with a didentate ligand (one pyridyl group removed) (Fig. 15, bottom) forms exclusively the bis(μ-oxo) dicopper(III) compound, according to its UV-VIS and rR spectroscopic properties (*82*).

B.2. Ligand Sterics

Stack and co-workers demonstrated the consequences of steric bulk surrounding the nitrogen chelator on the equilibrium using simple peralkylated ethylene and propylene diamine ligands (Fig. 16) (*75*). The didentate alkyl amine ligands preferentially adopted a bis(μ-oxo) structure when the nitrogen substituents were methyl groups. As the ligands increased steric demands with bulkier nitrogen substituents such as ethyl or *tertiary*-butyl groups, the equilibrium between peroxo and oxo compounds shifted towards peroxo. Side-on peroxo compounds are also preferred for diamines with a propylene $-CH_2-CH_2-CH_2-$ backbone vs. an ethylene $-CH_2-CH_2-$ backbone, in compounds with similar steric substituents, thus demonstrating a 5 vs. 6 membered chelate ring effect (see below).

The steric hindrance effect on the bis(μ-oxo) species is also manifested in the didentate ligands, $^{R1}Py1^{R2,R3}$ (Fig. 8, R1 = H or Me, R2 = Et, R3 = Bz-d$_2$). Although didentate ligands prefer Cu_2^{III}-(μ-oxo)$_2$ compounds, adding one methyl group on the 6-position of the pyridine moiety (changing R1 from H to Me) shifts the equilibrium from a bis(μ-oxo)compound in $[\{Cu^{III}(^HPy1^{Et,Bz\text{-}d2})\}_2(O^{2-})_2]^{2+}$ to a side-on peroxo dicopper(II) compound

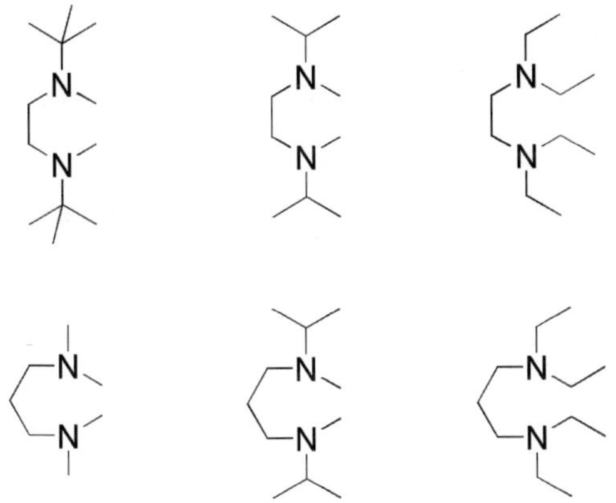

FIG. 16. Series of substituted didentate amine ligands.

in $[\{Cu^{II}(^{Me}Py1^{Et,Bz\text{-}d2})\}_2(O_2^{2-})]^{2+}$ presumably due to the added bulk from the methyl group (Fig. 8) (*83*).

B.3. Chelate Ring Size Effects in Tridentate Ligand Systems

We described above the effects of changing the chelate ring size in Cu^I/O_2 reactivity with tetradentate ligands, (Fig. 10), such that the longer armed $[Cu^I(TEPA)]^+$ complex is even unreactive towards dioxygen (*59*). Itoh and co-workers have observed chelate ring size variations for tridentate ligands where the effects are instead manifested in changes in the structures of copper-dioxygen adducts (*84*). The copper(I) complex of $^{Phe}L^{Pym2}$ forms a low-temperature dioxygen adduct with UV-VIS features indicative of a bis (μ-oxo) dicopper(III) adduct. In contrast, the copper(I) complex of $^{Phe}L^{Pye2}$ reacts with O_2 at low temperature to yield exclusively the side-on peroxo dicopper(II) adduct (Fig. 17) (*85*). Similar to what we observed for the tetradentate ligands, (see above) changing the alkyl linker between the central amine and the pyridine donors from a $-CH_2-$ to a $-CH_2-CH_2-$ group causes a shift towards positive potential ($E_{1/2}$) for the copper(I) complexes. Again, the chelate ring size to the coordinated copper(I) increases from a five membered ring to a six membered ring. This in effect causes a change in the ∠ N–Cu N ligand "bite" angle which, in turn, affects the geometry of the O_2 adduct. The bis(μ-oxo) dicopper(III) cores are better stabilized by more acute ∠ N–Cu–N angles due to the obtuse ∠ O–Cu–O angle in $Cu_2^{III}(O)_2$ in comparison with the side-on peroxo isomers (*86*).

As a last example, the uniformly five-membered chelate ring formed in Tolman's R_3-TACN ligands (Fig. 8) were compared with triazacyclodecane (TACD) and triazacyclododecane (TACDD) ligands (i.e., increasing the macrocycle by one $-CH_2-$, or three $-CH_2-$ groups, respectively (Fig. 18) (*87*). The copper(I) complex of the tridentate TACD contains two 5, and one 6 membered chelate ring, while that of TACDD binds Cu^I with three 6 membered chelate rings. As discussed above (Section III.A.2), the copper(I) reactivity of the isopropyl substituted TACN ligands forms mixtures

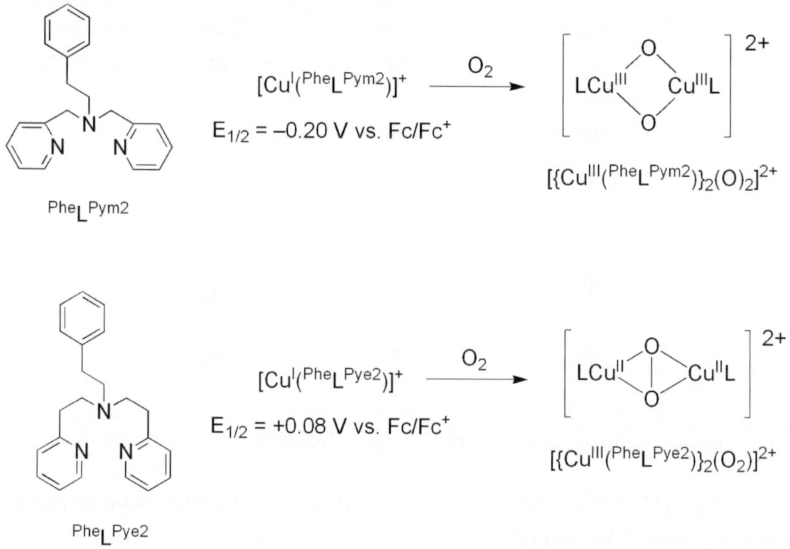

FIG. 17. Effect of chelate ring size in Cu^I/O_2 reactivity with tridentate ligands.

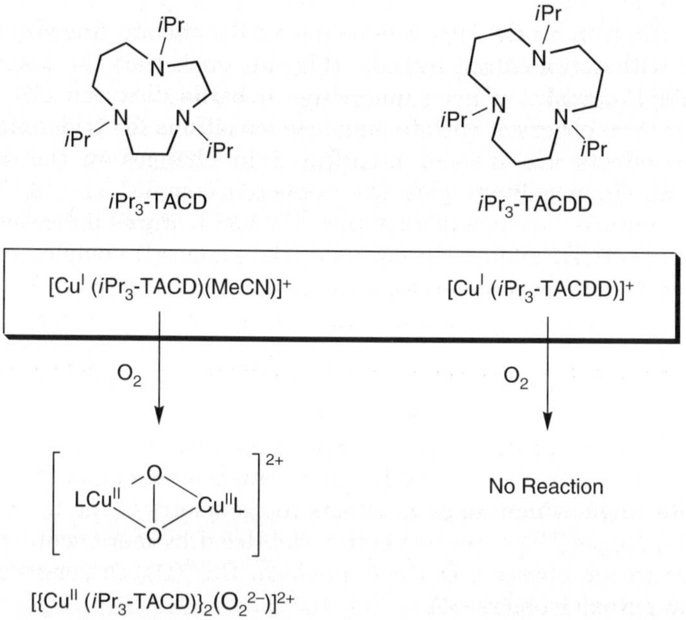

FIG. 18. Contrasting dioxygen reactivity with macrocyclic ligands.

of the side-on peroxo and bis(μ-oxo) compounds. However, with the TACD ligand, the Cu^I/O_2 reaction resulted in only a side-on $[\{Cu(i\text{Pr}_3\text{-TACD})\}_2(O_2)]^{2+}$ complex (Fig. 18) (*87*). The rationalization for the differential dioxygen reactivity is a greater distortion of the $[N_3–Cu–O_2–Cu–N_3]^{2+}$ core in the larger macrocycle (i.e., macrocycle rotation and/or smaller cone angles in order to accommodate the isopropyl groups' steric requirements). The complex with the three 6 membered chelate ring, $[Cu^I(TACDD)]^{1+}$ was unreactive towards dioxygen.

B.4. Type of Nitrogen Donor

The dioxygen chemistry of Cu^I complexes of polyamine ligands is very sensitive to the nature of the nitrogen donor. In the past in our own laboratories we typically used ligands containing pyridine units that capture the essence of an sp^2-hybridized nitrogen of a natural histidine imidazole donor ligand. Such a ligand was MePY2 (Fig. 8, R = H). The oxygenation product $[\{Cu(MePY2)\}_2O_2]^{2+}$ exhibits UV-VIS absorption features indicative of a side-on peroxide complex, but with a shoulder at 410 nm suggesting a small amount of isomeric dicopper(III) bis(μ-oxo) species in solution. The resonance Raman data confirmed both species were present with a $Cu_2^{III}(O)_2$ core vibration at 588 cm^{-1} and an unusually low side-on peroxide vibration at 730 cm^{-1} (*86*). Solid $[\{Cu(MePY2)\}_2O_2]^{2+}$ was analyzed by EXAFS and rR spectroscopies and best fits occur when an 80:20 side-on peroxo:bis(μ-oxo) ratio is considered (*88*). A crystal structure of the side-on peroxo isomer was solved illustrating a Cu…Cu distance of 3.45 Å and interperoxide distance of 1.67 Å, structural parameters which must represent an averaged structure. The minor dicopper(III) bis(μ-oxo) product was the first example of an alkyl/pyridyl ligand supporting a high-valent $Cu_2^{III}(O)_2$ species. Previously characterized bis(μ-oxo) dicopper(III) compounds were formed with copper(I) precursors containing (purely alkyl) substituted triazacyclononane ligands, or various peralkylated diamine ligands, as described above.

In general, if steric considerations permit it, copper(I) complexes of alkyl amine ligands tend to favor bis(μ-oxo) formation while Cu^I precursors of pyridyl containing ligands tend to favor the side-on peroxo isomer upon reaction with O_2. In fact, the initial examples of $Cu_2^{III}(O)_2$ compounds were prepared with alkyl amine ligand donors such as Tolman's substituted TACN and Stack's diamine ligands (see above) (*78,89–91*). Earlier copper-dioxygen reactivity studies with tridentate pyridyl ligands with our own Nn or XYL ligands (see below) only afforded side-on peroxo compounds. The rationalization is that alkyl nitrogen chelates are stronger σ-donors and can better stabilize the high-valent copper(III) cation.

However, a recent example from our group is in contrast to that intuition. The copper(I) complex of a sterically unhindered tridentate alkyl amine ligand, MeAN, (Fig. 19) preferentially adopts a side-on peroxide structure even in moderately coordinating solvents like THF. The UV-VIS absorption spectrum represents a typical side-on peroxide LMCT band at 360 nm ($\varepsilon = 22{,}000\ M^{-1}cm^{-1}$) with rR features at 721 cm^{-1} and 268 cm^{-1}, assigned as the O–O, and Cu…Cu vibrations. The unusually weak intraperoxide stretching frequency may be a consequence of added backbonding arising from the strong σ-donating alkyl amine ligands. However this example demonstrates that the bonding aspects of O_2 to copper(I) complexes are not yet completely understood and require further consideration.

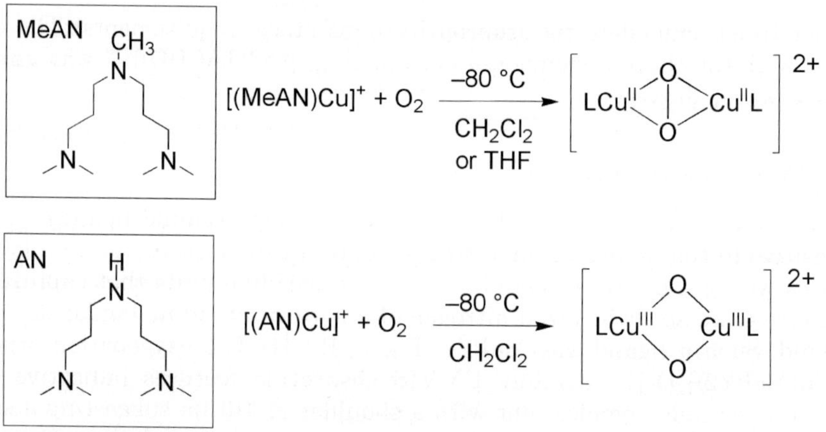

FIG. 19. Contrasting Cu^I/O_2 reactivity with analogous ligands.

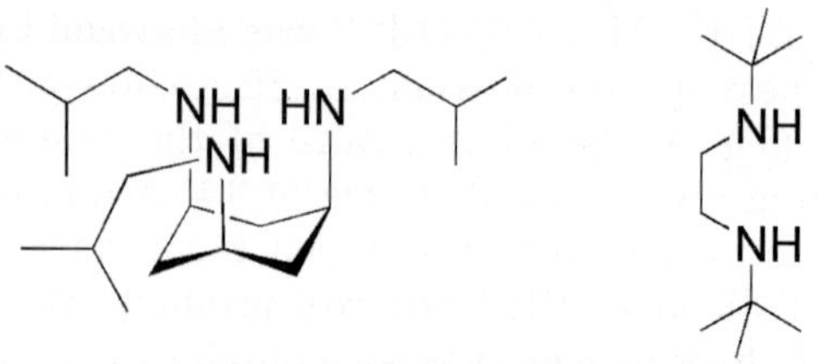

CHART 1.

A second variation in nitrogen donor ligands is the absence or presence of N–H bonds. While the majority of amine ligands in copper-dioxygen complexes contain tertiary amine ligands, there are now several examples of secondary amine ligands supporting copper-dioxygen adducts, including the derivatized TREN ligand ($R_1 = H$, $R_2 = Bn$) which led to a structurally characterized dicopper(II) end-on peroxo structure (see Section II.C). However, within the side-on peroxo/bis(μ-oxo) realm, the presence of an N–H moiety clearly imposes an effect on that equilibrium which still not fully understood, yet intriguing and worth mentioning. To date there are several secondary amine ligands that have been shown to support either $Cu_2^{II}(O_2)$ or $Cu_2^{III}(O)_2$ compounds. Stack and co-workers described an ethylenediamine ligand (Chart 1) that shifts the equilibrium from the typical bis(μ-oxo) dicopper (III) structure for related didentates towards the side-on peroxo dicopper(II) isomer (*92*). Masuda and co-workers describe a distinctive cyclohexane backed ligand that contains three secondary amine donors in a rigid *cis*-1,3,5 axial framework (Chart 1) (*93*). The ligand strongly binds Cu(I) and reacts with O_2 at low temperature to generate a bis(μ-oxo) complex in various solvents. Interestingly, the complex exhibits a Cu_2O_2 core vibration of $\nu = 570\,cm^{-1}$ ($^{18/16}\Delta = 26$), the lowest of all reported bis(μ-oxo) compounds (*1*). Whether or not this is due to the three secondary amine groups, is uncertain.

In our own group efforts, with analogous ligands, we have discovered drastic effects on the dioxygen reactivity of a copper(I) complex analogous to that of the ligand MeAN (Fig. 19). The tridentate amine ligands differ only by one substituent on the central nitrogen donor whereby a methyl group is replaced by a hydrogen atom. In CH_2Cl_2 at $-80°C$, $[Cu^I(MeAN)]^+$ reacts with O_2 forming solely a side-on $[\{Cu^{II}(MeAN)\}_2(O_2)]^{2+}$ complex, while the Cu^I complex of the secondary amine ligand, AN, forms only the high-valent $[\{Cu^{III}(AN)\}_2(\mu\text{-}O)_2]^{2+}$ species (Fig. 19) (*94*). Unlike $[\{Cu(MeAN)\}_2(O_2)]^{2+}$, the $[\{Cu(AN)\}_2(O)_2]^{2+}$ complex is susceptible to solvent effects and forms mixtures of $Cu^{II}_2(O_2)$ and $Cu^{III}{}_2(O)_2$ isomers in THF and acetone, and in diethyl ether the main dioxygen adduct is the dicopper(II) side-on peroxo species. The contrasting reactivity between the closely related $[\{Cu^{II}(MeAN)\}_2(O_2)]^{2+}$ and $[\{Cu^{III}(AN)\}_2(O)_2]^{2+}$ complexes calls attention to the sensitivity of every feature of an amine ligand. In secondary amine ligands there is a proton available for hydrogen bonding. Borovik and co-workers discovered a bis(μ-oxo) dicobalt(III) complex stable at room temperature due to the presence of four intramolecular hydrogen bonds (*95*). Save for a few examples, (*35,96*), the effect of hydrogen bonding has yet to be extensively explored in copper-dioxygen chemistry.

B.5. Electronic Effects

Until recently, the effect of ligand electronic variations upon the side-on peroxo/bis(μ-oxo) equilibrium had not been systematically investigated. By adding electron donating or withdrawing substituents to the *para*-position of a pyridine donor, we effectively altered the electron donating ability of the pyridine N-atom without inducing any other possibly conflicting ligand modifications such as sterics, denticity, or chelate ring size. The 4-R-pyridyl substituents, R = NMe_2, OMe, and Cl were incorporated into the tridentate ligand, R-MePY2, in each of the two pyridyl donors (Fig. 8) (*97*). This double substituent effect was manifested in the varying physical properties of the copper(I) complexes. The ν(C–O) frequency of the Cu^I-carbonyl adducts showed a gradual decrease in CO stretching frequency as the electron donating ability of R increased, as expected for having more Cu^I to CO back-bonding. Also, the Cu^{II}/Cu^I redox values for $[Cu^I(R\text{-}MePY2)(MeCN)_n]^+$ increased from –440 mV, to –360 mV, to –310 mV, to –270 mV vs. Fc/Fc^+ in DMF, for R = NMe_2, OMe, H, and Cl, respectively. In fact, the more positive potential of $[Cu(Cl\text{-}MePY2)(MeCN)]^+$ precludes the compound from reducing dioxygen and it forms no copper-dioxygen adduct. In contrast, the copper(I) compounds of R-MePY2 where R = NMe_2, OMe, and H, do activate O_2 and form mixtures of side-on peroxo and bis(μ-oxo) compounds, where the equilibrium lies largely in favor of the peroxo isomer. The equilibrium between the two species was affected by the pyridine *para*-substituents. Although not so clearly evident by the only modestly varying UV-VIS absorption spectra of the Cu_2O_2 complexes, rR studies confirm that the electron donating substituents on the ligands produce a greater relative amount of bis(μ-oxo) species. By scaling the rR intensity to the side-on

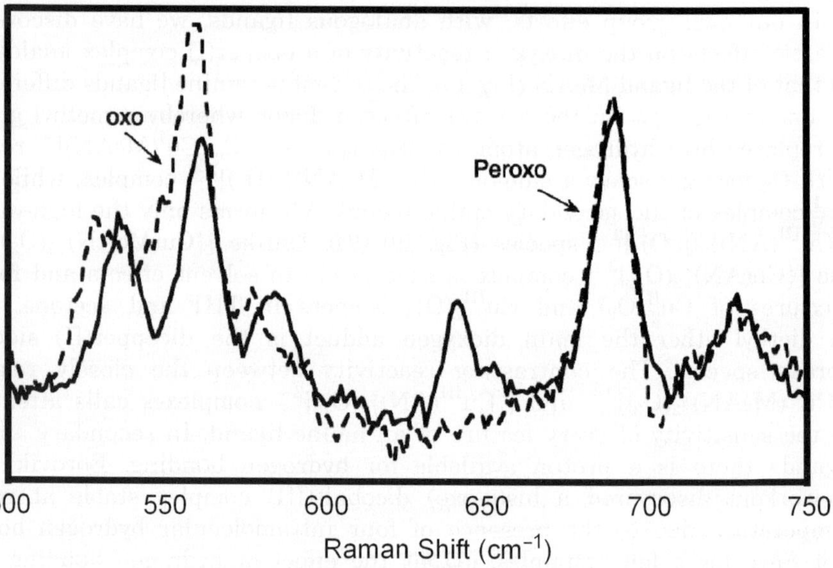

FIG. 20. rR spectroscopic evidence for a larger amount of μ-oxo complex $[\{Cu^{III}(L)\}_2(O)_2]^{2+}$ compared to peroxo species $[\{Cu^{II}(L)\}_2(O_2)]^{2+}$, L = H-MePY2 (solid line) or OMe-MePY2 (dotted line).

peroxide peak, it becomes clear that the OMe-derivative has a higher bis(μ-oxo) quantity present. (Fig. 20) And the NMe_2-derivative has an even higher relative amount of $Cu_2^{III}(O)_2$ (*64*).

The ligand electronic effect is more pronounced in a copper(I) complex with an analogous tridentate ligand, PYAN, whereby one of the pyridyl donors in MePY2 was replaced with a tertiary alkyl amine (Fig. 8). The $[Cu(PYAN)(MeCN)]^+$ complex not only reveals significant solvent effects on dioxygen reactivity (see above), but also illustrates dramatic electronic effects when the ligand is modified by a 4-R-pyridyl substituent (*98*). The copper(I) precursors of R-PYAN (R = NMe_2, OMe, H, and Cl) display similar trends in carbon monoxide backbonding and $E_{1/2}$ redox potentials as the R-MePY2 ligand series (i.e., the ν(C–O), and $E_{1/2}$ of the corresponding Cu^I complexes decrease as the ligand becomes a stronger donor by the R-substituent), although to a lesser extent. The decreased span is probably due to a single R-substituent vs. two R-substituents from each pyridine ring in R-MePY2. However, in contrast to the dioxygen reactivity of $[Cu(R\text{-}MePY2)]^+$, the equilibrium between the dicopper(II) side-on peroxo and dicopper(III) bis(μ-oxo) is drastically affected by the electronically modifying ligand substituent. According to the UV-VIS spectra of the oxygenated complexes, the complex with the most electron rich ligand, NMe_2-PYAN, forms exclusively $[\{Cu^{III}(NMe_2\text{-}PYAN)\}_2(O)_2]^{2+}$ in CH_2Cl_2 ($\lambda_{max} = 395$ nm), while the electron poor, or parent complex favor the formation of $[\{Cu^{II}(R\text{-}PYAN)\}_2(O_2)]^{2+}$ (where R = Cl or H) ($\lambda_{max} = 358$ nm) (Fig. 21). Therefore, the structure of the Cu_2O_2 complex of the R-PYAN ligand series can be completely tailored by pure electronic effects (*98*).

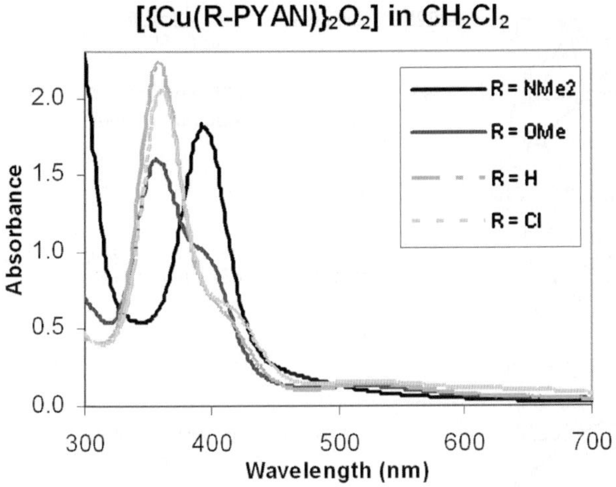

FIG. 21. Electronic effects affecting the side-on peroxo ($\lambda_{max} = 358$ nm) and bis(μ-oxo) ($\lambda_{max} = 395$ nm) equilibrium.

C. Tridentate Ligand Influences: Binucleating Ligand Constraints

Considerable ligand effects were observed when copper ions are constrained in some manner, such as possible Cu... Cu distance, by dinucleating ligands. The ligands Nn (Fig. 8), where the bis[2-(2-pyridyl)ethyl]amine (PY2) units are linked *via* a $(CH_2)_n$ spacer (n = 3–5), and their dicopper(I) complexes display different O_2 reactivity in terms of kinetics/thermodynamics, and structure and bonding descriptions. All dicopper(I) complexes $[Cu^I_2(Nn)]^{2+}$ reversibly bind O_2 at low temperature forming dioxygen adducts (2:1 Cu:O_2) with intense LMCT bands at 365 nm (N3) and 360 nm (N4 and N5) typical of side-on peroxo LMCT energies, but they also contain an additional intense band at 490 nm, 458 nm, and 423 nm (sh) for N3, N4, and N5, respectively (Fig. 22). The peroxo compounds of N3 and N4 were subjected to EXAFS which revealed short Cu... Cu distances of 3.2 Å and 3.4 Å, for $[Cu^{II}_2(N3)(O_2)]^{2+}$ and $[Cu^{II}_2(N4)(O_2)]^{2+}$, respectively. The rational for the structural differences lies in the geometric constraints of the supporting ligand, whereby the ligand containing the shortest methylene linker N3, enforces the shortest Cu... Cu distance.

To gain a better understanding of the consequences that the constraining ligands imposed on the bonding and structure of the O_2 adduct, and to determine the origin for the additional peak at ~423–490 nm, the $[Cu^{II}_2(Nn)(O_2)]^{2+}$ complexes were investigated with rR and correlated with electronic structure calculations. Resonance excitation into the newly observed LMCT band for $[Cu^{II}_2(Nn)(O_2)]^{2+}$ did not produce a characteristic bis(μ-oxo) peak at around 600 cm^{-1}. Instead, excitation profiles of those CT bands caused resonance enhancement of the O–O and Cu... Cu stretching vibrations indicating that the additional peak was indeed associated with the side-on peroxo complex and not a bis(μ-oxo) component. The origin of this new peak associated with a side-on peroxo is attributed to the "bent", or "butterfly"

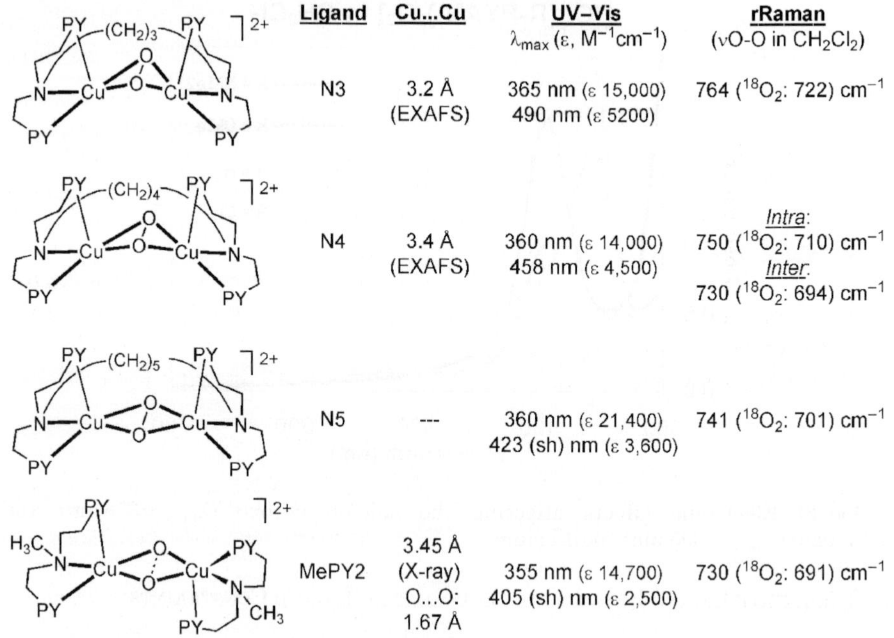

FIG. 22. Structural and spectroscopic properties of $[Cu^{II}_2(Nn)(O_2)]^{2+}$.

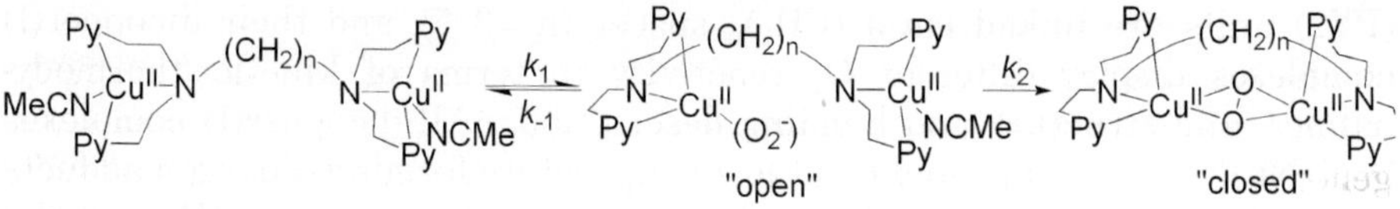

SCHEME 3.

structure of the dicopper-dioxygen core as commanded by the constraining ligands, and is assigned as the second component of the π^*_σ interaction. A clear trend was observed in the intramolecular peroxide stretching frequencies for the $[Cu^{II}_2(Nn)(O_2)]^{2+}$ series: ν(O–O) = 764 cm^{-1}, 750 cm^{-1}, and 741 cm^{-1}, for N5, N4, and N3, respectively. The trend continues with the mononuclear analogue, MePY2, whose side-on peroxo dicopper(II) complex has a peroxide stretching frequency of 730 cm^{-1} (see above).

The ligand constraints also show consequences in the kinetics and thermodynamics of O_2 binding in CH_2Cl_2 (*99*). At temperatures below < 213 K, kinetic analysis reveals an initial rate-determining reversible formation of a non-detectible superoxide adduct, followed by subsequent Cu(I) binding to form the "closed" side-on peroxo species (Scheme 3). However, at that temperature regime, definitive thermodynamic parameters can not be derived from kinetic measurements due to dimerization present at lower temperatures. At higher temperatures (253–283 K), the first reaction is far left-lying, and a pre-equilibrium superoxo species is implicated by the low, or negative activation enthalpies of formation of the side-on peroxo adduct.

Rate constants for the formation of the "closed" compounds, $k_{on,\,high}$, increase slightly as the ligand constraints ease (i.e., going from N3, to N4, to N5), and the equilibrium constant $K_{on} = (k_{on,\,high}/k_{off,\,high})$ for the side-on peroxide formation increase by two orders of magnitude in similar fashion. Thermodynamic parameters for K_{on} vary modestly as the methylene linker chain is increased: $\Delta H^{\circ}_{on} = 84 \pm 6$ kJ/mol (N3), 84 ± 1 kJ/mol (N4), and 81 ± 4 kJ/mol (N5), and $\Delta S^{\circ}_{on} = -297 \pm 22$ J/mol·K (N3), -270 ± 5 J/mol·K (N4), and -238 ± 15 J/mol·K (N5). The entropic factor decreases as the flexibility of the ligand increases. The kinetics for the intermolecular reaction of the mononuclear compound $[Cu(MePY2)(MeCN)]^+$ in CH_2Cl_2 are complex but the following general observation can be made: k_{on}, is considerably smaller than that for intramolecular $[Cu^{II}_2(Nn)(O_2)]^{2+}$ formation (*99*).

The changes in structure, spectroscopy and bonding, and kinetic and thermodynamic parameters for O_2-binding to the dicopper(I) complexes of the Nn ligand series is a pertinent example of the strong influence a ligand can impose on dioxygen reactivity. The ligand constraints established a new "butterfly" binding mode for side-on peroxo complexes with unique spectroscopic (UV-VIS) features owing to changes in molecular bonding. The spectral properties and O_2 reactivity parameters of $[Cu^{II}_2(Nn)(O_2)^{2-}]^{2+}$ are summarized in Fig. 22.

D. Summary of Ligand Effects on Copper-Dioxygen Reactivity

The previous discussion represents a sizeable but non-exhaustive section surveying of advances in copper-dioxygen reactivity and highlights the general trends in ligand features and consequences for the preference for various types of copper-dioxygen adducts. Monomeric Cu O_2 adducts are generally unstable intermediates on-route to the formation of dinuclear compounds. Some copper(II)-superoxo (or Cu^{III}-peroxo) mononuclear compounds can be stabilized *via* ligand design efforts. In dinuclear chemistry, it is evident that tetradentate ligands favor the formation of an end-on peroxo species, while tridentate and didentate ligands prefer side-on peroxo and/or bis(μ-oxo) formation. The latter two species can often be in a facile equilibrium which can be biased by external factors such as solvent, counterion, and temperature, and the specific nature of the tri-/bi-dentate ligand plays a large role in the structure of the resulting O_2 adducts. Generally speaking, sterically unhindered, didentate, strong σ-donating (alkyl nitrogen) ligands favor bis(μ-oxo) formation, and tridentate pyridine ligands or ligands with steric requirements can favor side-on peroxo formation. The examples of the various ligand effects described herein are merely trends for Cu^I/O_2 reactivity, and some examples that are in contrary to the aforementioned have also been documented. Sorting out the various influences in Cu^I/O_2 reactivity will be helpful for designing well characterized reactive Cu_nO_2 compounds which may be useful for investigating substrate reactivity and oxidation mechanisms in both synthetic systems and relevant biological enzymes.

IV. Ligand Effects on the Exogenous Substrate Reactivity of Cu_2O_2 Compounds

The ligand influences described above also have a fundamental impact on their oxidative chemistry. Changes in coordination number, redox properties, and steric accessibility, all affect the oxidative capability of the activated copper-O_2 complex. Here, we first highlight some initial observations from our group demonstrating reactivity differences from end-on, and side-on peroxo dicopper(II) complexes. We then discuss more recent mechanistic studies on a tyrosinase model system, and then present some very recent investigations on exogenous substrate reactivity by dicopper-dioxygen complexes from our laboratory.

A. End-on vs. Side-on Peroxo Reactivity – Initial Studies

Early after the detection and characterization of discrete Cu_2O_2 compounds such as the TMPA end-on peroxo and Nn side-on peroxo dicopper(II) compounds, we began to investigate their oxidative reactivity properties in order to see if we can mimic biological functionality and also to learn about the fundamental properties of each types of compounds. Analysis of the end-on peroxo dicopper(II) complexes, $[\{Cu(TMPA)\}_2(\mu\text{-}1,2\text{-}O_2)]^{2+}$ and $[Cu^{II}_2(XYL\text{-}O^-)(O_2)]^+$ (a terminal end-on (μ-1-O_2) compound formed from the low-temperature oxygenation of $[Cu^I_2(XYL\text{-}O^-)]^+$, see Fig. 23) (*100*), and their reaction with acids, organic peroxides, and substrates like 2,4-di-*tert*-butylphenol, and triphenylphosphane, gave a clear

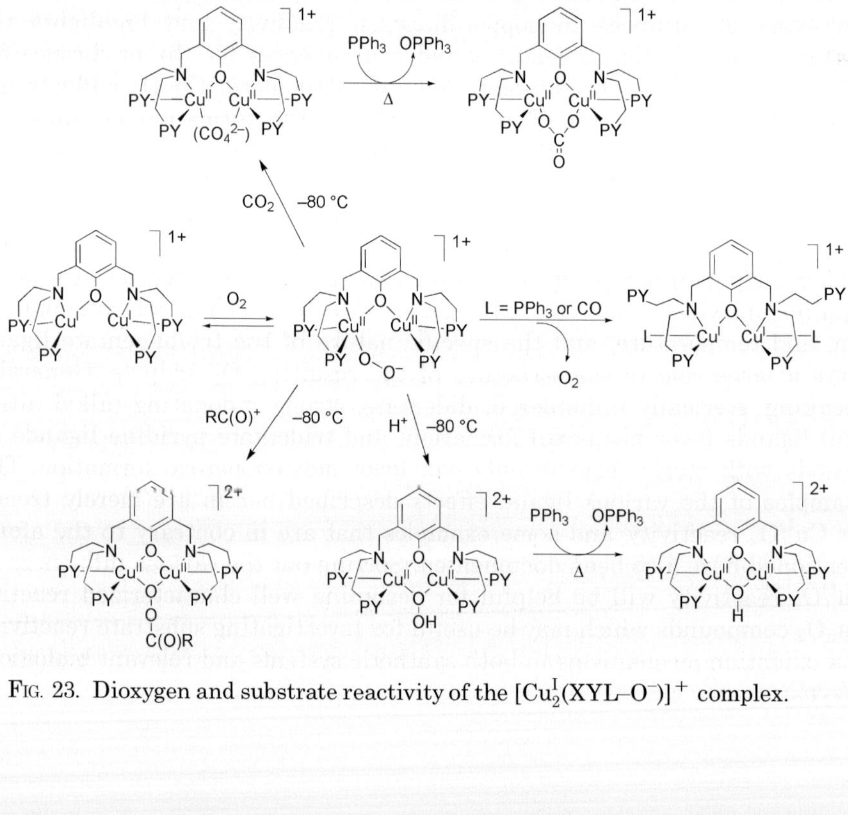

FIG. 23. Dioxygen and substrate reactivity of the $[Cu^I_2(XYL\text{–}O^-)]^+$ complex.

representation of their reactivity properties. Both of these end-on peroxo complexes were found to be basic, or nucleophilic, in nature according to the following reactivity trends (*101*). The stoichiometric reaction of $[Cu_2^{II}(XYL\text{-}O^-)(\mu\text{-}1\text{-}O_2)]^+$ with HBF_4 or HPF_6 led to $[Cu_2^{II}(XYL\text{-}O^-)(\mu\text{-}1,1\text{-}OOH)]^{2+}$ formation (*102*). Also, $[Cu_2^{II}(XYL\text{-}O^-)(O_2)]^+$ is acylated by addition of an acyl chloride generating the first structurally characterized percarboxylatodicopper(II) complex (*103*). Addition of excess H^+ to either $[Cu_2^{II}(XYL\text{-}O^-)(O_2)]^+$ or $[\{Cu^{II}(TMPA)\}_2(\mu\text{-}1,2\text{-}O_2)]^{2+}$ releases the peroxide ligands yielding H_2O_2 and the oxidized dicopper(II) complexes. Both end-on peroxo complexes react with CO_2 forming peroxycarbonato complexes which thermally decompose to the carbonato species with liberation of O_2(g) (Fig. 23). The reaction of the end-on Cu_2O_2 compounds with triphenylphosphane results in $[LCu^I(PPh_3)]^+$ and the release of molecular oxygen. Finally, the reaction with 2,4-di-*tert*-butylphenol as substrate results in acid/base chemistry where the phenol serves as a proton source. All these results are consistent with the nucleophilic nature of Cu_n^{II} end-on peroxo species (Fig. 24). However, these complexes can be electrophilically activated by addition of one equivalent of proton or CO_2 as the dicopper(II) hydroperoxo and peroxycarbonato products readily oxidize PPh_3 to $O{=}PPh_3$ (with warming). This chemistry is summarized in Fig. 23.

To compare and contrast, the same substrate survey was carried out on a dicopper(II) side-on peroxo complex supported by the dinucleating ligand N4 (Fig. 8, see Section III.C). $[Cu_2^{II}(N4)(O_2)]^{2+}$ exhibited contrastingly different reactivity than that described for the end-on peroxo complexes. It was found to be unreactive towards acids, acyl chlorides, and CO_2. Furthermore, unlike the nucleophilic end-on peroxo complexes, the side-on peroxo $[Cu_2^{II}(N4)(O_2)]^{2+}$ oxidized PPh_3 to $O{=}PPh_3$ (with warming) in good yields, and 2,4-di-*tert*-butylphenol was oxidatively dimerized, presumably by hydrogen atom abstraction (Fig. 24).

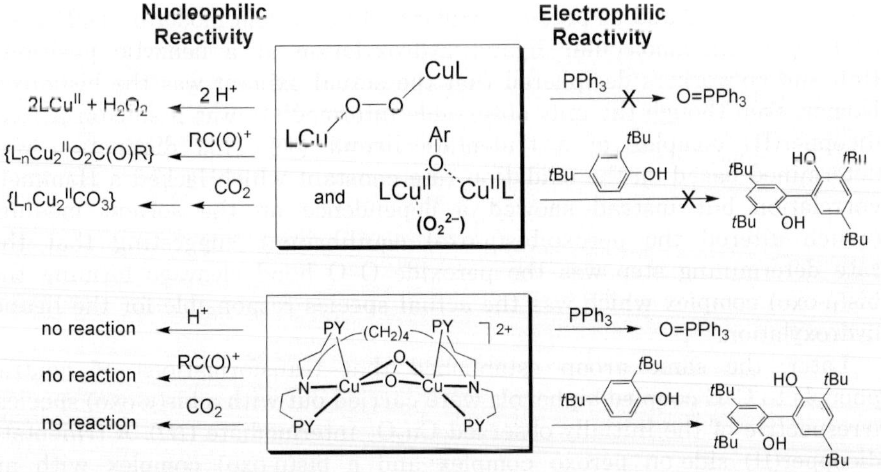

FIG. 24. Nucleophilic reactivity of end-on peroxo and electrophilic reactivity of side-on peroxo complexes.

This disparate reactivity exemplifies the key role that the ligand imposes on the subsequent reactivity. Copper complexes of tetradentate ligands form nucleophilic end-on peroxo compounds, and tridentate amine ligands generate electrophilic side-on peroxo dicopper(II) (or bis(μ-oxo) dicopper(III)) compounds.

B. Reactivity of Side-on Peroxo and Bis(μ-oxo) Complexes

It is now established that dicopper(II) side-on peroxo and dicopper(III) bis(μ-oxo) compounds are both electrophilic and can effect a wide range of substrate oxidations. The specific reaction depends on the particular ligand system and resulting O_2 adduct. Oxidative reactivity includes alcohol to aldehyde oxidations, and the oxidative coupling of phenols (*2*). Also, hydroxylase activity has been observed in both aromatic (*104*) or phenolate oxidations (cresolase activity) (*85,92,105,106*), as well as in aliphatic (*82*) hydrocarbons at benzylic positions, akin to DβH reactivity. Oxidative *N*-dealkylations, as occur in PAM, by Cu_2O_2 complexes take place in intramolecular (*107*) (ligand oxidation) and intermolecular (*97*) (substrate) reactions. Finally, Cu_2O_2 compounds can also carry out oxygen atom transfer reactions to phosphanes and sulfides (*108*), and benzylamine oxidation to benzonitrile has also been documented (*92*).

It is thought that bis(μ-oxo)dicopper(III) compounds are good for carrying out hydrogen atom abstraction type reactions (*107*), while side-on peroxo dicopper(II) compounds are better for oxygen atom insertion reactions (*75,109,110*). However, there are several examples where these generalizations do not hold true. Tolman and co-workers observed intramolecular aromatic hydroxylation by a bis(μ-oxo) dicopper(III) complex, though the mechanism of oxidation is ill-defined (*104*). Also, it is often the case where the actual species responsible for the oxidation is obscured due to the facile equilibrium between $[Cu^{II}_2(O_2)]^{2+}$ and $[Cu^{III}_2(O)_2]^{2+}$ isomers (*97*), and both species should be considered as possible oxidants. In a system modeling DβH reactivity (i.e., intramolecular ligand hydroxylation at a benzylic position), Itoh and co-workers deciphered that the actual oxidant was the bis(μ-oxo) isomer, even though the only observable intermediate was a side-on peroxo dicopper(II) complex of a tridentate ligand (*80*). This distinction was determined based on the oxidation rate constant which lacked a Hammett correlation but instead showed a dependence on the solvent medium (which altered the peroxo/bis(μ-oxo) equilibrium), suggesting that the rate determining step was the peroxide O–O bond cleavage forming the bis(μ-oxo) complex which was the actual species responsible for the ligand hydroxylation.

Later, the same group established that transformations of neutral phenols to C–C coupled biphenols were carried out with a bis(μ-oxo) species, irrespective of the initially observed Cu_2O_2 intermediate (*111*). A tridentate dicopper(II) side-on peroxo complex and a bis(μ-oxo) complex with an analogous didentate ligand, $^{H}PY1^{Et,Bn\text{-}d2}$ (Fig. 8, R1 = H, R2 = Et, R3 = Bn-d_2), reacted with various substituted phenols *via* a similar proton-coupled electron transfer mechanism (see Section IV.D), with the only difference being that the oxidation rate constants for the side-on peroxo reaction were two orders of magnitude smaller than those of the reaction by the dicopper (III) bis(μ-oxo) complex. It was concluded that the bis(μ-oxo) dicopper(III) complex was the actual oxidant responsible for the phenol oxidations, even in cases where only a side-on peroxo isomer was observed. Despite these advances, the reactivity tendencies and mechanisms of side-on peroxo and bis(μ-oxo) complexes are complex and still require further understanding.

C. Aromatic Hydroxylation

Early on in our laboratories we successfully carried out tyrosinase-like activity with a dicopper complex $[Cu^{I}_{2}(XYL\text{-}H)]^{2+}$ whereby an arene C–H bond is rapidly (within seconds) hydroxylated under very mild conditions (i.e., −80°C, ≤1 atm O_2) (*112*) (Fig. 25, R = H). A side-on peroxo structure was confirmed for the dioxygen intermediate using stopped flow UV-VIS and spectroscopic probing employing derivatized ligands that slowed the hydroxylation step and allowed direct probing of a reactive intermediate (*109,113*). Manometry experiments demonstrated a Cu:O_2 stoichiometry of 2:1, and using labeled $^{18}O_2$ proved that the hydroxyl O-atom comes from molecular oxygen.

We have gathered compelling evidence supporting the mechanism of hydroxylation to be one where the reaction occurs *via* an electrophilic attack on the juxtaposed aryl group. Stopped flow kinetic measurements of the oxidation of substituted $[Cu^{I}_{2}(R\text{-}XYL\text{-}H))]^{2+}$ show that $k_{hydroxyl}$ (Fig. 25) increases as R becomes more electron donating; i.e., stabilizing

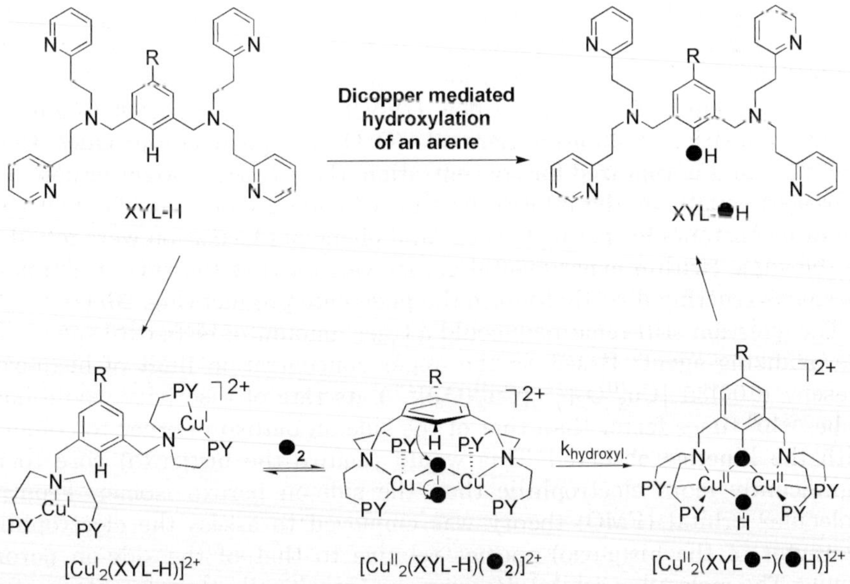

FIG. 25. Tyrosinase-like activity by a synthetic dicopper(II) dioxygen complex.

a carbocation intermediate associated with aromatic electrophilic substitution ($\rho \approx -2.1$) (*109*). Furthermore, no kinetic deuterium isotope effect is observed on $k_{hydroxyl}$ when using $[Cu_2^I(R\text{-}XYL\text{-}D)]^{2+}$ as the substrate complex (*114*), clearly eliminating a hydrogen atom abstraction type mechanism. Substitution to a methyl group in the hydroxylation 2-position, $[Cu_2^I(H\text{-}XYL\text{-}CH_3)]^{2+}$ results in aromatic hydroxylation accompanied by a 1,2 migration of the methyl group, reminiscent to the "NIH shift" observed by iron hydroxylases involved with electrophilic oxidations (*115,116*).

More recently, this aromatic hydroxylation reaction has been examined in greater detail by rR spectroscopy and with a nitro-substituted XYL-H derivative. The relative rate of the hydroxylation reaction of $[Cu_2^I(NO_2\text{-}XYL\text{-}H)]^{2+}$ is diminished compared to formation of the peroxo intermediate $[Cu_2^{II}(NO_2\text{-}XYL\text{-}H)(O_2)]^{2+}$, making the latter accessible to spectroscopic analysis (*109*). The oxygenated product $[Cu_2^{II}(NO_2\text{-}XYL\text{-}H)(O_2)]^{2+}$ has UV-VIS absorption bands at 358 nm ($\varepsilon = 20{,}000\ M^{-1}cm^{-1}$) and 530 nm ($\varepsilon = 1{,}200\ M^{-1}cm^{-1}$), similar to other side-on peroxo dicopper(II) complexes, but it also has an additional intense peak at 435 nm ($\varepsilon = 5{,}000\ M^{-1}cm^{-1}$). Despite the tendency for certain side-on peroxo compounds to be in equilibrium with isomeric bis(μ-oxo) compounds (see above), which also have LMCT transitions near ~400–420 nm, this new peak is not due to the presence of a bis(μ-oxo) species. With resonance excitation, no symmetric $Cu_2^{III}(O)_2$ core stretching mode was observed near 600 cm^{-1} (*117*). Instead, a rR profile of the 435 nm absorption band shows that it is associated with the 277 cm^{-1} peak associated assigned as the side-on peroxo Cu…Cu stretch. As determined by comparison with a known bis(μ-oxo) standard, the amount of residual $Cu_2^{III}(O)_2$ possibly in equilibrium was determined to be 0.13 % (0.0013) relative to the amount of Cu_2^{II}-(side-on-O_2) present.

The kinetics of hydroxylation were monitored by rR comparing the decay of the intensity of the peroxide stretch to the increasing C–O stretch of the newly forming phenol, and were found to be consistent with the kinetics previously measured by stopped flow UV-VIS spectroscopy (*109*). Compared to the intensities of standard peroxide (O–O) and phenolate product (C–O) stretches, and normalized for concentration, the relative amount of peroxide decreases equally as the relative amount of phenol increases. The fact that the rate constants for peroxide decay and phenolate formation were found to be the same (within experimental error), verifies that the side-on peroxide species is reacting directly to form the phenolate product (Fig. 26) (*117*).

The question still remained: could a trace amount of bis(μ-oxo) species be the oxidizing agent? Based on the upper concentration limit of bis(μ-oxo) present, (0.0013:1 $[Cu_2^{III}O_2]^{2+}$:$[Cu_2^{II}(O_2)]^{2+}$), its rate of reactivity would have to be $> 10^{0}$ times faster than that of the side-on peroxo in order to coincide with the kinetics observed. This would require the bis(μ-oxo) core to be significantly more electrophilic than the side-on peroxo isomer. Frontier molecular orbital (FMO) theory was employed to assess the electrophilic character of the bis(μ-oxo) species relative to that of the side-on peroxo isomer. The molecular orbital descriptions (specifically the percentage of π_σ^* character) and relative energies of the relevant LUMO's of the side-on peroxo

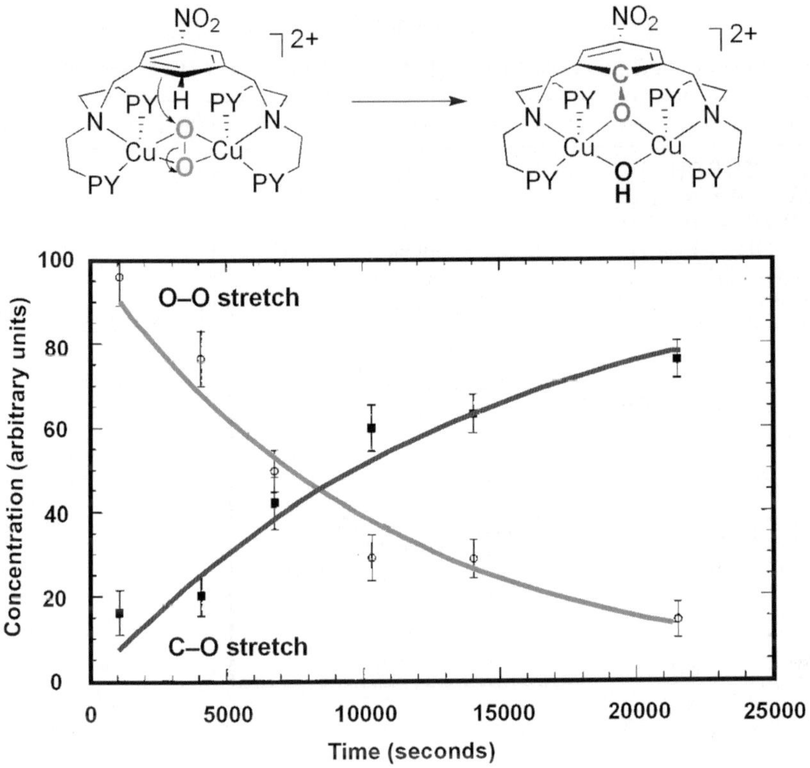

FIG. 26. Concentration of the side-on peroxide $[Cu^{II}_2(NO_2\text{-XYL-H})(O_2)]^{2+}$ and phenolate product $[Cu^{II}_2(NO_2\text{-XYL-O}^-)(OH)]^{2+}$ as determined by the intensities of ν(O–O) and ν(C–O) as a function of time.

and bis(μ-oxo) cores appear to be similar as determined either by density functional theory (*42*) or *ab initio* calculations (*45*). It was concluded that each species should be roughly equally activated for electrophilic aromatic hydroxylation. Also, considering the additional charge donation from the copper to the (O^{2-}) in the $Cu^{III}_2(O^{2-})_2$ isomer, (i.e., more negative charge on the oxide ligands) the bis(μ-oxo) species would be relatively more disfavored for electrophilic attack as compared to the side-on peroxo structure (*117,118*). Since theoretical calculations failed to describe the bis(μ-oxo) as having enhanced electrophilic character, the species would not react several orders of magnitude faster than the side-on peroxo species; thus, the latter dicopper(II) peroxo complex was assigned as the species responsible for the aromatic hydroxylation of the XYL ligand.

D. Intermolecular Oxidations

To complement our studies with previous tridentate dinucleating ligands such as the XYL and Nn based ligands, we turned our efforts to the tridentate mononucleating ligands, R-MePY2 (Fig. 8) for further insights into copper-dioxygen activation and reactivity. The resulting Cu_2O_2 compounds

TABLE I

SUBSTRATE OXIDATIONS OF $[\{Cu(R\text{-}MePY2)\}_2O_2]^{2+}$ IN CH_2Cl_2 AT −80°C

Substrate	H	OMe	NMe_2
THF → THF-OH	67 %	82 %	84 %
DMA → MA + CH_2O	69 %	83 %	96 %
$PhCH_2OH$ → PhCHO	72 %	88 %	100 %
Ph_2CHOH → Ph_2CO	79 %	92 %	100 %
MeOH → CH_2O	31 %	38 %	37 %
TEMPOH → TEMPO•	100 %	100 %	100 %

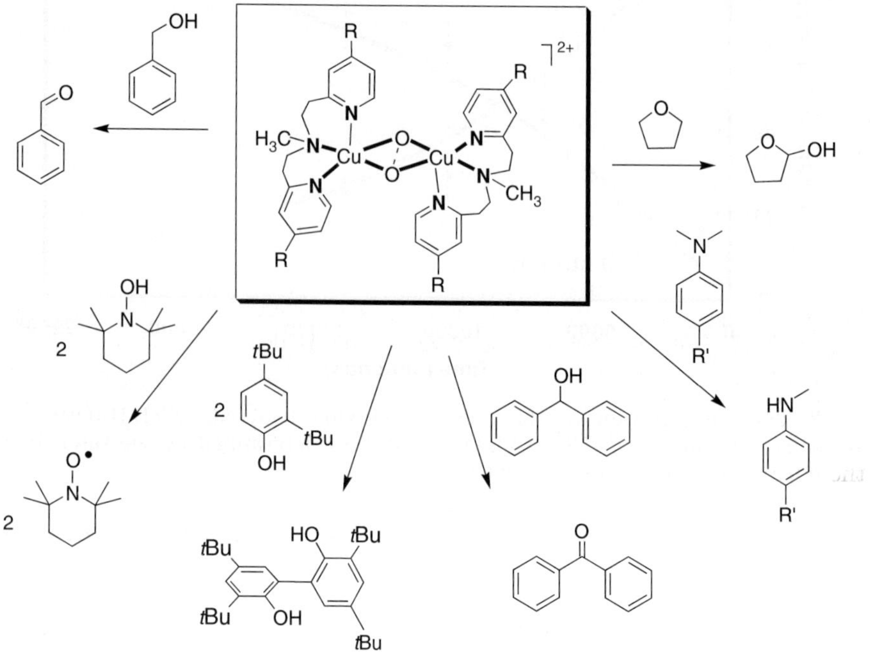

FIG. 27. Exogenous substrate reactivity of $[\{Cu(R\text{-}MePY2)\}_2(O)_2]^{2+}$.

are highly efficient oxidants showing unprecedented substrate reactivity (Table I, Fig. 27) (*97*). Benzyl alcohol and benzhydrol are efficiently converted to benzaldehyde and benzophenone, respectively. Intramolecular oxidative *N*-dealkylations had previously been observed with Cu_2O_2 compounds (*107*), but to our knowledge, our system was the first to carry out this same reaction in an *intermolecular* fashion with an *exogenous* substrate. *N,N*-dimethylaniline (DMA) was converted to *N*-methylaniline (MA) and formaldehyde. The stoichiometric reaction occurred efficiently in moderate to high yields depending on the R-substituent on the ligand.

Tetrahydrofuran (THF) oxidation was also observed and documented for the first time for Cu_2O_2 reactivity. The reaction was clean giving mainly 2-hydroxy-tetrahydrofuran (THF-OH) as the oxidized product. When using

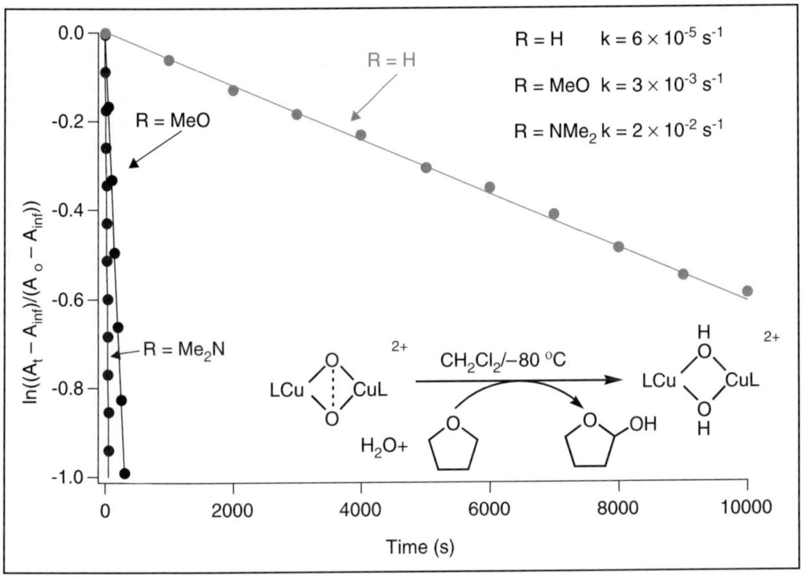

FIG. 28. Differences in THF oxidation rates for the $[\{Cu^{II}(R\text{-}MePY2)\}_2(O_2)]^{2+}$ series in CH_2Cl_2.

labeled $^{18}O_2$ as the dioxygen source, the hydroxylated product contains >80 % THF-^{18}OH confirming the oxygen atom is derived from dioxygen and the Cu_2O_2 species. THF oxidations by $[\{Cu(R\text{-}MePY2)\}_2O_2]^{2+}$ exhibited dramatic ligand effects as seen by enhanced (~1500 fold) reaction rates as the ligand is made more electron donating with the 4-R substituent on the pyridyl group (Fig. 28).

E. Mechanistic Studies

A current focus in our laboratories has been to uncover mechanistic details concerning substrate oxidations by dicopper-dioxygen compounds. We have recently been involved with comprehensive mechanistic studies on the oxidation of dimethylaniline (DMA) by $[\{Cu(R\text{-}MePY2)\}_2(O_2)]^{2+}$ with an emphasis on the precise distinction between a hydrogen atom abstraction mechanism (herein considered a "concerted" electron transfer-proton transfer mechanism-ETPT) and a proton-coupled electron transfer mechanism, specifically where a rate limiting electron transfer step is followed by a proton transfer step (ET/PT) (Scheme 4). In order to distinguish the different mechanisms, we have studied linear free energy correlations, deuterium kinetic isotope effects, and employed mechanistic probes; we also considered thermodynamic factors in favoring an ETPT vs. an ET/PT mechanism.

E.1. Linear Free Energy Correlations

To differentiate between a concerted ETPT mechanism and a sequential ET/PT pathway we analyzed the oxidation rates of *para*-substituted

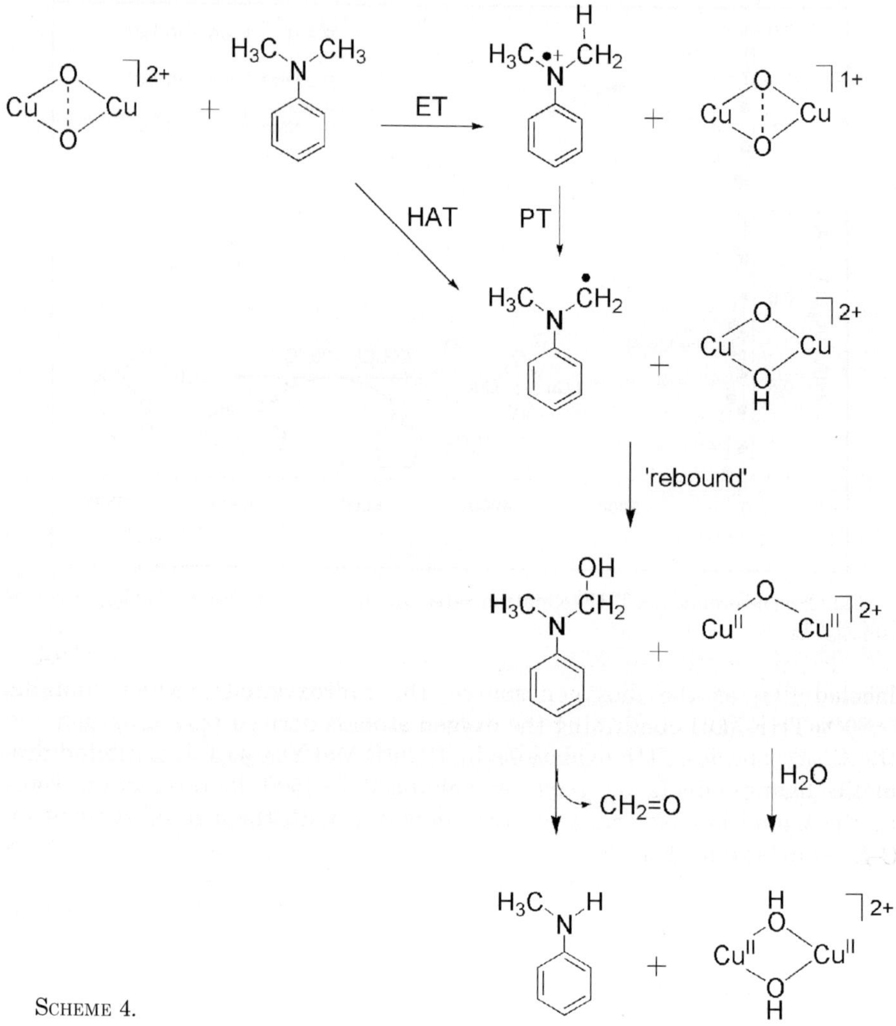

SCHEME 4.

N,N-dimethylanilines which have varying redox potentials. The slope (ρ value) of a correlation plot of log k_{ox} vs. the Hammett σ^+ parameter can give clues regarding the mechanism of oxidation. A ρ value near zero could suggest a ΔE-independent ETPT mechanism while a negative value such as $\rho = -2$ more likely could correspond to a rate limiting ET pathway. In between values suggest varying degrees of proton coupled electron transfer mechanisms.

The rate of reaction of R′-DMA (R = MeO, Me, H, and CN) by $[\{Cu(R\text{-}MePY2)\}_2O_2]^{2+}$ was analyzed by competition reactions between R′-DMA vs. H-DMA to get a relative rate constant, k_{rel}. The reactions by $[\{Cu(H\text{-}MePY2)\}_2O_2]^{2+}$ showed a strong dependence on the substrate σ^+ value with a $\rho = -2.1$ suggesting the oxidation of DMA occurs *via* a rate limiting electron transfer process. However, as the Cu_2O_2 compound becomes less oxidizing based on its electron donating ligand substituent, the results of the linear free energy correlation become inconclusive. R′DMA oxidations by $[\{Cu(OMe\text{-}MePY2)\}_2O_2]^{2+}$ and $[\{Cu(NMe_2\text{-}MePY2)\}_2O_2]^{2+}$ give a ρ value of -0.99 and -0.47, respectively; values intermediate than those for an ETPT mechanism and an ET/PT mechanism (*119*).

E.2. Deuterium Kinetic Isotope Effects

Kinetic isotope effects (KIE) have long been used to study hydrogen atom abstraction reactions. Intricate deuterium kinetic isotope studies have been useful in deducing the mechanism of DMA oxidations by Cytochrome P450 (*120,121*). Similar studies with the complexes $[\{Cu(R\text{-}MePY2)\}_2O_2]^{2+}$ have also presented valuable insight. Kinetic isotope effects should be negligible for rate limiting ET mechanisms, and significant for ETPT mechanisms. The value of the KIE (depending on σ^+, or $E^\circ_{1/2}$) gives an idea of the reversibility of the ET step in a coupled ET/PT mechanism, (i.e., the sequential PT step will have an effect on σ^+). Also, a comparison of the magnitude of the kinetic isotope effect in an intramolecular (KIE_{intra}) reaction and an intermolecular reaction (KIE_{inter}), i.e., using *N*-trideuteromethyl-*N*-methylaniline (R′-DMA-d_3) for KIE_{intra}, and *NN*,-bis (trideuteromethyl)aniline (R′-DMA-d_6) vs. *N,N*-dimethylaniline (R′-DMA) for KIE_{inter}, has proven to be a powerful probe for deciphering between a ETPT or ET mechanism (*122*). The probe is based on the different outcomes of KIE_{intra} and KIE_{inter} in rate limiting ETPT or ET/PT pathways. In the case of a concerted ETPT mechanism, the KIE is dependent on the C–H vs. C–D bond dissociation energy; thus the KIE will be observed equally in the inter- and intramolecular deuterium isotope experiment ($KIE_{intra} \approx KIE_{inter}$) (Fig. 29). However, in a rate limiting ET process, the intramolecular reaction can lead to discrimination in the subsequent PT step, and should therefore have a $KIE_{intra} > 1$, while the KIE_{inter} should be close to unity since only the ET event will dictate the product distribution. Hence, an ET/PT mechanism can be identified if $KIE_{intra} > KIE_{inter}$ (Fig. 30).

For experiments involving R′-DMA oxidation by $[\{Cu(H\text{-}MePY2)\}_2O_2]^{2+}$ at $-80°C$ in CH_2Cl_2, the kinetic isotope effects are small in magnitude and KIE_{intra} is greater than KIE_{inter}, suggesting a rate-limiting ET process (Table II) (*123*). Furthermore, the kinetic isotope effects in the intramolecular reactions of the different R′-DMA substrates shows an increase as σ^+ (and $E_{1/2}$) become more negative, eventually reaching an asymptote. The increase in KIE_{intra} is due to better H vs. D discrimination in the proton transfer step made possible by the stabilization of the *N*-radical cation by the electron donating R′ group. In contrast, the intermolecular KIE profiles show no such dependence on the R′DMA σ^+ value. This suggests that the subsequent PT step has little influence on the *N*-methylaniline product, and that the reaction is pre-determined by the mostly irreversible ET step (Fig. 30).

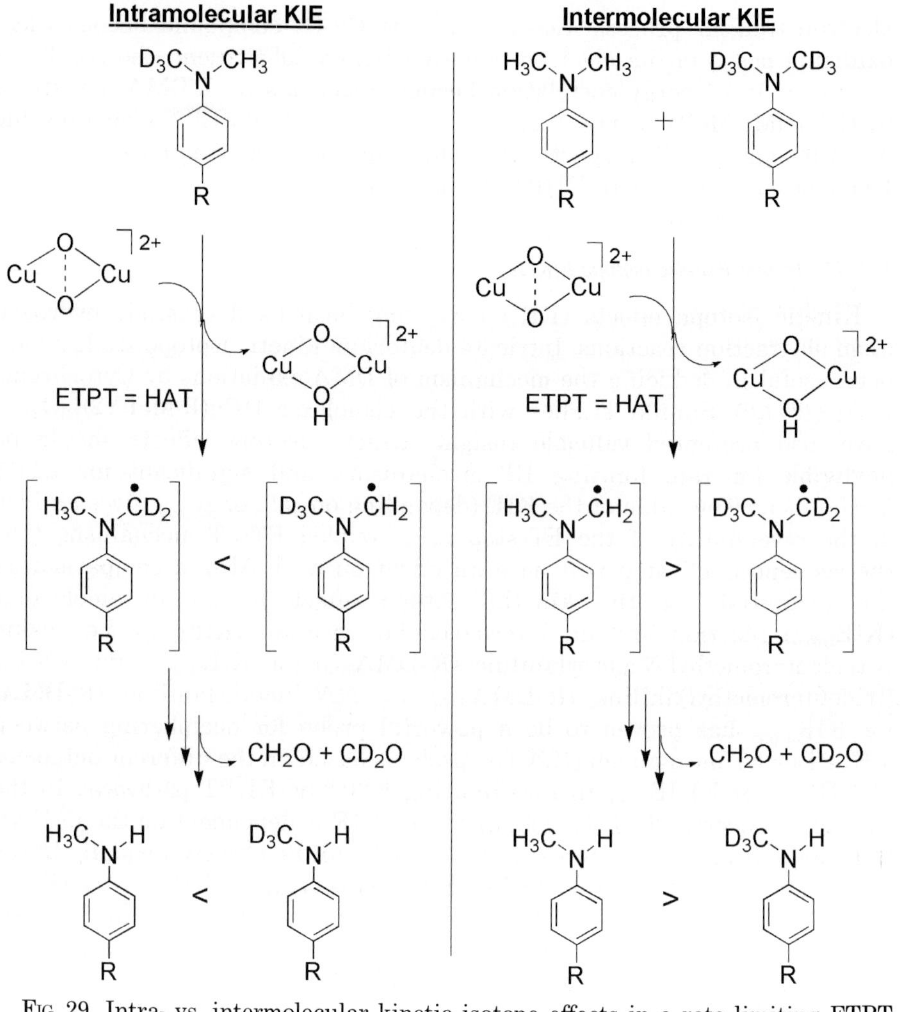

FIG. 29. Intra- vs. intermolecular kinetic isotope effects in a rate limiting ETPT mechanism showing $KIE_{intra} \approx KIE_{inter}$. See text.

TABLE II

KIE_{intra} VS. KIE_{inter} FOR THE *N*-DEALKYLATION OF R′-DMA BY $[\{Cu(R\text{-}MePY2)\}_2O_2]^{2+}$

	$[\{Cu(H\text{-}MePY2)\}_2O_2]^{2+}$		$[\{Cu(OMe\text{-}MePY2)\}_2O_2]^{2+}$		$[\{Cu(NMe_2\text{-}MePY2)\}_2O_2]^{2+}$		$E_{1/2}$[a]
	KIE_{intra}	KIE_{inter}	KIE_{intra}	KIE_{inter}	KIE_{intra}	KIE_{inter}	
MeO-DMA	4.7 (*8*)	1.7 (*6*)	7.5 (*2*)	2.3 (*4*)	7.3 (*4*)	2.7 (*2*)	0.53
Me-DMA	4.6 (*6*)	2.3 (*3*)	5.0 (*7*)	3.0 (*2*)	5.8 (*8*)	3.8 (*2*)	0.72
H-DMA	4.1 (*6*)	2.4 (*6*)	6.1 (*7*)	2.7 (*6*)	12.0 (*9*)	11.4 (*15*)	0.92
CN-DMA	2.6 (*8*)	2.1 (*8*)	13.9 (*11*)	13.1 (*19*)	14.9 (*7*)	15.0 (*4*)	1.21

[a]In CH_2Cl_2 at room temperature (vs. SCE).

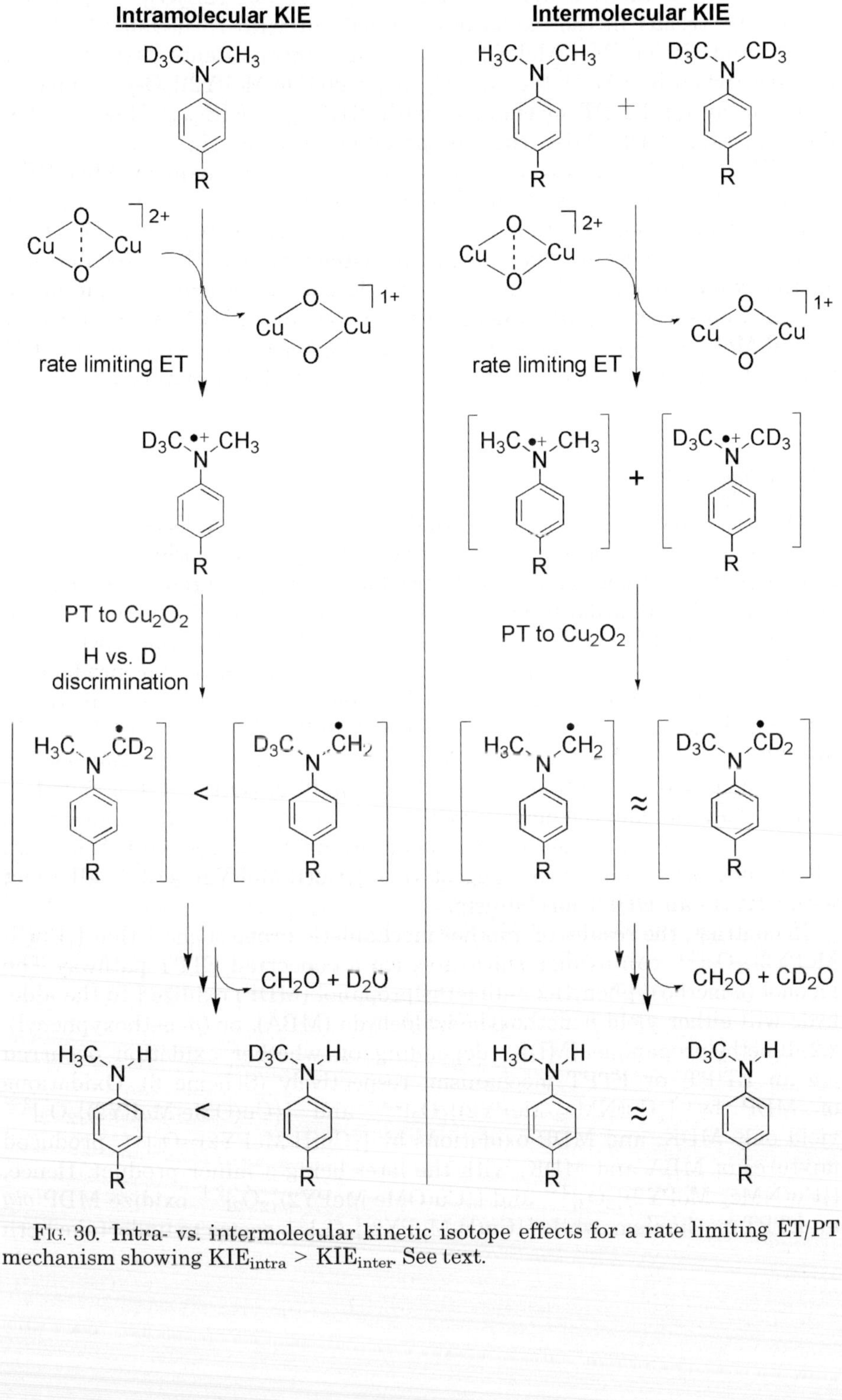

FIG. 30. Intra- vs. intermolecular kinetic isotope effects for a rate limiting ET/PT mechanism showing $KIE_{intra} > KIE_{inter}$. See text.

In contrast to the R′-DMA oxidations by $[\{Cu(H\text{-}MePY2)\}_2O_2]^{2+}$, the kinetic isotope trends in the oxidations carried out with $[\{Cu(OMe\text{-}MePY2)\}_2O_2]^{2+}$ and $[\{Cu(NMe_2\text{-}MePY2)\}_2O_2]^{2+}$ are inconsistent. Oxidation of R′-DMA (when R′ = MeO, Me, and H) by $[\{Cu(OMe\text{-}MePY2)\}_2O_2]^{2+}$ appears to occur by an ET/PT mechanism with $KIE_{intra} > KIE_{inter}$. However, for the oxidation of CN-DMA, the isotope effects become significantly larger, and $KIE_{intra} \approx KIE_{inter}$, strongly indicating an ETPT mechanism. When $[\{Cu(NMe_2\text{-}MePY2)\}_2O_2]^{2+}$ is used as the oxidizing agent, the KIE_{inter} and KIE_{intra} profiles suggest that the reaction with MeO-DMA and Me-DMA proceed *via* an ET/PT mechanism, and a switch from ET/PT to ETPT occurs for the oxidation of H-DMA and CN-DMA. Hence, as the energy required for the ET step increases, when using weaker oxidants (R = NMe_2 or OMe in $[\{Cu(R\text{-}MePY2)\}_2O_2]^{2+}$) and substrates with higher redox potentials (R = CN in R′-DMA), the reaction proceeds *via* a concerted ETPT mechanism.

E.3. Mechanistic Probes

The use of mechanistic probes can distinguish between ET/PT and ETPT oxidation pathways based on the relative stability of the cationic or radical intermediates, respectively, which produce different product outcomes. *N*-cyclopropyl-*N*-methylaniline (CPA) is one such probe which can undergo *N*-dealkylation yielding either *N*-methylaniline (MA), or *N*-cyclopropylaniline (CA), where MA or ring opened products can be the major product in an ET/PT mechanism, and CA is the dominant product of an ETPT reaction (Scheme 5). The reaction of $[\{Cu(H\text{-}MePY2)\}_2O_2]^{2+}$ and $[\{Cu(OMe\text{-}MePY2)\}_2O_2]^{2+}$ with CPA results in 50–60 % MA and other ring opened compounds. After oxidation by $[\{Cu(NMe_2\text{-}MePY2)\}_2O_2]^{2+}$, CA is found in only ~2 % yield, and the major product was *N*-methylquinolinium cation – a ring opened product of the CPA *N*-radical cation formed from an ET pathway. The results with this probe suggest that $[\{Cu(R\text{-}MePY2)\}_2O_2]^{2+}$ all react with CPA *via* an ET/PT mechanism.

In contrast, the results of another mechanistic probe suggest that $[\{Cu(R\text{-}MePY2)\}_2O_2]^{2+}$ can oxidize substrates *via* a concerted ETPT pathway. The alcohol (*p*-methoxyphenyl)-2,2-dimethylpropanol (MDP) oxidized to the aldehyde will either yield *p*-methoxybenzaldehyde (MBA), or (*p*-methoxyphenyl)-2,2-dimethylpropanone (MDK) depending on whether oxidation occurred *via* an ET/PT, or ETPT mechanism, respectively (Scheme 6). Oxidations of MDP by $[\{Cu(NMe_2\text{-}MePY2)\}_2O_2]^{2+}$ and $[\{Cu(OMe\text{-}MePY2)\}_2O_2]^{2+}$ yield only MDK, and MDP oxidations by $[\{Cu(H\text{-}MePY2)\}_2O_2]^{2+}$ produced mixtures of MBA and MDK, with the later being a minor product. Hence, $[\{Cu(NMe_2\text{-}MePY2)\}_2O_2]^{2+}$ and $[\{Cu(OMe\text{-}MePY2)\}_2O_2]^{2+}$ oxidize MDP *via* an ETPT mechanism, while $[\{Cu(H\text{-}MePY2)\}_2O_2]^{2+}$ seems to undertake both

ET/PT and ETPT pathways. Perhaps, it is easier to oxidize the CMA aniline probe by one electron compared the alcohol MDP substrate. Thus, the differential oxidative pathways taken by $[\{Cu(R\text{-}MePY2)\}_2O_2]^{2+}$ for CMA and MDP vary depending on the strength of the Cu_2O_2 oxidant, and the relative ease of the one electron oxidation of the substrate.

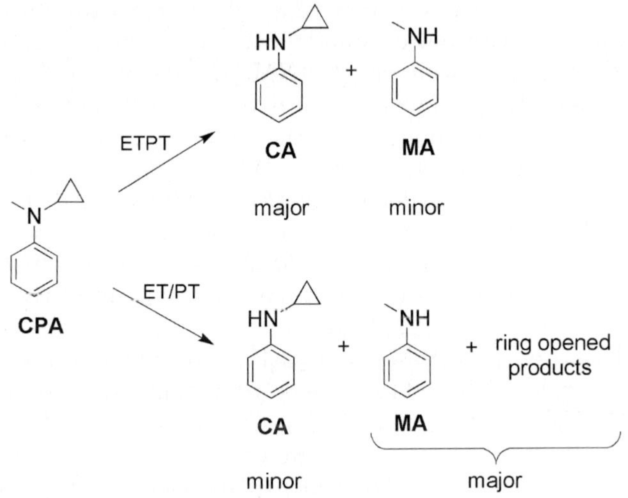

SCHEME 5.

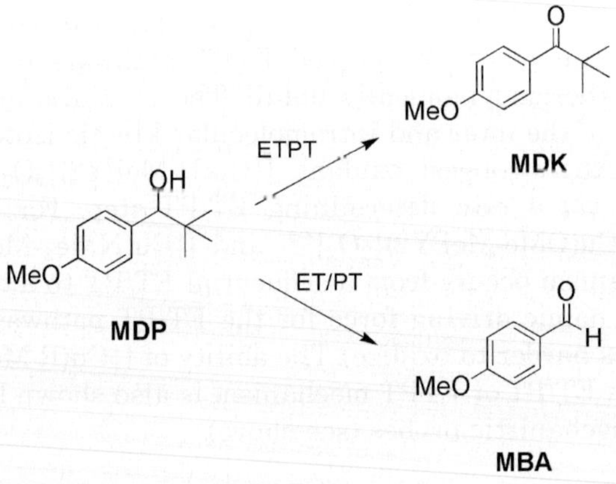

SCHEME 6.

E.4. Thermodynamic Considerations

To try to tie things together, we sought to determine the redox potentials of each $[\{Cu(R\text{-}MePY2)\}_2O_2]^{2+}$ oxidant, seeking an understanding of the preferential ETPT and ET/PT pathways from a thermodynamic perspective. In CH_2Cl_2, the low temperature (–80°C) cyclic voltammograms of $[\{Cu(H\text{-}MePY2)\}_2O_2]^{2+}$ and $[\{Cu(OMe\text{-}MePY2)\}_2O_2]^{2+}$ show quasi-reversible behavior with $E_{1/2} \sim 550$ and ~ 400 mV vs. $Fc^{+/0}$, respectively, presumably corresponding to the $[Cu_2O_2]^{2+}/[Cu_2O_2]^{+}$ couple (*124*). As expected, the complex with a better electron donating substituent exhibits a decreased redox potential as it can better stabilize the more highly charged $Cu_2(O_2)$ core. The reduction potential of $[\{Cu(NMe_2\text{-}MePY2)\}_2O_2]^{2+}$ could not be determined from cyclic voltammetry but it was estimated by "spectroelectrochemical titrations" (i.e., monitoring the presence/decay of the dicopper-dioxygen core spectrophotometrically). While, $[\{Cu(OMe\text{-}MePY2)\}_2O_2]^{2+}$ was readily reduced by ethylferrocene ($E_{1/2} = 380$ mV vs. SCE in CH_2Cl_2), $[\{Cu(NMe_2\text{-}MePY2)\}_2O_2]^{2+}$ was unreactive towards ethylferrocene, but did react with dimethylferrocene ($E_{1/2} = 230$ mV vs. SCE in CH_2Cl_2). Hence, $[\{Cu(NMe_2\text{-}MePY2)\}_2O_2]^{2+}$ is the least powerful oxidant of the series, with $E_{1/2} < 400$ mV vs. SCE in CH_2Cl_2 (*124*).

The proclivity of $[\{Cu(R\text{-}MePY2)\}_2O_2]^{2+}$ to undergo coupled ET/PT reactions was explored by adding weak reductants with proton sources. Addition of $[Fe^{II}(phen)_2(NO\text{-}phen)](ClO_4)_2$ ($E_{1/2} = 1.3$ V vs. SCE in CH_2Cl_2) to $[\{Cu(NMe_2\text{-}MePY2)\}_2O_2]^{2+}$ at –80°C in CH_2Cl_2 did not produce a redox reaction. However, addition of $[Fe^{II}(phen)_2(NO\text{-}phen)](ClO_4)_2$ together with a proton source such as $NH_4(B(C_6F_5)_4)$ produces $[Fe^{III}(phen)_2(NO\text{-}phen)]^{3+}$ and $[\{Cu(NMe_2\text{-}MePY2)\}_2(OH)_2]^{2+}$ indicating that a proton coupled electron transfer (PCET) reaction occurred. Curiously, addition of $[Fe^{II}(phen)_2(NO\text{-}phen)](ClO_4)_2$ with a weaker acid such as acetic acid did not promote any PCET reaction. These findings suggest that the complexes $[\{Cu(R\text{-}MePY2)\}_2O_2]^{2+}$ can react *via* coupled ETPT pathways if an ET pathway is prohibitively thermodynamically uphill. These results are in agreement with the results of the inter and intramolecular kinetic isotope effects (vide supra) whereby the strongest oxidant, $[\{Cu(H\text{-}MePY2)\}_2O_2]^{2+}$ reacts with dimethylaniline *via* a rate determining ET/PT step. For the less powerful oxidants, $[\{Cu(OMe\text{-}MePY2)\}_2O_2]^{2+}$ and $[\{Cu(NMe_2\text{-}MePY2)\}_2O_2]^{2+}$, a switch in mechanism occurs from a sequential ET/PT to a concerted ETPT, as the thermodynamic driving force for the ET/PT pathway increases (i.e., R′-DMA becomes harder to oxidize). The ability of $[\{Cu(R\text{-}MePY2)\}_2O_2]^{2+}$ to undergo either a ET/PT or ETPT mechanism is also shown by the reactivity profiles of the mechanistic probes (see above).

E.5. Substrate Binding Requirement

Previously, Itoh and co-workers presented the importance of substrate coordination to the Cu_2O_2 core in intermolecular oxidative reactivity. The reaction of *p*-substituted neutral phenols with a side-on peroxo dicopper(II) complex supported by a tridentate ligand results in oxidative coupling forming biphenol compounds (*111*). Instead, the reaction with deprotonated phenols, results in aromatic hydroxylation, whereby phenolate coordination is a key step the oxygenation chemistry (*110*). The same group also showed that the oxidation of thioether compounds by the bis(μ-oxo) dicopper(III) compound with the ligand $^{H}PY1^{Et,Bz\text{-}d2}$ (Fig. 8) first involved substrate coordination to the metal center followed by the oxygen atom transfer from the $Cu_2^{III}(O)_2$ to the thioether substrate (*108*).

In our own efforts, we also observe the prerequisite of substrate binding in order for a reaction to occur. Analysis of the rates of oxidation of THF by $[\{Cu(NMe_2\text{-}MePY2)\}_2O_2]^{2+}$ also imply a substrate binding step prior to oxidation. With varying THF concentrations, plots of k_{obs} vs. [THF] show linear behavior indicating a second-order oxidation reaction. Variable temperature Eyring analysis show that the reaction occurs with a very low-activation enthalpy, $\Delta H^{*} = 15.5$ kJ/mol, indicative of a left-lying pre-equilibrium (perhaps substrate binding) prior to the oxidation step. Clear evidence for substrate binding to $[\{Cu(R\text{-}MePY2)\}_2O_2]^{2+}$ is observed in the kinetic traces of dimethylaniline oxidations. Plots of k_{obs} vs. [DMA] show saturation behavior at higher DMA concentrations; the binding constants are weak ranging from 850 to 10 M^{-1} at low temperatures (−85 to −65°C) (Fig. 31).

The substrate binding prerequisite for oxidation is also supported by the types of substrates that can, and can not, be oxidized by $[\{Cu(R\text{-}MePY2)\}_2O_2]^{2+}$. THF, with a C–H bond dissociation energy (BDE)

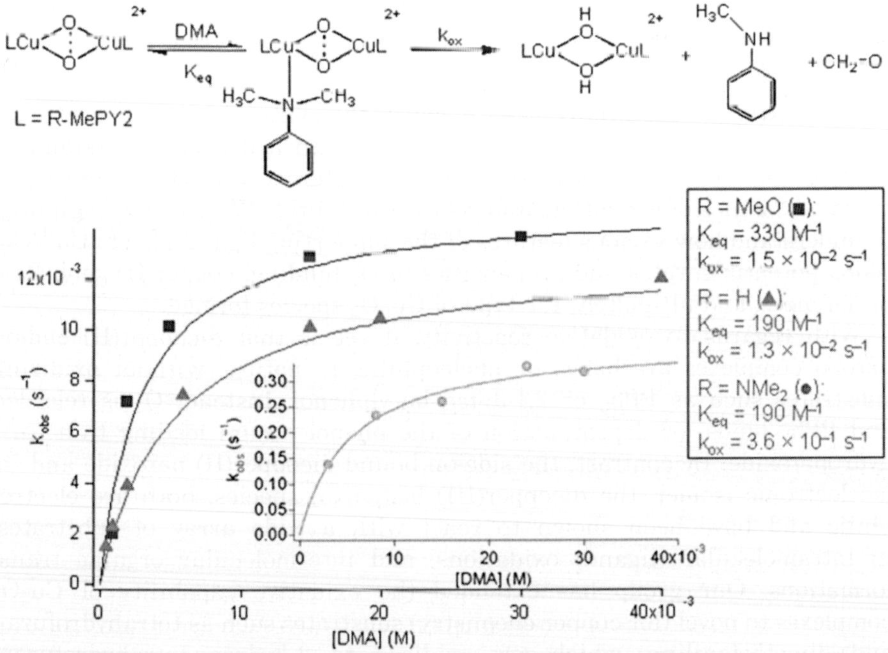

FIG. 31. Oxidation of DMA by $[\{Cu(R\text{-}MePY2)\}_2O_2]^{2+}$ showing saturation behavior.

of 385 kJ/mol gets readily oxidized, yet $[\{Cu(R\text{-}MePY2)\}_2O_2]^{2+}$ is unreactive towards toluene even though it has a lower BDE of 372 kJ/mol. The substrate 9,10-dihydroanthracene has an even lower C–H BDE but it is not oxidized by $[\{Cu(R\text{-}MePY2)\}_2O_2]^{2+}$ (*124*), while substrates with higher BDE's (such as benzyl alcohol) do get oxidized. A common feature of all oxidizable substrates by $[\{Cu(R\text{-}MePY2)\}_2O_2]^{2+}$ is the presence of a heteroatom that can potentially coordinate, serving to put the oxidizable C–H moiety into close proximity to the Cu_2O_2 center. Consistent with this notion is the observation that $[\{Cu(R\text{-}MePY2)\}_2O_2]^{2+}$ are unreactive towards the sterically modified 2,5-dimethyl-tetrahydrofuran.

E.6. Summary of $[\{Cu(R\text{-}MePY2)\}_2O_2]^{2+}$ Substrate Oxidations

A wide range of substrates are oxygenated by $[\{Cu(R\text{-}MePY2)\}_2O_2]^{2+}$, and mechanistic investigations have provided considerable insights. Only substrates with an "accessible" heteroatom were oxidized demonstrating the requirement of substrate binding prior to oxidation. Both ET/PT and ETPT = HAT mechanisms appear to take place based on the mechanistic criteria (linear free energy correlations, KIE's, and studies with probes). ETPT occurs when ET becomes too difficult thermodynamically according to the redox potentials of the oxidant and/or the substrate involved. Finally, as substrate binding appears to be a key step in the oxidation reaction, it is probably more correct to say that ET oxidations by $[\{Cu(R\text{-}MePY2)\}_2O_2]^{2+}$-can be considered "inner-sphere electron transfer" oxidations.

V. Summary and Outlook

Significant advances have been made in the synthesis and characterization of discrete copper-dioxygen structures, with a deeper understanding of their spectroscopic origins (i.e., bonding interactions), the kinetics and thermodynamics of formation, and relative stabilities. We are now beginning to understand how various features of the supporting ligand affects Cu^{II}/Cu^{I} redox potentials, rates and propensities of O_2 binding, copper(II) geometric preferences, and ultimately, the type of Cu_nO_2 species formed.

With regards to oxidative reactivity, it seems that dicopper(II) end-on peroxo complexes are basic, or nucleophilic, in nature, without oxidizing substrates such as PPh_3 or 2,4-di-*tert*-butylphenol. Instead, O_2 is released and PPh_3 binds, or deprotonation of the phenol occurs leading to a Cu^{II}-hydroperoxide. By contrast, the side-on bound dicopper(II) peroxide and its isoelectronic isomer, the dicopper(III) bis(μ-oxo) species, both are electrophilic and have been shown to react with a wide array of substrates, in intramolecular (ligand) oxidations, and intermolecular organic transformations. Our group has extended the oxidative capability of Cu_2O_2 complexes to novel (for copper chemistry) substrates such as tetrahydrofuran and dimethylaniline, which are oxidized to 2-hydroxy-tetrahydrofuran, and methylaniline plus formaldehyde, respectively.

Furthermore, we have been actively investigating the mechanism of substrate oxidations. For the tyrosinase-like hydroxylation of an arene reaction by a dicopper(II) side-on peroxo complex $[Cu^{II}_2(NO_2\text{-XYL})(O_2)]$, we have unambiguously confirmed that the mechanism involves a type of electrophilic aromatic substitution by the side-on peroxo moiety. In intermolecular oxidations by $[\{Cu(R\text{-MePY2})\}_2O_2]^{2+}$, comprehensive mechanistic studies including linear free energy correlations, deuterium kinetic isotope effects, and the utilization of mechanistic probes indicate that the oxidant can react *via* both a concerted ETPT mechanism, or a sequential ET/PT pathway, where the thermodynamics of the ET limited reaction can cause a switch in mechanism to the ETPT reaction. We also draw attention to the importance of the substrate coming into close proximity to the $[\{Cu(R\text{-MePY2})\}_2O_2]^{2+}$ oxidant in order for a reaction to take place.

In aspiring to design robust, yet powerful synthetic oxidant reagents or catalysts, and for a thorough understanding of copper-mediated oxidations in enzymatic systems, further studies of copper-dioxygen structure and reactivity are still warranted. There are ample questions yet to be unraveled. Can we generate mononuclear copper(II)-superoxide complexes and probe to see if they are capable of hydrogen atom abstraction reactivity as proposed for DβH and PHM? What are the inherent differences in the reactivities of the (often isoenergetic) dicopper(II) side-on peroxo and dicopper(III) bis(μ-oxo) species and what role, if any, does the latter play in biological systems? Finally, have we yet discovered all of the possible Cu_nO_2 structural and reactivity types, or could there be novel structures that have yet to be isolated?

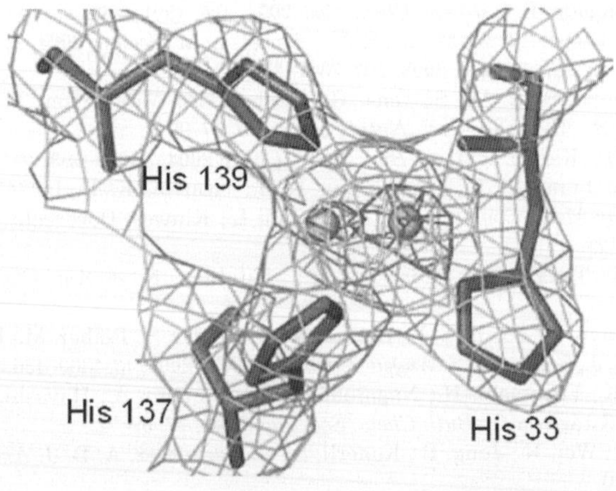

FIG. 32. A dicopper site in particulate methane monooxygenase.

A recent, long sought X-ray crystal structure of the particulate copper-dependent form of methane monooxygenase (pMMO) hints of the possibility – or even probably – of new as yet undiscovered Cu_nO_2 structures and reactivity types. The structure reveals three copper ions in coordination modes atypical for copper mediated dioxygen processing enzymes (*125*). A dicopper moiety is observed with a Cu... Cu separation of only 2.6 Å, and an unprecedented (in biology) low coordination number for copper ion. Histidines 137 and 139 are in coordinating distance to one copper while the other can coordinate to both the terminal amino- and γ-nitrogen of His 33 (Fig. 32). A second mononuclear copper site and a zinc site (in the as-isolated crystal) are also possible catalytic centers. This structure describes a new potential site or sites for copper-based oxidations and opens new ideas for synthetic modeling investigations in terms of copper ligation and O_2-reactivity.

References

1. Mirica, L. M.; Ottenwaelder, X.; Stack, T. D. P. *Chem. Rev.* **2004**, *104*, 1013–1045.
2. Lewis, E. A.; Tolman, W. B. *Chem. Rev.* **2004**, *104*, 1047–1076.
3. Hazes, B.; Magnus, K. A.; Bonaventura, C.; Bonaventura, J.; Dauter, Z.; Kalk, K. H.; Hol, W. G. J. *Protein Sci.* **1993**, *2*, 597–619.
4. Magnus, K. A.; Hazes, B.; Ton-That, H.; Bonaventura, C.; Bonaventura, J.; Hol, W. G. J. *Proteins: Struct. Funct. Genet.* **1994**, *19*, 302–309.
5. Cuff, M. E.; Miller, K. I.; van Holde, K. E.; Hendrickson, W. A. *J. Mol. Biol.* **1998**, *278*, 855–870.
6. Gerdemann, C.; Eicken, C.; Krebs, B. *Acc. Chem. Res.* **2002**, *35*, 183–191.
7. Solomon, E. I.; Sundaram, U. M.; Machonkin, T. E. *Chem. Rev.* **1996**, *96*, 2563–2605.
8. Klinman, J. P. *Chem. Rev.* **1996**, *96*, 2541–2561.
9. Eipper, B. A.; Stoffers, D. A.; Mains, R. E. *Annu. Rev. Neurosci.* **1992**, *15*, 57–85.
10. Prigge, S. T.; Kolhekar, A. S.; Eipper, B. A.; Mains, R. E.; Amzel, L. M. *Science* **1997**, *278*, 1300–1305.
11. Evans, J. P.; Ahn, K.; Klinman, J. P. *J. Biol. Chem.* **2003**, *278*, 49691–49698.
12. Evans, J. P.; Ahn, K.; Klinman, J. P. *J. Biol. Chem.* **2004**, *279*, 5048.
13. Chen, P.; Solomon, E. I. *Proc. Natl. Acad. Sci. USA* **2004**, *101*, 13105–13110.
14. Chen, P.; Solomon, E. I. *J. Am. Chem. Soc.* **2004**, *126*, 4991–5000.
15. Prigge, S. T.; Eipper, B. A.; Mains, R. E.; Amzel, L. M. *Science* **2004**, *304*, 864–867.
16. Whittaker, J. W. *Chem. Rev.* **2003**, *103*, 2347–2363.
17. Ito, N.; Phillips, S. E. V.; Stevens, C.; Ogel, Z. B.; McPherson, M. J.; Keen, J. N.; Yadav, K. D. S.; Knowles, P. F. *Nature* **1991**, *350*, 87–90.
18. Hatcher, L. Q.; Karlin, K. D. *J. Biol. Inorg. Chem.* **2004**, *9*, 669–683.
19. Zhang, C. X.; Liang, H.-C.; Humphreys, K. J.; Karlin, K. D. In "*Catalytic Activation of Dioxygen by Metal Complexes*"; Ed. Simandi L.; Kluwer: Dordrecht, The Netherlands, **2003**, pp. 79–121.
20. Fujusawa, K.; Tanaka, M.; Moro-oka, Y.; Kitajima, N. *J. Am. Chem. Soc.* **1994**, *116*, 12079–12080.
21. Schatz, M.; Raab, V.; Foxon, S. P.; Brehm, G.; Schneider, S.; Reiher, M.; Holthausen, M. C.; Sundermeyer, J.; Schindler, S. *Angew. Chem. Int. Edit.* **2004**, *43*, 4360–4363.
22. Komiyama, K.; Furutachi, H.; Nagatomo, S.; Hashimoto, A.; Hayashi, H.; Fujinami, S.; Suzuki, M.; Kitagawa, T. *Bull. Chem. Soc. Jap.* **2004**, *77*, 59–72.
23. Karlin, K. D.; Wei, N.; Jung, B.; Kaderli, S.; Zuberbühler, A. D. *J. Am. Chem. Soc.* **1991**, *113*, 5868–5870.

24. Karlin, K. D.; Wei, N.; Jung, B.; Kaderli, S.; Niklaus, P.; Zuberbühler, A. D. *J. Am. Chem. Soc.* **1993,** *115*, 9506–9514.
25. Karlin, K. D.; Kaderli, S.; Zuberbühler, A. D. *Acc. Chem. Res.* **1997,** *30*, 139–147.
26. Weitzer, M.; Schindler, S.; Brehm, G.; Schneider, S.; Hormann, E.; Jung, B.; Kaderli, S.; Zuberbühler, A. D. *Inorg. Chem.* **2003,** *42*, 1800–1806.
27. Chen, P.; Root, D. E.; Campochiaro, C.; Fujisawa, K.; Solomon, E. I. *J. Am. Chem. Soc.* **2003,** *125*, 466–474.
28. Spencer, D. J. E.; Aboelella, N. W.; Reynolds, A. M.; Holland, P. L.; Tolman, W. B. *J. Am. Chem. Soc.* **2002,** *124*, 2108–2109.
29. Aboelella, N. W.; Lewis, E. A.; Reynolds, A. M.; Brennessel, W. W.; Cramer, C. J.; Tolman, W. B. *J. Am. Chem. Soc.* **2002,** *124*, 10660–10661.
30. Osako, T.; Nagatomo, S.; Tachi, Y.; Kitagawa, T.; Itoh, S. *Angew. Chem. Int. Ed.* **2002,** *41*, 4325–4328.
31. Ohtsu, H.; Itoh, S.; Nagatomo, S.; Kitagawa, T.; Ogo, S.; Watanabe, Y.; Fukuzumi, S. *Inorg. Chem.* **2001,** *40*, 3200–3207.
32. Yamaguchi, S.; Nagatomo, S.; Kitagawa, T.; Funahashi, Y.; Ozawa, T.; Jitsukawa, K.; Masuda, H. *Inorg. Chem.* **2003,** *42*, 6968–6970.
33. Kodera, M.; Kita, T.; Miura, I.; Nakayama, N.; Kawata, T.; Kano, K.; Hirota, S. *J. Am. Chem. Soc.* **2001,** *123*, 7715–7716.
34. Fujii, T.; Naito, A.; Yamaguchi, S.; Wada, A.; Funahashi, Y.; Jitsukawa, K.; Nagatomo, S.; Kitagawa, T.; Masuda, H. *Chemical Communications* **2003**, 2700–2701.
35. Yamaguchi, S.; Wada, A.; Nagatomo, S.; Kitagawa, T.; Jitsukawa, K.; Masuda, H. *Chem. Lett.* **2004,** *33*, 1556–1557.
36. Wada, A.; Harata, M.; Hasegawa, K.; Jitsukawa, K.; Masuda, H.; Mukai, M.; Kitagawa, T.; Einaga, H. *Angew. Chem. Int. Ed.* **1998,** *37*, 798–799.
37. Jacobson, R. R.; Tyeklár, Z.; Karlin, K. D.; Liu, S.; Zubieta, J. *J. Am. Chem. Soc.* **1988,** *110*, 3690–3692.
38. Tyeklár, Z.; Jacobson, R. R.; Wei, N.; Murthy, N. N.; Zubieta, J.; Karlin, K. D. *J. Am. Chem. Soc.* **1993,** *115*, 2677–2689.
39. Baldwin, M. J.; Ross, P. K.; Pate, J. E.; Tyeklár, Z.; Karlin, K. D.; Solomon, E. I. *J. Am. Chem. Soc.* **1991,** *113*, 8671–8679.
40. Kitajima, N.; Fujisawa, K.; Fujimoto, C.; Moro-oka, Y.; Hashimoto, S.; Kitagawa, T.; Toriumi, K.; Tatsumi, K.; Nakamura, A. *J. Am. Chem. Soc.* **1992,** *114*, 1277–1291.
41. Halfen, J. A.; Mahapatra, S.; Wilkinson, E. C.; Kaderli, S., Young Jr., V. G.; Que Jr., L.; Zuberbühler, A. D.; Tolman, W. B. *Science* **1996,** *271*, 1397–1400.
42. (a) DuBois, J. L.; Mukherjee, P.; Stack, T. D. P.; Hedman, B.; Solomon, E. I.; Hodgson, K. O. *J. Am. Chem. Soc.* **2000,** *122*, 5775–5787; (b) Henson, M. J.; Mukherjee, P.; Root, D. E.; Stack, T. D. P.; Solomon, E. I. *J. Am. Chem. Soc.* **1999,** *121*, 10332–10345.
43. Holland, P. L.; Cramer, C. J.; Wilkinson, E. C.; Mahapatra, S.; Rodgers, K. R.; Itoh, S.; Taki, M.; Fukuzumi, S.; Que Jr., L.; Tolman, W. B. *J. Am. Chem. Soc.* **2000,** *122*, 792–802.
44. Que Jr., L.; Tolman, W. B. *Angew. Chem. Int. Ed.* **2002,** *41*, 1114–1137.
45. Cramer, C. J.; Smith, B. A.; Tolman, W. B. *J. Am. Chem. Soc.* **1996,** *118*, 11283–11287.
46. Bérces, A. *Inorg. Chem.* **1997,** *36*, 4831–4837.
47. Flock, M.; Pierloot, K. *J. Phys. Chem. A.* **1999,** *103*, 95–102.
48. Messerschmidt, A. *Adv. Inorg. Chem.* **1993,** *40*, 121–185.
49. Enguita, F. J.; Marcal, D.; Martins, L. O.; Grenha, R.; Henriques, A. O.; Lindley, P. F.; Carrondo, M. A. *Journal of Biological Chemistry* **2004,** *279*, 23472–23476.
50. Hakulinen, N.; Kiiskinen, L.-L.; Kruus, K.; Saloheimo, M.; Paananen, A.; Koivula, A.; Rouvinen, J. *Nature Structural Biology* **2002,** *9*, 601–605.
51. Cole, A. P.; Root, D. E.; Mukherjee, P.; Solomon, E. I.; Stack, T. D. P. *Science* **1996,** *273*, 1848–1850.
52. Mirica, L. M.; Stack, T. D. P. *Inorg. Chem.* **2005,** *44*, 2131–2133.
53. Karlin, K. D.; Gan, Q.-F.; Farooq, A.; Liu, S.; Zubieta, J. *Inorg. Chim. Acta* **1989,** *165*, 37–39.
54. Karlin, K. D.; Gan, Q.-F.; Farooq, A.; Liu, S.; Zubieta, J. *Inorg. Chem.* **1990,** *29*, 2549–2551.

55. Karlin, K. D.; Gan, Q.-F.; Tyeklár, Z. *Chem. Commun.* **1999**, 2295–2296.
56. Reim, J.; Krebs, B. *Angew. Chem. Int. Ed. Engl.* **1994**, *33*, 1969–1971.
57. Reim, J.; Werner, R.; Haase, W.; Krebs, B. *Chem. Eur. J.* **1998**, *4*, 289–298.
58. Meyer, F.; Pritzkow, H. *Angew. Chem. Int. Ed.* **2000**, *39*, 2112–2115.
59. Schatz, M.; Becker, M.; Thaler, F.; Hampel, F.; Schindler, S.; Jacobson, R. R.; Tyeklár, Z.; Murthy, N. N.; Ghosh, P.; Chen, Q.; Zubieta, J.; Karlin, K. D. *Inorg. Chem.* **2001**, *40*, 2312–2322.
60. Karlin, K. D.; Hayes, J. C.; Shi, J.; Hutchinson, J. P.; Zubieta, J. *Inorg. Chem.* **1982**, *21*, 4106–4108.
61. Ambundo, E. A.; Deydier, M.-V.; Grall, A. J.; Aguera-Vega, N.; Dressel, L. T.; Cooper, T. H.; Heeg, M. J.; Ochrymowycz, L. A.; Rorabacher, D. B. *Inorg. Chem.* **1999**, *38*, 4233–4242.
62. Rorabacher, D. B. *Chem. Rev.* **2004**, *104*, 651–697.
63. Zhang, C. X.; Kaderli, S.; Costas, M.; Kim; E.-i.; Neuhold, Y.-M.; Karlin, K. D.; Zuberbühler, A. D. *Inorg. Chem.* **2003**, *42*, 1807–1824.
64. Henson, M. J.; Vance, M. A.; Zhang, C. X.; Liang, H.-C.; Karlin, K. D.; Solomon, E. I. *J. Am Chem. Soc.* **2003**, *125*, 5186–5192.
65. Halfen, J. A.; Young, J. V. G.; Tolman, W. B. *J. Am. Chem. Soc.* **1996**, *118*, 10920–10921.
66. Bol, J. E.; Driessen, W.; Ho, R. Y. N.; Maase, B.; Que, J. L.; Reedijk, J. *Angew. Chem. Int. Ed. Engl.* **1997**, *36*, 998–1000.
67. Börzel, H.; Comba, P.; Katsichtis, C.; Kiefer, W.; Lienke, A.; Nagel, V.; Pritzkow, H. *Chem. Eur. J.* **1999**, *5*, 1716–1721.
68. Börzel, H.; Comba, P.; Hagen, K. S.; Kerscher, M.; Pritzkow, H.; Schatz, M.; Schindler, S.; Walter, O. *Inorg. Chem.* **2002**, *41*.
69. Becker, M.; Heinemann, F. W.; Schindler, S. *Chem. Eur. J.* **1999**, *5*, 3124–3129.
70. He, C.; DuBois, J. L.; Hedman, B.; Hodgson, K. O.; Lippard, S. J. *Angew. Chem. Int. Ed.* **2001**, *40*, 1484–1487.
71. Hayashi, H.; Fujinami, S.; Nagatomo, S.; Ogo, S.; Suzuki, M.; Uehara, A.; Watanabe, Y.; Kitagawa, T. *J. Am. Chem. Soc.* **2000**, *122*, 2124–2125.
72. Mizuno, M.; Hayashi, H.; Fujinami, S.; Furutachi, H.; Nagatomo, S.; Otake, S.; Uozumi, K.; Suzuki, M.; Kitagawa, T. *Inorg. Chem.* **2003**, *42*, 8534–8544.
73. Tachi, Y.; Aita, K.; Teramae, S.; Fumito, T.; Naruta, Y.; Fukuzumi, S.; Itoh, S. *Inorg. Chem.* **2004**, *43*, 4558–4560.
74. Cahoy, J.; Holland, P. L.; Tolman, W. B. *Inorg. Chem.* **1999**, *38*, 2161–2168.
75. Mahadevan, V.; Henson, M. J.; Solomon, E. I.; Stack, T. D. P. *J. Am. Chem. Soc.* **2000**, *122*, 10249–10250.
76. Liang, H.-C.; Henson, M. J.; Hatcher, L. Q.; Vance, M. A.; Zhang, C. X.; Lahti, D.; Kaderli, S.; Sommer, R. D.; Rheingold, A. L.; Zuberbüehler, A. D.; Solomon, E. I.; Karlin, K. D. *Inorg. Chem.* **2004**, *43*, 4115–4117.
77. Stack, T. D. P. *Dalton Trans.* **2003**, 1881–1889.
78. Mahadevan, V.; DuBois, J. L.; Hedman, B.; Hodgson, K. O.; Stack, T. D. P. *J. Am. Chem. Soc.* **1999**, *121*, 5583–5584.
79. These results will be published in detail.
80. Itoh, S.; Nakao, H.; Berreau, L. M.; Kondo, T.; Komatsu, M.; Fukuzumi, S. *J. Am. Chem. Soc.* **1998**, *120*, 2890–2899.
81. Osako, T.; Tachi, Y.; Taki, M.; Fukuzumi, S.; Itoh, S. *Inorg. Chem.* **2001**, *40*, 6604–6609.
82. Itoh, S.; Taki, M.; Nakao, H.; Holland, P. L.; Tolman, W. B.; Que Jr., L.; Fukuzumi, S. *Angew. Chem. Int. Ed.* **2000**, *39*, 398–400.
83. Taki, M.; Teramae, S.; Nagatomo, S.; Tachi, Y.; Kitagawa, T.; Itoh, S.; Fukuzumi, S. *J. Am. Chem. Soc.* **2002**, *124*, 6367–6377.
84. Osako, T.; Ueno, Y.; Tachi, Y.; Itoh, S. *Inorg. Chem.* **2003**, *42*, 8087–8097.
85. Itoh, S.; Kumei, H.; Taki, M.; Nagatomo, S.; Kitagawa, T.; Fukuzumi, S. *J. Am. Chem. Soc.* **2001**, *123*, 6708–6709.
86. Pidcock, E.; DeBeer, S.; Obias, H. V.; Hedman, B.; Hodgson, K. O.; Karlin, K. D.; Solomon, E. I. *J. Am. Chem. Soc.* **1999**, *121*, 1870–1878.

87. Lam, B. M. T.; Halfen, J. A.; Young Jr., V. G.; Hagadorn, J. R.; Holland, P. L.; Lledos, A.; Cucurull-Sanchez, L.; Novoa, J. J.; Alvarez, S.; Tolman, W. B. *Inorg. Chem.* **2000**, *39*, 4059–4072.
88. Obias, H. V.; Lin, Y.; Murthy, N. N.; Pidcock, E.; Solomon, E. I.; Ralle, M.; Blackburn, N. J.; Neuhold, Y.-M.; Zuberbühler, A. D.; Karlin, K. D. *J. Am. Chem. Soc.* **1998**, *120*, 12960–12961.
89. Mahapatra, S.; Halfen, J. A.; Wilkinson, E. C.; Pan, G.; Cramer, C. J.; Que Jr., L.; Tolman, W. B. *J. Am. Chem. Soc.* **1995**, *117*, 8865–8866.
90. Mahapatra, S.; Halfen, J. A.; Wilkinson, E. C.; Pan, G.; Wang, X.; Young, J. V. G.; Cramer, C. J.; Que, J. L.; Tolman, W. B. *J. Am. Chem. Soc.* **1996**, *118*, 11555–11574.
91. Mahadevan, V.; Hou, Z.; Cole, A. P.; Root, D. E.; Lal, T. K.; Solomon, E. I.; Stack, T. D. P. *J. Am. Chem. Soc.* **1997**, *119*, 11996–11997.
92. Mirica, L. M.; Vance, M.; Rudd, D. J.; Hedman, B.; Hodgson, K. O.; Solomon, E. I.; Stack, T. D. P. *J. Am. Chem. Soc.* **2002**, *124*, 9332–9333.
93. Arii, H.; Saito, Y.; Nagatomo, S.; Kitagawa, T.; Funahashi, Y.; Jitsukawa, K.; Masuda, H. *Chem. Lett.* **2003**, *32*, 156–157.
94. Liang, H.-C.; Zhang, C. X.; Henson, M. J.; Sommer, R. D.; Hatwell, K. R.; Kaderli, S.; Zuberbüehler, A. D.; Rheingold, A. L.; Solomon, E. I.; Karlin, K. D. *J. Am. Chem. Soc.* **2002**, *124*, 4170–4171.
95. Larsen, P. L.; Parolin, T. J.; Powell, D. R.; Hendrich, M. P.; Borovik, A. S. *Angew. Chem. Int. Edit.* **2003**, *42*, 85–89.
96. Yamaguchi, S.; Wada, A.; Funahashi, Y.; Nagatomo, S.; Kitagawa, T.; Jitsukawa, K.; Masuda, H. *European Journal of Inorganic Chemistry* **2003**, 4378–4386.
97. Zhang, C. X.; Liang, H.-C.; Kim; E.-i.; Shearer, J.; Helton, M. E.; Kim, E.; Kaderli, S.; Incarvito, C. D.; Zuberbühler, A. D.; Rheingold, A. L.; Karlin, K. D. *J. Am. Chem. Soc.* **2003**, *125*, 634–635.
98. Hatcher, L. Q.; Vance, M. A.; Sarjeant, A. N.; Solomon, E. I.; Karlin, K. D. **2005**, submitted for publication.
99. Liang, H.-C.; Karlin, K. D.; Dyson, R.; Kaderli, S.; Jung, B.; Zuberbühler, A. D. *Inorg. Chem.* **2000**, *39*, 5884–5894.
100. Karlin, K. D.; Cruse, R. W.; Gultneh, Y.; Farooq, A.; Hayes, J. C.; Zubieta, J. *J. Am. Chem. Soc.* **1987**, *109*, 2668–2679.
101. Paul, P. P.; Tyeklár, Z.; Jacobson, R. R.; Karlin, K. D. *J. Am. Chem. Soc.* **1991**, *113*, 5322–5332.
102. Karlin, K. D.; Ghosh, P.; Cruse, R. W.; Farooq, A.; Gultneh, Y.; Jacobson, R. R.; Blackburn, N. J.; Strange, R. W.; Zubieta, J. *J. Am. Chem. Soc.* **1988**, *110*, 6769–6780.
103. Ghosh, P.; Tyeklár, Z.; Karlin, K. D.; Jacobson, R. R.; Zubieta, J. *J. Am. Chem. Soc.* **1987**, *109*, 6889–6891.
104. Holland, P. L.; Rodgers, K. R.; Tolman, W. B. *Angew. Chem. Int. Ed.* **1999**, *38*, 1139–1142.
105. Santagostini, L.; Gullotti, M.; Monzani, E.; Casella, L.; Dillinger, R.; Tuczek, F. *Chem. Eur. J.* **2000**, *6*, 519–522.
106. Battaini, G.; De Carolis, M.; Monzani, E.; Tuczek, F.; Casella, L. *Chem. Commun.* **2003**, 726–727.
107. Mahapatra, S.; Halfen, J. A.; Tolman, W. B. *J. Am. Chem. Soc.* **1996**, *118*, 11575–11586.
108. Taki, M.; Itoh, S.; Fukuzumi, S. *J. Am. Chem. Soc.* **2002**, *124*, 998–1002.
109. Karlin, K. D.; Nasir, M. S.; Cohen, B. I.; Cruse, R. W.; Kaderli, S.; Zuberbühler, A. D. *J. Am. Chem. Soc.* **1994**, *116*, 1324–1336.
110. Itoh, S.; Kumei, H.; Taki, M.; Nagatomo, S.; Kitagawa, T.; Fukuzumi, S. *J. Am. Chem. Soc.* **2001**, *123*, 6708–6709.
111. Osako, T.; Ohkubo, K.; Taki, M.; Tachi, Y.; Fukuzumi, S.; Itoh, S. *J. Am. Chem. Soc.* **2003**, *125*, 11027–11033.
112. Karlin, K. D.; Gultneh, Y.; Hayes, J. C.; Cruse, R. W.; McKown, J. W.; Hutchinson, J. P.; Zubieta, J. *J. Am. Chem. Soc.* **1984**, *106*, 2121–2128.

113. Karlin, K. D.; Cruse, R. W.; Haka, M. S.; Gultneh, Y.; Cohen, B. I. *Inorg. Chim. Acta.* **1986**, *125*, L43–L44.
114. Cruse, R. W.; Kaderli, S.; Karlin, K. D.; Zuberbühler, A. D. *J. Am. Chem. Soc.* **1988**, *110*, 6882–6883.
115. Karlin, K. D.; Cohen, B. I.; Jacobson, R. R.; Zubieta, J. *J. Am. Chem. Soc.* **1987**, *109*, 6194–6196.
116. Nasir, M. S.; Cohen, B. I.; Karlin, K. D. *J. Am. Chem. Soc.* **1992**, *114*, 2482–2494.
117. Pidcock, E.; Obias, H. V.; Zhang, C. X.; Karlin, K. D.; Solomon, E. I. *J. Am. Chem. Soc.* **1998**, *120*, 7841–7847.
118. Solomon, E. I.; Chen, P.; Metz, M.; Lee, S. K.; Palmer, A. E. *Angew. Chem. Int. Ed.* **2001**, *40*, 4570–4590.
119. Shearer, J.; Zhang, C. X.; Hatcher, L. Q.; Karlin, K. D. *J. Am. Chem. Soc.* **2003**, *125*, 12670–12671.
120. Karki, S. B.; Dinnocenzo, J. P.; Jones, J. P.; Korzekwa, K. R. *J. Am. Chem. Soc.* **1995**, *117*, 3657–3664.
121. Goto, Y.; Watanabe, Y.; Fukuzumi, S.; Jones, J. P.; Dinnocenzo, J. P. *J. Am. Chem. Soc.* **1998**, *120*, 10762–10763.
122. Baciocchi, E.; Lanzalunga, O.; Lapi, A.; Manduchi, L. *J. Am. Chem. Soc.* **1998**, *120*, 5783–5787.
123. All KIE values are within the semiclassical limit for ETPT and ET/PT reactions at −80°C (k_H/k_D could reach a maximum of 23.1).
124. Shearer, J.; Zhang, C. X.; Zakharov, L. N.; Rheingold, A. L.; Karlin, K. D. *J. Am. Chem. Soc.* **2005**, *127*, 5469–5483.
125. Lieberman, R. L.; Rosenzweig, A. C. *Nature* **2005**, *434*, 177–182.

BIOMIMETIC OXIDATIONS BY DINUCLEAR AND TRINUCLEAR COPPER COMPLEXES

GIUSEPPE BATTAINI[a], ALESSANDRO GRANATA[a], ENRICO MONZANI[a], MICHELE GULLOTTI[b] and LUIGI CASELLA[a,*]

[a]Dipartimento di Chimica Generale, Università di Pavia, Via Taramelli 12, 27100 Pavia, Italy

[b]Dipartimento di Chimica Inorganica, Metallorganica ed Analitica, Università di Milano, Istituto ISTM-CNR, Via Venezian 21, 20133 Milano, Italy

I. Copper Proteins Containing Dinuclear and Trinuclear Metal Clusters

A. Introduction

Copper is widely employed as metal cofactor in proteins and enzymes that perform dioxygen binding, activation, and reduction processes *(1–6)*; provide general overviews describing the catalytic reactions and structural features of the protein active sites. The activities of copper enzymes using dioxygen as a substrate can be broadly classified as oxidase, monooxygenase, or dioxygenase, depending on the stoichiometry of the catalytic reactions:

oxidase:

$$4\,S + O_2 + 4\,H^+ \longrightarrow 4\,S^+ + 2H_2O \qquad (1)$$

$$2\,SH_2 + O_2 \longrightarrow 2\,S + 2H_2O \qquad (2)$$

ADVANCES IN INORGANIC CHEMISTRY
VOLUME 58 ISSN 0898-8838 / DOI: 10.1016/S0898-8838(05)58005-X

$$SH_2 + O_2 \longrightarrow S + H_2O_2 \quad (3)$$

$$RCH_2NH_2 + O_2 + H_2O \longrightarrow RCHO + H_2O_2 + NH_3 \quad (4)$$

monooxygenase:

$$SH_2 + O_2 \longrightarrow SO + H_2O \quad (5)$$

$$S + DH_2 + O_2 \longrightarrow SO + D + H_2O \quad (6)$$

dioxygenase:

$$S + O_2 \longrightarrow SO_2 \quad (7)$$

where S, or SH_2 is a substrate molecule and DH_2 an external reductant. In oxidase reactions, dioxygen can be either reduced to water, e.g., by cytochrome *c* oxidase (reaction 1, where S is reduced cytochrome *c*) and laccase (reaction 2), or dihydrogen peroxide, e.g., by galactose oxidase (reaction 3) and copper amine oxidase (reaction 4). Oxidase activities essentially involve electron and proton transfers from the substrate to O_2, which can be two-electron reduced to H_2O_2, or four-electron reduced to H_2O; when an oxygen atom is inserted into the substrate, as in the amine oxidase reaction 4, it comes from water and not from O_2. By contrast, in monooxygenase and dioxygenase reactions, one, or two oxygen atoms from O_2 are inserted into the substrate, respectively. Monooxygenase reactions usually require an external reducing cosubstrate, as in the case of dopamine β-hydroxylase (reaction 6, where DH_2 is ascorbic acid), while for tyrosinase also these reducing equivalents are provided by the substrate.

Copper proteins and enzymes can be classified into structural types, according to the nuclearity of their active sites, which can be mononuclear, dinuclear, or polynuclear, but in general protein function is not directly correlated with the nuclearity of the active site. For instance, mononuclear copper enzymes perform oxidase, monooxygenase, and dioxygenase activities. However, in most cases the reactivity of these copper centers is assisted by redox organic cofactors, resulting from posttranslational modifications of endogenous amino acids (*7*), as in galactose oxidase (*8*) and amine oxidase (*2,9*), or additional distant copper centers, as in dopamine β-hydroxylase (*10*) and peptidylglycine α-hydroxylating monooxygenase (*11*). Cytochrome *c* oxidase, on the other hand, contains a much more complex enzyme machinery, with both copper and heme centers (*12*). Quercetin 2,3-dioxygenase is the only known mononuclear dioxygenase (*13*). The proteins of interest in the context of this review are those containing electronically coupled, dinuclear, and trinuclear copper centers and performing oxidase, or oxygenase activities. The best known representative members of the former family of proteins are hemocyanin, catechol oxidase and tyrosinase, and those of the latter family are ascorbate oxidase, laccase, and ceruloplasmin. The basic properties of these proteins are the necessary reference for any attempt to reproduce in biomimetic systems key aspects of the protein structure and reactivity.

B. Dinuclear Copper Proteins and Enzymes

B.1. Hemocyanin

Hemocyanins are large, multisubunit dioxygen transporting proteins found in the hemolymph of many invertebrate species of the phyla of molluscs and arthropods (*14*). The subunits of molluscan hemocyanins contain functional units with a molecular weight of about 50,000 Da, each of which contains a dioxygen binding dicopper center. Arthropodal hemocyanins occur as hexamers, or multihexamers of subunits with a molecular weight of about 75,000 Da. As shown by the comparison of various X-ray crystal structures of the proteins from *Panulirus interruptus* (*15*), *Limulus polyphemus* (*16*), and *Octopus* (*17*), the active site of the hemocyanins of the two families is quite similar and is buried within the α-helices of a four helix bundle motif. These helices provide the six histidine ligands for the copper centers, which are three-coordinated in the deoxy form and five-coordinated, with approximate square-pyramidal geometry, in the oxy form (Fig. 1). Dioxygen is bound in a side-on μ-η^2:η^2 mode, and both the O–O distance (~1.4 Å) and resonance Raman $\nu(O_2)$ stretching frequency (~750 cm^{-1}) indicate the peroxo nature of the bound O_2 residue, which results from an oxidative addition of O_2 to the dicopper(I) centers of the deoxy form of the protein to give a peroxodicopper(II) species. Other characteristic spectroscopic signatures for oxy-hemocyanin are the optical LMCT bands at 345 ($\varepsilon \sim 20{,}000\ M^{-1}cm^{-1}$) and 570 nm ($\varepsilon \sim 1000\ M^{-1}cm^{-1}$), responsible for the intense blue color of the protein, and lack of EPR features and diamagnetism, due to strong magnetic coupling between the two copper(II) centers.

Although reversible binding of dioxygen is the normal activity of hemocyanins, these proteins also exhibit weak catalase (*18,19*) and catecholase (*20*) activities (only molluscan hemocyanins in the latter case), as shown by studies contributed also by our group. These activities involve the intermediate formation of a non-physiological, Met form of the protein, with oxidized Cu(II) centers, which we have shown to contain a di-μ-hydroxo bridge (*21*). The catecholase activity of molluscan hemocyanin is of particular relevance here because the mechanism is completely different from those

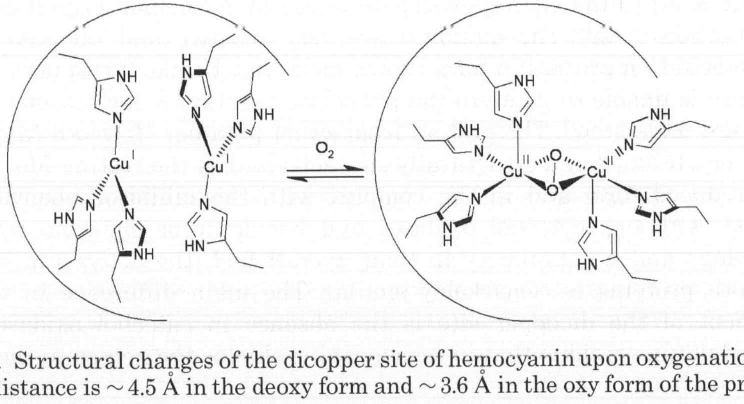

Fig. 1. Structural changes of the dicopper site of hemocyanin upon oxygenation, the Cu–Cu distance is ~4.5 Å in the deoxy form and ~3.6 Å in the oxy form of the protein.

of catechol oxidase and tyrosinase. For structural reasons, the active site of hemocyanins is inaccessible to molecules of the size of a phenol, or a catechol (*15–17*). This prevents the formation of a ternary complex between the dicopper center, dioxygen and the substrate, where O_2 and the phenolic derivative are simultaneously bound to the metal centers, that is the necessary condition to promote the enzymatic activities of catechol oxidase and tyrosinase (see below). Therefore, in the case of hemocyanin the catecholase activity proceeds through a complex mechanism involving monoelectronic steps and the transformation of catechol into a semiquinone radical species, quinone being generated by dismutation of two molecules of semiquinone (*20*). The control of the active-site accessibility is an important means for regulating the activity of hemocyanin, and in fact it has been shown that some monophenolase (reaction 8) and catecholase (reaction 9) activities can be introduced into the protein from *Euripelma californium* after limited proteolysis with trypsin and chymotripsin (*22,23*). This treatment cleaves the *N*-terminal peptide containing the highly conserved phenylalanine residue (Phe49 in *L. polyphemus*) shielding the access channel to the dinuclear copper site and thus makes binding of the substrate possible.

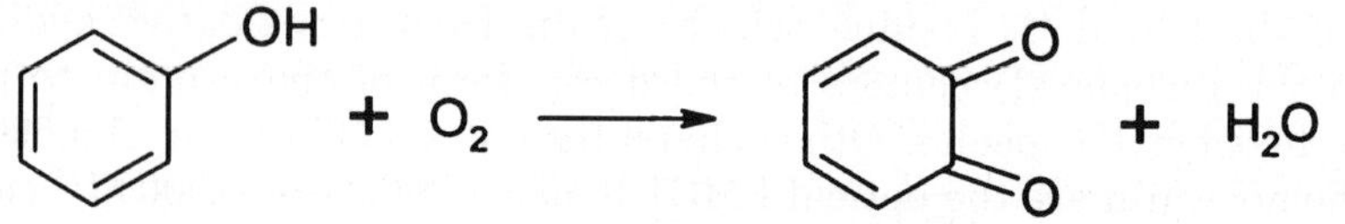

(8)

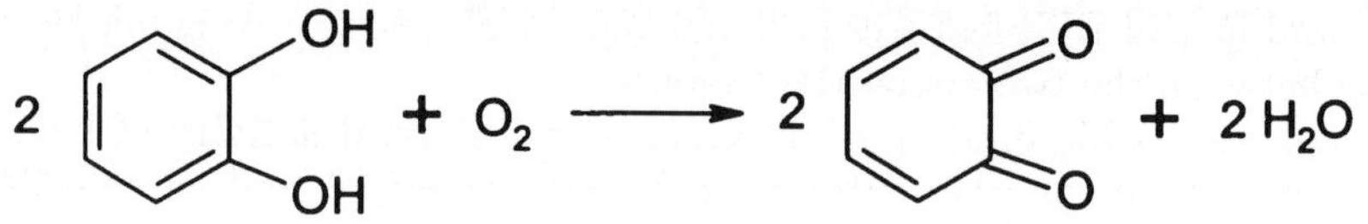

(9)

B.2. Catechol Oxidase

Catechol oxidase, also known as *ortho*-diphenol oxidase, or even polyphenol oxidase (EC 1.10.3.1), is an ubiquitous plant enzyme which catalyzes the two-electron oxidation of a broad range of catecholic substrates (like caffeic acid) to the corresponding quinones by molecular oxygen according to reaction 9 (*24*). The quinones are very reactive and undergo a rapid polymerization process to form brown melanins. Unlike tyrosinase, catechol oxidase is unable to catalyze the phenolase reaction 8, for reasons that are still not understood. The protein from sweet potatoes (*Ipomoea batatas*) has been crystallized and structurally characterized in the resting Met form, in the reduced form and in the complex with the inhibitor phenylthiourea (*25,26*). Although catechol oxidases and hemocyanins show no significant homology and no similarity in their overall fold, the active-site structure of both proteins is remarkably similar. The main difference in the environment of the dicopper site is the absence in catechol oxidase of the phenylalanine residue that in hemocyanins shields the access to the metals.

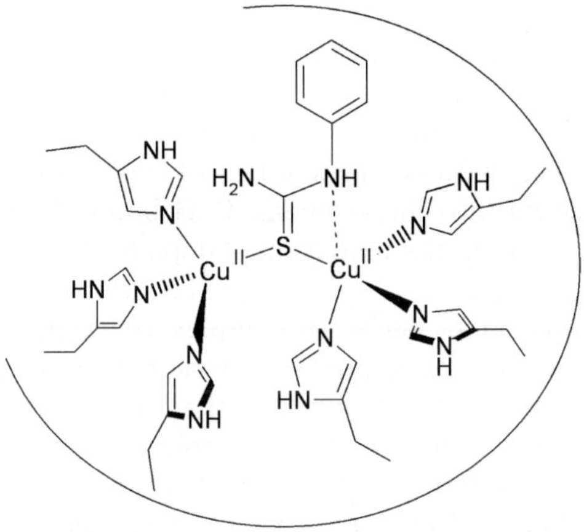

FIG. 2. Binding of phenylthiourea to the Met form of catechol oxidase. The amide nitrogen of the inhibitor interacts weakly (Cu-N distance is 2.6 Å) with one of the copper centers (CuB).

Interestingly, superposition of the structures of catechol oxidase complexed with phenylthiourea with that of *L. polyphemus* oxyhemocyanin shows that the aromatic ring of phenyltiourea aligns perfectly with the phenyl ring of Phe49. The inhibitor is bound to both copper(II) ions with the sulfur atom acting as a bridging ligand and one of the nitrogen atoms of the thiourea providing a weak bond to one of the coppers (Fig. 2). This replaces the μ-hydroxo bridge present in the structure of the Met form of the protein.

Another interesting feature in the structure of *I. batatas* catechol oxidase is a covalent thioether bond between the C_ε atom of the imidazole ring of one of the histidine ligands (His109) and the sulfur atom of a cysteine residue (Cys92). A similar posttranslational modification is actually present in *Octopus* hemocyanin (*17*) and probably in other molluscan hemocyanins (*27*), and in some fungal tyrosinases (*28*). The absence of such a Cys-His bridge in arthropodal hemocyanins and, for instance, in human tyrosinase means that the thioether modification does not play any direct role in the functional activity of the proteins, but it adds structural constraint around the copper center (CuA). The spectroscopic data of catechol oxidases are similar to those of the corresponding forms of hemocyanin and tyrosinase (*29,30*). The oxy form can be obtained by treatment of Met catechol oxidase with excess dihydrogen peroxide. Though, this oxy form is less stable than that of hemocyanin and tyrosinase: it cannot be fully produced and spontaneously decays, with partial enzyme inactivation, also because the enzyme exhibits some catalase activity. The instability of the oxy form may be a reason for the lack of monooxygenase reactivity of this enzyme towards monophenols (*26*), but it has also been suggested that access to one of the coppers (CuA), which is essential for binding of the phenol, is blocked in the structure of *I. batatas* catechol oxidase (*31*).

B.3. Tyrosinase

Tyrosinase (EC 1.14.18.1) is the enzyme mostly responsible for the formation of brown melanin pigments (*32*), since it catalyzes the energetically demanding, initial step of a complex series of reactions involving the insertion of an oxygen atom into the aromatic C–H bond of tyrosine (reaction 8). Tyrosinase is also active in the oxidation of dopa to dopaquinone (reaction 9). Melanin formation depends on the fast evolution of dopaquinone to dopachrome and the subsequent series of complex reactions undergone by this key intermediate, which also involve participation of amino acids like cysteine (*33*). We have recently shown that the quinone intermediates can be trapped when the slow reacting fluorophenols are used as substrates (*34*). The monophenolase reaction is often shown to proceed through the initial formation of dopa followed by oxidation to dopaquinone (*35*), but formation of dopa *in vitro* occurs through an indirect non-enzymatic mechanism (*33*). The evidence for stepwise conversion of phenol to catechol and from catechol to quinone is indeed weak and is based on the detection of nM amounts of catechol released in the oxidation of 4-*tert*-butylphenol, which is a very poor substrate for tyrosinase and lacks the charged amino and carboxylate functions on the alkyl side chain (*35*). In a recent paper, Itoh *et al.* used a borate buffer at pH 9.0 to trap the catechol resulting from the enzymatic oxidation of the phenol (*36*). However, the excess of external reductant used to sustain the enzymatic catalysis is probably responsible for the reduction of quinone directly formed in the reaction. Catechol oxidation is more than one order of magnitude faster than phenol oxygenation and it is difficult to stop the reaction at the level of a bound catechol. In addition, there might be structural effects on the enzyme, since the activity at pH 9.0 is markedly reduced. Native tyrosinase occurs in the Met form, which is inactive against phenols; the enzyme is thus usually activated *in vitro* by a catechol, which converts the enzyme to the deoxy form. Deoxy-tyrosinase readily binds dioxygen and the resulting oxy form initiates the catalytic cycle (Scheme 1). The catecholase catalytic cycle is more complex, because it involves the oxidation of two molecules of substrate per molecule of dioxygen reduced (Scheme 2). Here one catechol molecule is oxidized by the Met form and one by the oxy form of the enzyme. In these schemes, binding of catechol to the Met form of tyrosinase has been represented to occur in the η^2:η^1 bridging mode, rather than the commonly adopted μ-1,4 bridging mode (*1*), as is suggested by evidence from inhibitor binding studies to half-Met *S. antibioticus* tyrosinase (*37*), spectroscopic studies on biomimetic catecholate-dicopper(II) complexes (*38*), and by the structure of the catechol oxidase complex with phenylthiourea (*25*). Like hemocyanin and catechol oxidase, also tyrosinase exhibits a weak catalase activity (*39,40*).

Although the sructure of tyrosinase is not known, evolutionary correlations (*41*), sequence homology (*28*), and spectroscopic similarities (*6*) with hemocyanin, particularly molluscan hemocyanin, and catechol oxidase indicate a common active site with six histidine ligands bound to the two copper ions. Thus, for instance, the electronic spectrum of oxy-tyrosinase exhibits

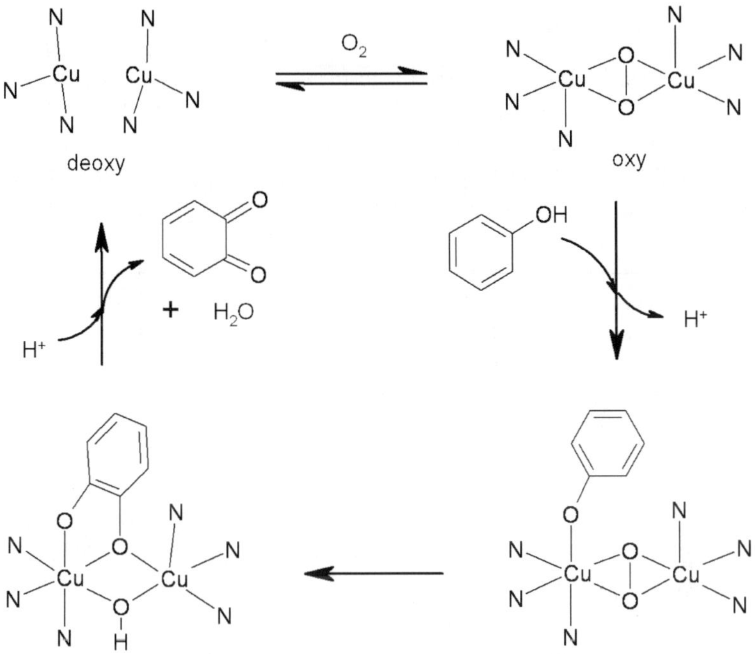

SCHEME 1. The monophenolase catalytic cycle of tyrosinase (adapted from (*44*)).

LMCT bands at 345 ($\varepsilon = 18{,}000\,M^{-1}cm^{-1}$) and 590 nm ($\varepsilon = 1200\,M^{-1}cm^{-1}$), and the resonance Raman spectrum a peak at 755 cm^{-1}, which are characteristic signatures of the μ-η^2:η^2 binding mode of dioxygen. In spite of intensive efforts, little is known about the catalytic intermediates occuring in the reaction cycles of Scheme 1 and 2. Recent mechanistic studies using *p*-substituted phenols have shown that the phenol hydroxylation by the μ-η^2:η^2-peroxodicopper(II) species occurs through an electrophilic aromatic substitution mechanism with a Hammett reaction constant $\rho = -2.4$ (*36*). This mechanism is basically confirmed by theoretical computations, according to which the energy barrier for the monophenolase reaction (14.4 kcal mol^{-1}) is determined by cleavage of the peroxo O–O bond (*42*). Though, the ternary active species of Cu_2/dioxygen/phenolate emerging from these computations contains an unprecedented mixed-valent Cu(II)–Cu(I) center with dioxygen bound as a bridging peroxo radical. The phenol hydroxylation would then assume a radical character that is in contrast with experimental evidence (*43*). We have recently shown with studies at variable temperature that the monophenolase and catecholase reactions of tyrosinase are characterized by $\Delta H^{\neq}$ values of 62 ± 8 kJ mol^{-1}, which are little dependent on substrate and reaction types (*44*). This indicates that the key event controlling enzyme activity is common to the transition states involved in the monophenolase and catecholase reactions, and this necessarily leads to the conclusion that peroxide O–O bond cleavage is the main determinant for the evolution of the ternary complexes of the enzymatic reactions.

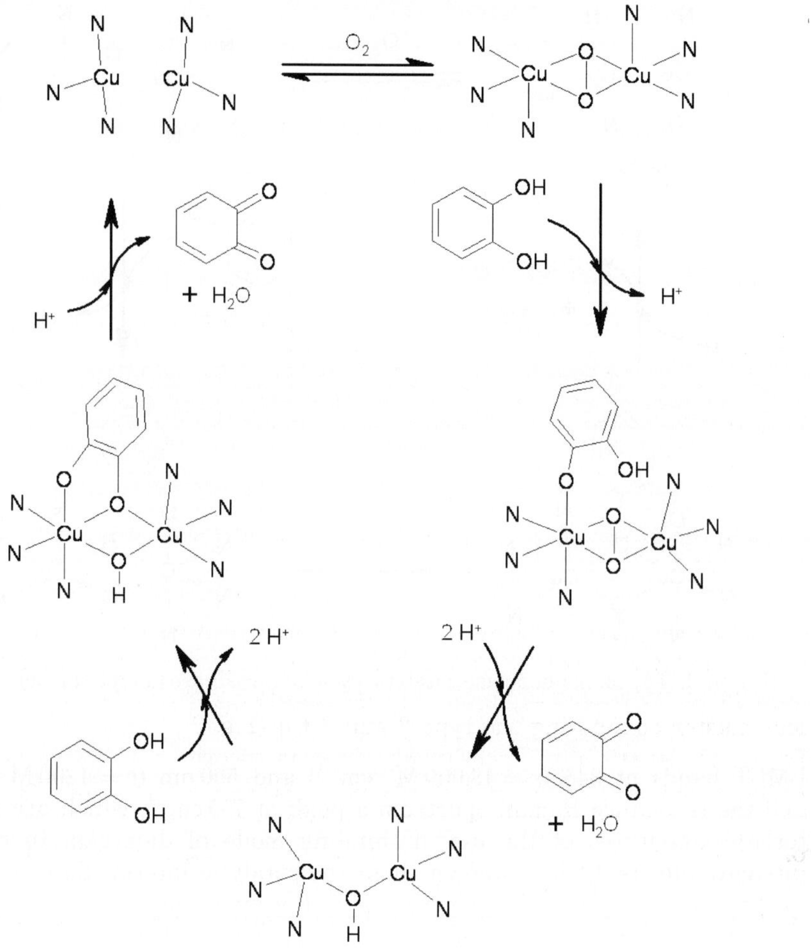

SCHEME 2. The catecholase catalytic cycle of tyrosinase (adapted from (*44*)).

The $\Delta H^{\neq}$ values are also fully consistent with the theoretical calculations mentioned above (*42*).

C. Multicopper Oxidases

The "blue" multicopper oxidases constitute a heterogeneous family of enzymes from different sources (*1*). In addition to the well characterized members of this family, ascorbate oxidase (*45,46*), laccase (*47,48*), and ceruloplasmin (*49,50*), all from higher organisms, two other proteins have attracted much recent interest: Fet3p, which is involved in iron uptake in yeast (*51*), and CueO, which is required for copper homeostasis in *Escherichia coli* (*52*). The characteristic reactivity of these enzymes is the one-electron oxidation of four substrate equivalents coupled to the four-electron reduction of dioxygen to water (*1*). These processes occur at a catalytic unit constituted by four copper atoms classified according to their spectroscopic properties in

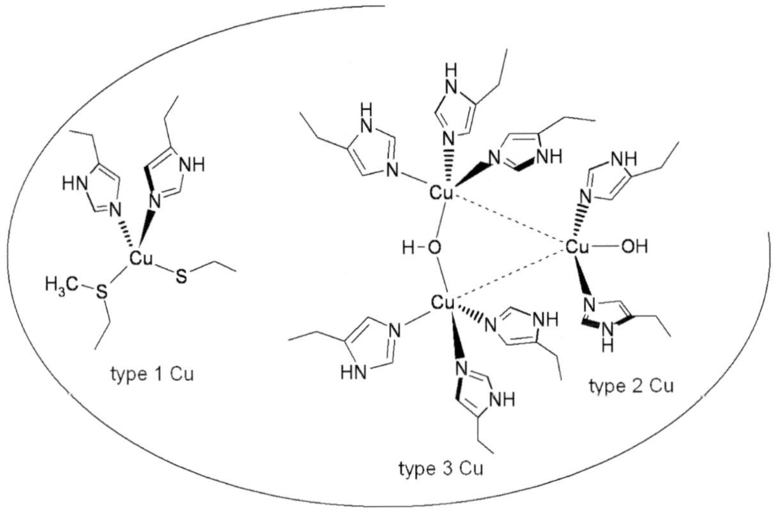

FIG. 3. Structure of the active site of multicopper oxidases. Type 1 Cu in fungal laccase lacks the methionine ligand.

the +2 redox state as type 1 (or blue) Cu, type 2 (or normal) Cu, and type 3 (or coupled dinuclear) Cu sites (*53*). The structure of the active site actually contains a mononuclear type 1 Cu separated by about 12.5 Å from a trinuclear cluster comprising the type 2 and 3 Cu (Fig. 3). The ligands of type 1 Cu are the usual set of His, Cys, His, Met, in ascorbate oxidase and ceruloplasmin, while in the fungal laccase structures the Met ligand is missing and type 1 Cu is three-coordinated (*47,48*). The trinuclear cluster has a total of eight histidine ligands subdivided by three, three and two between the pair of type 3 Cu and the type 2 Cu, respectively. An important feature is that type 1 Cu is connected through bonds to the trinuclear cluster, because the Cys ligand is adjacent to two of the type 3 His ligands. This has functional importance because the role of type 1 Cu is to transfer electrons from the substrate to the trinuclear cluster, which is the site of dioxygen binding and reduction. It should also be noted that ceruloplasmin contains two additional type 1 Cu centers, the function of which is unknown. Sequence analysis and evolutionary relationships show that multicopper oxidases are multidomain proteins evolved from a common ancestor monodomain protein homologous with cupredoxins, the electron transfer proteins containing a single type 1 Cu center (*54*), by a cascade of domain duplication events (*55*). While type 1 Cu has an intradomain location, the trinuclear type 2/type 3 Cu cluster occurs at an interdomain locus.

The catalytic cycle of multicopper oxidases is very complex because the four redox active Cu centers need to store reducing equivalents coming from four substrate molecules and transfer the electrons to dioxygen. From spectroscopic studies and fast kinetic experiments carried out on laccase it has been shown that reaction of fully reduced enzyme with O_2 produces two intermediates, labeled as "peroxy" intermediate and "native" intermediate (Scheme 3). The peroxy intermediate has been trapped in a derivative of

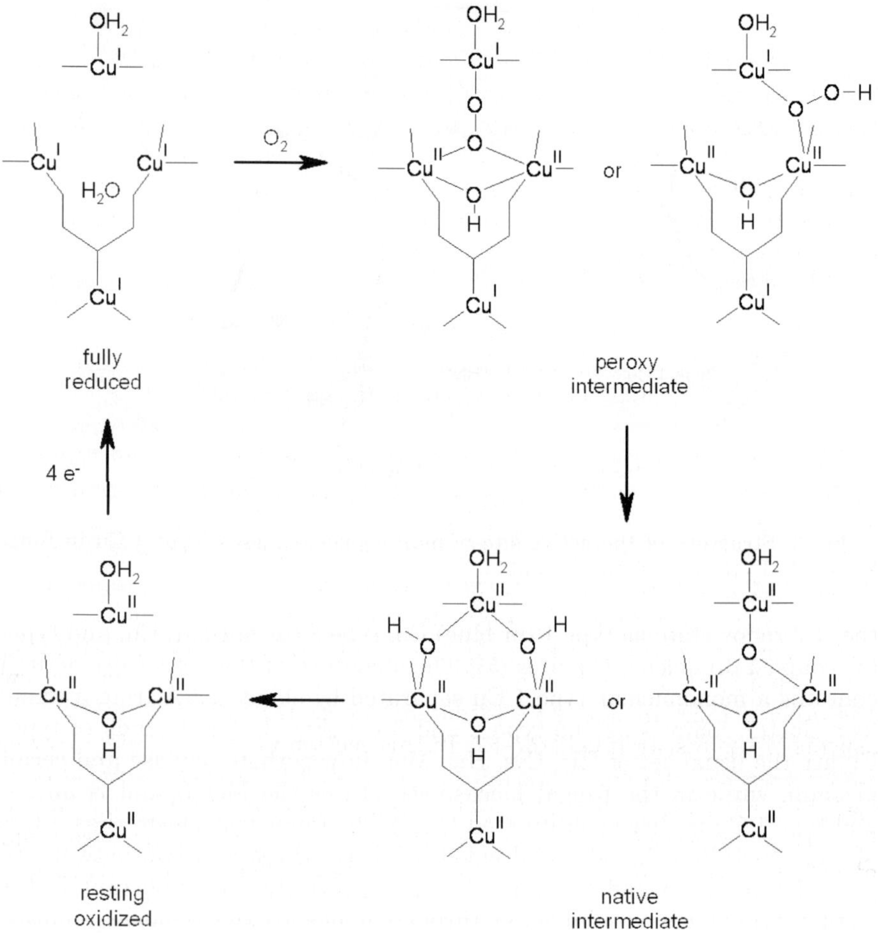

SCHEME 3. Mechanism of reduction of O_2 in a sequence of two-electron steps by multicopper oxidases (adapted from (*57*)).

laccase where type 1 Cu has been replaced with the redox inactive mercuric ion (*56*). The native intermediate is instead observed using native laccase (*1,57,58*). The peroxy intermediate of type 1 Hg laccase decays to resting type 1 Hg laccase through a species similar to the native intermediate. The structures of the two intermediates has not been completely clarified; the peroxy intermediate contains a bridged peroxo, or hydroperoxo residue within the trinuclear cluster, while the native intermediate contains a fully oxidized cluster in which all three Cu ions are bridged by the oxygen product. This structure differs from that of the resting enzyme where type 2 Cu is not bridged to type 3 Cu. An essential aspect of this mechanism is a fast electron transfer from the substrate to the enzyme. For this reason, the substrate must bind close to type 1 Cu. The location of the substrate binding site is indicated by an empty cleft near type 1 Cu in the structure of ascorbate oxidase (*45*), and by the structure of *Trametes versicolor* laccase, which contains a molecule of 2,5-xylidine, an enzyme inducer in the fungus culture, bound in a large cavity near type 1 Cu and actually hydrogen bonded to one of the His ligands (*47*). We have addressed the problem of the substrate binding site in ascorbate oxidase by performing NMR relaxation studies and docking calculations on the enzyme complexes with phenolic compounds, which act as competitive inhibitors against ascorbate (*59,60*). Also in this case we have obtained indication that the bound inhibitor is hydrogen bonded to one of the His ligand of type 1 Cu.

II. Biomimetic Oxidations by Dinuclear Complexes as Tyrosinase and Catechol Oxidase Models: General Aspects

The chemistry of copper containing enzymes has stimulated an enormous amount of research activity in the field of coordination chemistry over the years. This activity has different aims: (i) design of structural models for the metal active sites, (ii) understanding the mechanism of formation and spectroscopic characteristics of Cu_nO_2 complexes, (iii) finding correlations between structural/electronic properties and reactivity of Cu_nO_2 complexes, and (iv) building of biomimetic catalysts to promote chemical transformations of potential synthetic and industrial interest. Several excellent reviews and other articles in this issue extensively cover the literature in all major aspects of the research field (*61–69*). In this section we will therefore concentrate on the systems developed in our group, and the related systems developed by other groups, that perform biomimetic reactions of relevance for the activity of tyrosinase and catechol oxidase. The type of copper mediated oxidations of interest here can be summarized as follows: (a) oxygen atom insertion into aromatic C–H bonds of the ligands supporting the complexes, (b) genuine monophenolase reactions on exogenous phenolic substrates, and (c) catecholase activity on diphenolic substrates. In the latter case, we will only deal with those systems performing catalytic conversion of catechols into quinones.

Copper-dioxygen complexes are necessary intermediates in the biomimetic oxidations performed by type 3 Cu models. Several structural motifs for synthetic Cu_nO_2 complexes exist and the main types are shown in Scheme 4. Their mechanism of formation, structural characteristics, spectroscopic and thermodynamic properties, and reactivity have been recently thoroughly reviewed (*62,63*). In general, the reaction between a Cu(I) complex and dioxygen initially forms a 1:1 Cu/O_2 adduct that, in the absence of sterically hindered ligands, rapidly evolves to the thermodynamically more stable 2:1 Cu/O_2 adduct. According to the usual formalism, these processes can be described by the following equilibria:

$$LCu^{I} + O_2 \rightleftharpoons LCu^{II}O_2^- \tag{10}$$

$$LCu^{II}O_2^- + LCu^{I} \rightleftharpoons LCu^{II}O_2^{2-}Cu^{II}L \quad (\text{or } LCu^{III}(O^{2-})_2Cu^{III}L) \tag{11}$$

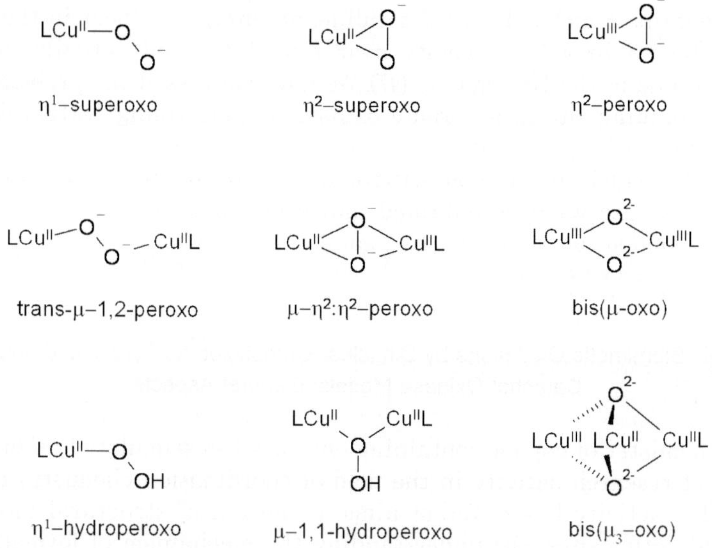

SCHEME 4. Structures of the main types of copper-dioxygen complexes; some related copper-hydroperoxide complexes are also included.

Adducts of 1:1 Cu/O_2 stoichiometry formulated as $LCu^{III}O_2^{2-}$ also exist (*70*), but do not seem to be relevant in biology. Also, of the two types of isoelectronic complexes with 2:1 Cu/O_2 stoichiometry, only the μ-η^2:η^2-peroxodicopper(II) species has been characterized as the oxy form of type 3 Cu proteins (see Section I). It is perhaps interesting to emphasize that the initial identification of this type of Cu_2O_2 complex as the binding mode of dioxygen in oxyhemocyanin came from the structural characterization by Kitajima *et al.* of the μ-η^2:η^2-peroxodicopper(II) complex obtained from Cu(I) complexes with sterically demanding tris(pyrazolyl)borate ligands (*71*). On the other hand, it cannot be excluded that bis(μ-oxo)dicopper(III) species could be transiently formed during enzyme activity, upon cleavage of the O–O bond in the side-on peroxo species (*23*). Actually, at least for some Cu_2O_2 complexes, the μ-η^2:η^2-peroxo and bis(μ-oxo) forms can be interconverted, since the energy barrier to the interconversion is small (*62,63*). From the synthetic point of view, formation of a given type of Cu-dioxygen adduct depends on the denticity of the ligand employed, its steric demand and the nature of the donor atoms.

In Sections III and IV we will thus focus on the model monooxygenase and oxidase reactions promoted by mononuclear and dinuclear complexes which make use of the ligands collected in Fig. 4. In principle, a suitable model complex for the type 3 Cu sites should contain two copper ions each coordinated to three nitrogen donors, and since the natural ligands are imidazole groups, most of the ligands in Fig. 4, and in the more extensive lists reported in (*62,63*), contain several heterocyclic nitrogen donors. For practical reasons, the synthesis of pyridine-, or benzimidazole-containing

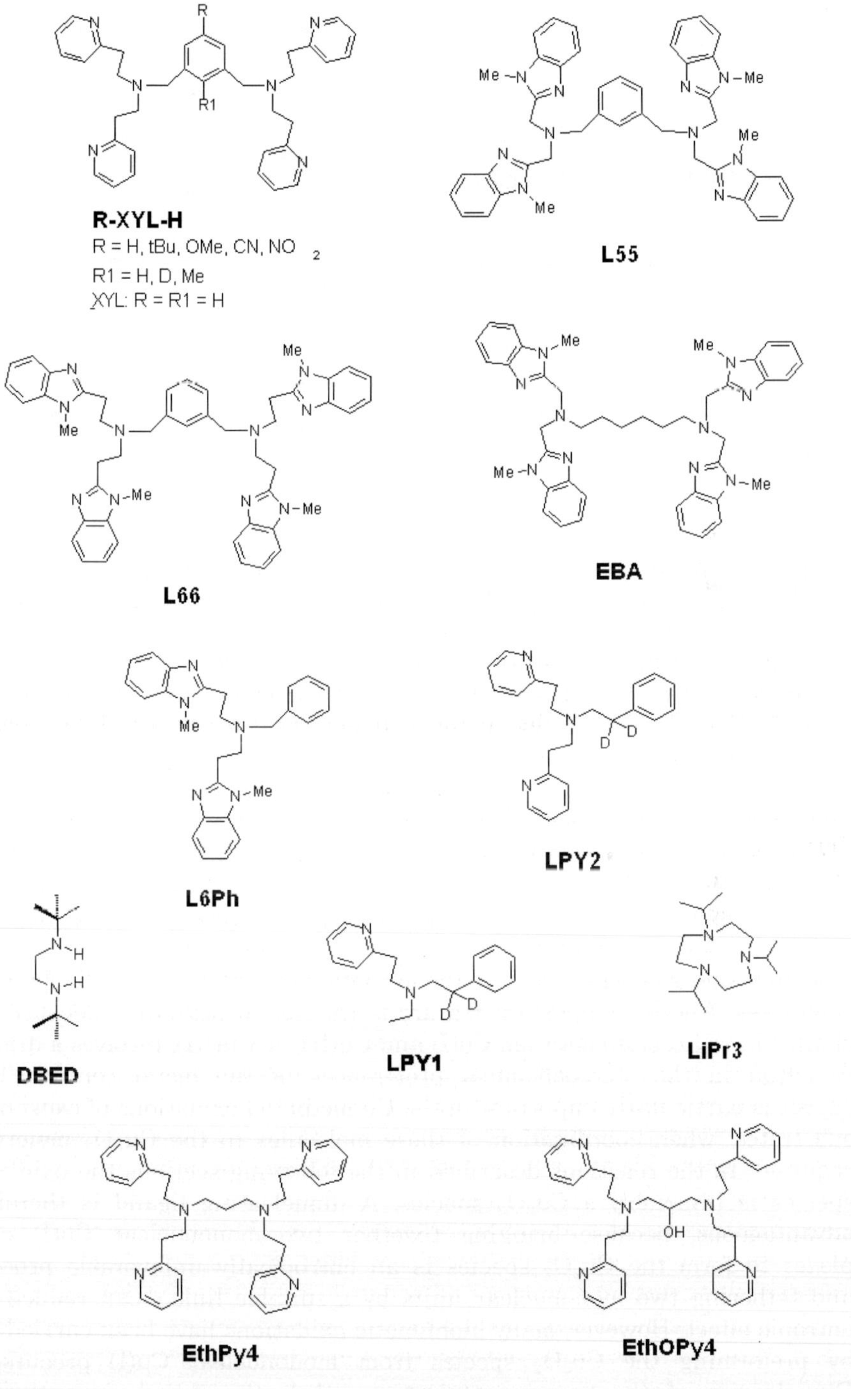

FIG. 4. Ligands used for biomimetic copper complexes of tyrosinase and catechol oxidase.

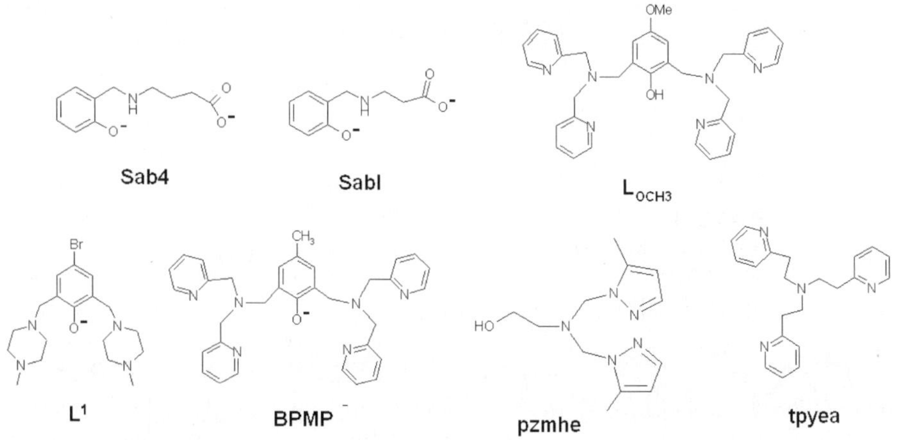

FIG. 4. Continued.

ligands is much easier than that of the parent imidazole derivatives, which are therefore much less frequently employed. Aliphatic amine donors, generally tertiary amine groups, normally occur in the linking units between heterocyclic residues; however, chelating ligands with entirely aliphatic amine donor sets, usually carrying bulky substituents, have been also used recently. The identity of the nitrogen donors affects the redox properties of the copper centers and this is, *inter alia*, an important parameter to take into account when the properties of the biomimetic complexes are considered. For instance, aliphatic amine ligands are strong σ-donors and tend to stabilize the Cu(III) oxidation state, and therefore, in the reaction between Cu(I) complexes and O_2, to form Cu_2O_2 adducts of bis(μ-oxo)dicopper(III) type (unless the presence of bulky substituents prevent a close Cu...Cu approach) (*63*). On the other hand, polydentate ligands rich in heterocyclic nitrogen donors generally lead to the formation of μ-η^2:η^2-peroxodicopper(II) complexes. Another important feature is the ligand flexibility, because the change in redox state between Cu(I) and Cu(II), or Cu(III), involves a drastic variation in the stereochemical preferences of the metal centers. This aspect is particularly important in the Cu mediated oxidations of exogenous substrates, when coordination of these molecules to the Cu_nO_2 moiety is required. In the reactions described in the following sections the oxidizing species is invariably a Cu_2O_2 species. A dinucleating ligand is therefore advantageous, because bringing together two mononuclear Cu(I) complexes to form the Cu_2O_2 species is an entropically unfavorable process, and tethering two mononuclear units by a suitable linker can reduce the entropic effect. However, many biomimetic oxidations have been carried out by preforming the Cu_2O_2 species from mononuclear Cu(I) precursors. Dinuclearity of the copper catalyst, on the other hand, is extremely important in *catalytic* reactions, because it can considerably increase both the thermodynamic and chemical stability of the catalyst, and the kinetics of the process.

III. Phenolase Activity

A. Intramolecular Ligand Oxidations

Except for a few cases (*72–76*), the reaction of dioxygen with Cu(I) complexes generally results, even at relatively low temperature, in the oxidation of the metal ion and intra- or intermolecular processes which usually cause decomposition *via* ligand oxidation. The activation of dioxygen in regiospecific ligand hydroxylation reactions has provided fundamental insight into the mechanism of C–H bond activation at Cu_nO_2 centers and the reactivity of the Cu_nO_2 species. Several mononuclear and dinuclear Cu(I) complexes have been reported which give rise to specific ligand hydroxylations by reaction with O_2 (*61,62,64,66,77,78*). However, only a few of them are particularly relevant to the mechanistic understanding of the phenolase activity of tyrosinase, because their reactions involve an arene residue (*79–91*). In addition, although the number of characterized μ-η^2:η^2-peroxo, or bis(μ-oxo) dicopper species which represent the active intermediates in these reactions is limited, the emerging reactivity pattern displays clearly different behavior for the two types of complexes. In fact, in all cases in which the reactions have been shown, or are likely to be due, to a bis(μ-oxo)dicopper(III) moiety, this species appears to be competent in H atom abstraction and radical type oxidations which are not typical of tyrosinase catalyzed reactions (*67,92–107*).

Particularly interesting in this context have been the dinuclear Cu(I) complexes which react with dioxygen to give Cu_nO_2 adducts and easily undergo hydroxylation at a *m*-xylyl moiety present in the ligand. The most detailed mechanistic studies of this intramolecular hydroxylation have been performed by Karlin and coworkers on the copper complexes derived from the R-XYL-H family of ligands (Fig. 4) (*61,64,79,108–111*). The reaction of $[Cu_2(\text{R-XYL-H})]^{2+}$ complexes with dioxygen at room temperature gives rapidly rise to the hydroxylated complexes $[Cu_2(\text{R-XYL-O})(OH)]^{2+}$ (Scheme 5). Both the oxygen atoms inserted into the product derive from molecular oxygen, as it has been demonstrated with experiments using $^{18}O_2$. Kinetic and spectroscopic studies carried out at low temperature identified the adducts $[Cu_2(\text{R-XYL-H})O_2]^{2+}$ as the reactive intermediates. These intermediates are relatively stabilized by electron withdrawing substituents, such as $R = NO_2$. In this case, Raman spectroscopic data unambiguously established the μ-η^2:η^2-peroxo nature of the bound dioxygen residue (*111*). The mechanism proposed for the xylyl hydroxylation by the peroxo intermediate involves electrophilic attack to the arene π system (Scheme 5). The carbocation formed by electrophilic attack by the peroxo species regains its aromaticity by intramolecular proton transfer. The lack of kinetic isotope effect on the hydroxylation reaction suggests that the initial attack on the aromatic moiety represents the rate-determining step. Interestingly, the oxygenation of the parent copper(I) complex containing a methyl substituent on the xylyl ring position where the attack of the peroxo intermediate occurs, $[Cu_2(\text{R-XYL-Me})]^{2+}$, produces a 1,2 methyl migration (NIH shift),

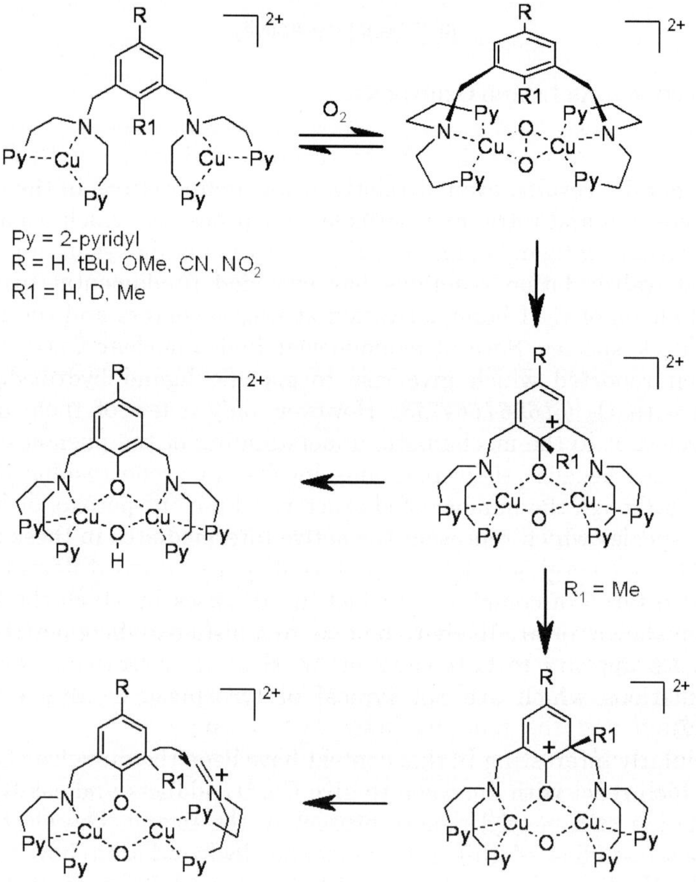

SCHEME 5. Reversible oxygenation of xylyl dicopper(I) complexes with pyridine arms and mechanism of ligand hydroxylation and copper mediated NIH shift (1,2-migration) reactions.

i.e., a carbocation intermediate, which further supports the electrophilic nature of the reaction (Scheme 5) (*112,113*).

Consistent with the electrophilic character of the reaction of the μ-η^2:η^2-peroxo species, the reaction rate constants decrease with the electron withdrawing strength of the substituent in the position *para* to the C–H bond undergoing hydroxylation. The Hammett constant ρ obtained for this reaction is therefore negative, −2.1 at 193 K, but small compared with other electrophilic aromatic substitutions (*109*). The relatively low ρ value appears to be related to the ideal orientation and close proximity of the peroxo group to the aromatic ring (*61*), that can justify the lack of sensitivity towards the change of the ring substituent. Other piece of evidence supports this explanation, such as the small negative volume of activation, which reflects minimal geometrical rearrangement in the transition state (*110*). Theoretical calculations and an orbital model to describe the reaction mechanism have also been reported (see below) (*111,114*).

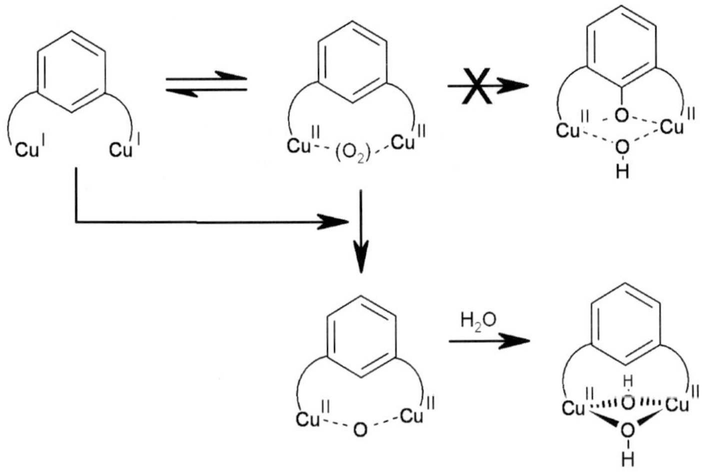

SCHEME 6. Pattern of reactivity of the dioxygen complexes of xylyl dicopper(I) complexes with heterocyclic arms different from pyridines.

Besides the complexes with R-XYL-H, the reactivity of numerous dicopper(I) complexes of related xylyl-containing dinucleating ligands with dioxygen has been described. It is not completely clear why replacement of the pyridine donors with other *N*-heterocycles prevents the ligand hydroxylation and the reaction between the Cu^{I}_{2} species and O_2 invariably yields products resulting from four-electron reduction of O_2, e.g., the bis(hydroxo)dicopper(II) complexes (*85,107,115–120*). This is the case of the dicopper(I) complexes with the ligands L55 (*107*) and L66 (*120*), and the related ligand EBA (*119*), studied by our group, which bear benzimidazole donor groups in the chelating arms. These Cu(I) complexes are able to activate molecular oxygen, as shown by their reactions with external substrates (Section III.B), therefore it is likely that the reactive oxygen intermediates which initially form rapidly react with the copper(I) species present in solution, forming oxodicopper(II), or bis(hydroxo)dicopper(II) complexes (Scheme 6) (*107,119*).

Stereochemical issues are crucial in the ligand hydroxylation reaction. Using frontier molecular orbital theory the pathway for the arene hydroxylation performed by the μ-η^2:η^2-peroxodicopper(II) complex $[Cu_2(R\text{-}XYL\text{-}H)(O_2)]^{2+}$ has been hypothesized (*111*). Although structural data for this intermediate are not available, a conceivable model has been drawn from the atomic coordinates of the complex $[Cu_2(R\text{-}XYL\text{-}O)(OH)]^{2+}$. As shown in Fig. 5, the configuration of the N3CuOOCuN3 unit containing approximately square-pyramidal copper(II) centers, involves a staggered disposition of the two tridentate nitrogen donors and a *trans* arrangement of the axial ligands. The peroxide O–O atoms and the arene C2 and C5 atoms lie approximately on the same line, with the arene ring plane twisted with respect to the Cu_2O_2 plane (by $\sim 35^{\circ}$). Under these conditions, the peroxide π^*_{σ} component of the LUMO can easily overlap with the arene HOMO driving the arene hydroxylation by $[Cu_2(R\text{-}XYL\text{-}H)(O_2)]^{2+}$ at position 2 of the

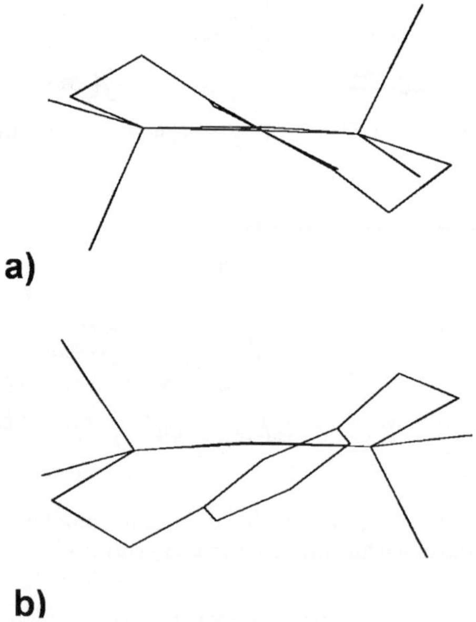

FIG. 5. Structural models of the μ-η^2:η^2-peroxo complexes $[Cu_2(R\text{-}XYL\text{-}H)(O_2)]^{2+}$ (a) and $[Cu_2(L66)(O_2)]^{2+}$ (b) obtained from the atomic coordinates of $[Cu_2(R\text{-}XYL\text{-}O)(OH)]^{2+}$, and $[Cu_2(L66\text{-}O)(OH)]^{2+}$, respectively.

xylyl ring. On the other hand, despite the μ-η^2:η^2-peroxo complex $[Cu_2(L66)(O_2)]^{2+}$ exhibits similar features to $[Cu_2(R\text{-}XYL\text{-}H)(O_2)]^{2+}$ (*121*), it does not undergo the ligand hydroxylation. The recently reported coordinates of the complex $[Cu_2(L66\text{-}O)(OH)]^{2+}$ (*90*) can be used to apply the same structural model of $[Cu_2(L66)(O_2)]^{2+}$, as for $[Cu_2(R\text{-}XYL\text{-}H)(O_2)]^{2+}$ (Fig. 5). It is interesting to note that although the structures of the two complexes are very similar in the copper coordination units and in their staggered *trans*-arrangement, the arene ring of the xylyl linker is markedly tilted with respect to the Cu_2O_2 plane in $[Cu_2(L66)(O_2)]^{2+}$. The five-membered rings of the benzimidazole chelating arms (implying different bite angles than the six-membered ring pyridines) and their steric hindrance are probably responsible for the geometry difference between the two complexes. Therefore, the superimposition of the dioxygen LUMO and the arene HOMO is disrupted in $[Cu_2(L66)(O_2)]^{2+}$, and the hydroxylation through the proposed pathway cannot take place.

The reaction of the dicopper(II) complex of XYL with H_2O_2 also undergoes hydroxylation at position 2 of the xylyl ring to form $[Cu_2(XYL\text{-}O)(OH)]^{2+}$ (*122*). The kinetic data are pH dependent and intriguingly indicate a second-order dependence on the dicopper(II) complex, suggesting that a tetracopper intermediate must be formed. The (μ-1,1-hydroperoxo)dicopper(II) complex which is initially formed may interact with a second dicopper(II) complex, acting as a Lewis acid to increase the nucleophilicity of the terminal OH group, thus allowing the remaining oxygen atom to react as an electrophile (*61*). In the related L66 system, the reactivity of $[Cu_2(L66)]^{4+}$ with dihydrogen

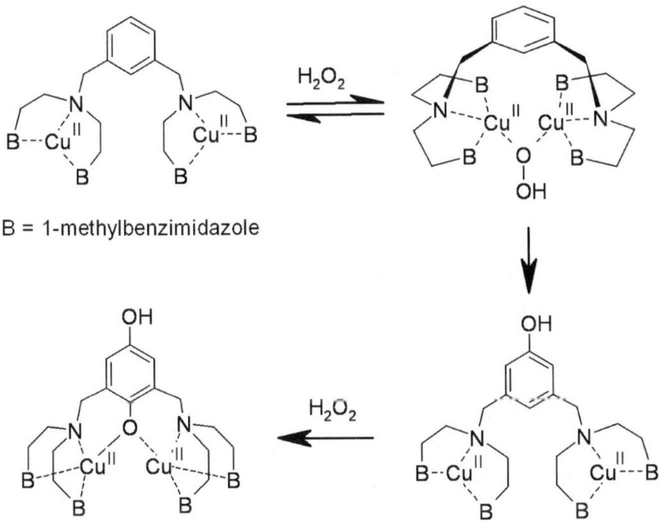

SCHEME 7. Double ligand hydroxylation undergone by $[Cu_2(L66)]^{4+}$ upon reaction with dihydrogen peroxide.

peroxide is completely different. This reaction, in fact, produces a double arene hydroxylation yielding the μ-(4-hydroxy)phenolate (μ-hydroquinone) complex $[Cu_2(HO\text{-}L66\text{-}O)]^{3+}$ (*90*). Using labeled $H_2{}^{18}O_2$, the incorporation in the product of both oxygen atoms from the peroxide has been demonstrated. This extremely facile reaction occurs in two different steps (Scheme 7), which follow first-order behavior with respect to the dicopper complexes and saturation behavior with respect to dihydrogen peroxide, respectively. The first step is slower and proceeds through a dicopper(II)-hydroperoxide complex, which evolves to the monohydroxylated intermediate. Thermodynamic data ($\Delta H^{\neq} = 39.1 \pm 0.9$ kJ mol^{-1}, $\Delta S^{\neq} = -115.0 \pm 2.4$ J K^{-1} mol^{-1}) indicate a highly constrained transition state that evolves through a concerted bond-making and bond-breaking process accompanying the oxygen transfer. The second step is about 25 times faster at room temperature ($\Delta H^{\neq} = 77.8 \pm 1.6$ kJ mol^{-1}; $\Delta S^{\neq} = -14.0 \pm 0.4$ J K^{-1} mol^{-1}) which can be explained considering that the ligand after the first hydroxylation is considerably electron richer. The reduced absolute entropic value associated with the second hydroxylation is conceivably associated to a less constrained hydroperoxo intermediate. The binding constants of the peroxide extrapolated from saturation kinetics are very similar, indicating that H_2O_2 binds similarly to the starting complex and the monohydroxylated intermediate, and that the two species must have the same ionic charge. Notably, for the first hydroxylation, the pH dependence of the rate has an asymmetric bell shape with a maximum corresponding to the conditions where the monohydroxo species $[Cu_2(L66)(OH)]^{3+}$ is maximized. This facilitates the binding of dihydrogen peroxide to the dicopper(II) complex as a hydroperoxide and maintains sufficient conformational mobility to allow an efficient second aromatic hydroxylation.

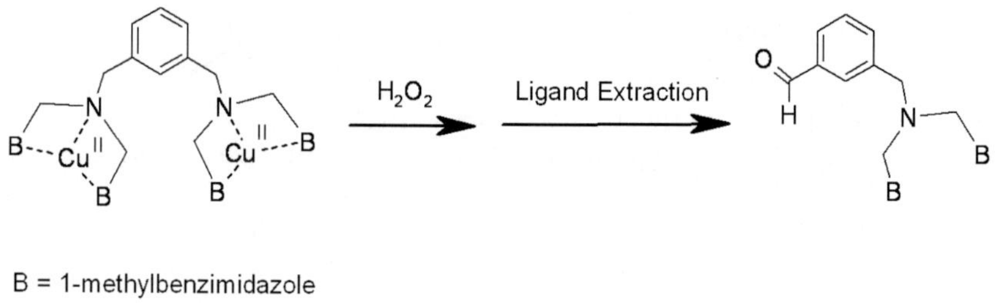

SCHEME 8. Ligand oxidative degradation undergone by $[Cu_2(L55)]^{4+}$ upon reaction with dihydrogen peroxide.

The hydroperoxo complex of $[Cu_2(L66)]^{4+}$ has been spectroscopically characterized at low temperature. Its spectral features consist of a moderately intense LMCT band at 342 nm ($\Delta\varepsilon \approx 8600\,M^{-1}cm^{-1}$) and two weaker bands at 444 nm ($\Delta\varepsilon \approx 850\,M^{-1}cm^{-1}$) and 610 nm ($\Delta\varepsilon \approx 400\,M^{-1}cm^{-1}$). This pattern of LMCT bands differs from those exhibited by mononuclear copper(II)-hydroperoxo complexes and suggests a bridging η^1:η^1 binding mode of the ligand (*90*). The hydroperoxo intermediate was also characterized by EPR spectroscopy, where a broad and unresolved signal at $g \approx 2.12$ is consistent with a weakly antiferromagnetically coupled dicopper(II) center by a single bridging ligand. It has been proposed that the disposition of the bridging hydroperoxo moiety can lead to a direct overlap between the arene HOMO and the peroxide σ^* component. Taking into account the orbital overlap, this reasonably explains the regiochemistry of the first hydroxylation at position 5 of the xylyl residue.

The structural difference imposed by the smaller metal chelate rings in the complex $[Cu_2(L55)]^{4+}$, five-membered instead of six-membered as in $[Cu_2(L66)]^{4+}$, determine a completely different reactivity behavior with dihydrogen peroxide (*107*). The complex $[Cu_2(L55)]^{4+}$, undergoes a ligand hydroxylation reaction in the presence of H_2O_2, but this occurs at a benzylic position and with much slower rate (Scheme 8). The involvement of a hydroperoxo complex of $[Cu_2(L55)]^{4+}$ has been suggested for this reaction, though no intermediate has been characterized yet.

B. REACTIONS ON EXOGENOUS PHENOLIC SUBSTRATES

The reactions of μ-η^2:η^2-peroxodicopper(II) and bis(μ-oxo)dicopper(III) complexes with exogenous substrates are particularly relevant for the understanding of enzymes catalysis. Comparative studies using different Cu_2O_2 complexes have revealed notable changes in reactivity with external substrates and provided mechanistic insight into synthetic and biological oxidation processes. Among the large number of synthetic copper(I) complexes producing relatively stable adducts of μ-η^2:η^2-peroxo, μ-1,2-peroxo, or bis(μ-oxo) types by reaction with O_2 at low temperature (*61–63,66,77,78,123*), many undergo ligand hydroxylation, or biologically less relevant reactions like radical oxidations, or hydrogen atom abstractions that normally lead to C–C coupling products. The reactions with phenols, or phenolates have attracted particular attention because of their relevance to the most peculiar catalytic transformation promoted by tyrosinase (*64,124–133*). Formation of *ortho*-catechol, or *ortho*-quinone by phenol oxidation reactions mediated by copper(I) complexes and dioxygen has been documented in a number of cases (*134–137*), but direct information on the reaction mechanism and spectroscopic data on the Cu_2O_2 intermediates involved are generally lacking.

The few characterized copper-dioxygen adducts capable of mediating regiospecific hydroxylation of an exogenous phenol in the *ortho* position have been reported to be only μ-η^2:η^2-peroxodicopper(II) species (*121,138–140*). These include the complexes derived from the dinucleating ligand L66 and from the mononucleating L6Ph, LPy2, and DBED. For structural reasons (*63*), these ligands induce the formation of μ-η^2:η^2-peroxo species, while the bis(μ-oxo) isomers are non-detectable in solution.

The complex $[Cu_2(L66)]^{2+}$ is the first one for which it has been possible to reproduce the phenolase activity of tyrosinase (*121*). The complex reacts with dioxygen at low temperature yielding an adduct with spectral characteristics (LMCT bands at 362 nm, $\varepsilon = 15{,}000\ M^{-1}cm^{-1}$, 455 nm, $\varepsilon = 2000\ M^{-1}cm^{-1}$ and 550 nm, $\varepsilon = 900\ M^{-1}cm^{-1}$) and resonance Raman pattern (760 cm^{-1}, shifted to 719 cm^{-1} with $^{18}O_2$) which unambiguously identify it as a side-on μ-η^2:η^2-peroxodicopper(II) species (Scheme 9). The dioxygen binding constant ($K = 1.4 \times 10^4\ M^{-1}$ at 195 K) is about three orders of magnitude smaller than that reported for the oxygenation of the parent $[Cu_2(XYL)]^{2+}$ complex (*61*). The difference may be due to the larger size of the benzimidazole groups which hinder the conformational rearrangement of the Cu(I) complex upon

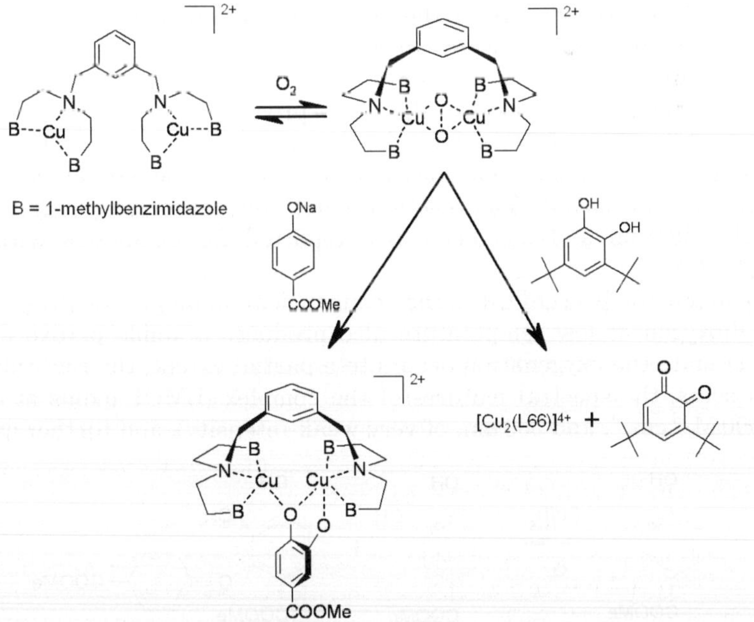

SCHEME 9. Reversible oxygenation, and phenol hydroxylation and catechol oxidation promoted by the complex $[Cu_2(L66)]^{2+}$ at low temperatures.

formation of the bridged peroxodicopper(II) species. As discussed in Section III.A. $[Cu_2(XYL)]^{2+}$ does not form a stable dioxygen adduct and undergoes a fast ligand hydroxylation, while the peroxo complex of $[Cu_2(L66)]^{2+}$ is a reactive species that can be used towards exogenous substrates.

In fact, when 4-carbomethoxy-phenolate is allowed to react with $[Cu_2(L66)(O_2)]^{2+}$ at –60°C in acetone the reaction takes place with replacement of the 362 nm peroxo band with a weaker band near 340 nm, which is characteristic of the corresponding catecholate adduct of $[Cu_2(L66)]^{4+}$ (Scheme 9), previously reported by our group (*136*). The conversion of the phenolate to catecholate (in 40 % yield with respect of the dioxygen adduct, using a stoichiometric amount of 4-carbomethoxy-phenolate with respect to the complex) has been also proven by workup of the reaction mixture. The oxidation of 3,5-di-*tert*-butyl-catechol to the quinone by $[Cu_2(L66)(O_2)]^{2+}$ was also studied at low temperature (Scheme 9). In this case, the decrease of the copper-peroxide band at 362 nm and the development of the quinone band at 420 nm follow the same time profile ($k_{obs} = 2.8 \times 10^{-3}\,s^{-1}$ and $2.6 \times 10^{-3}\,s^{-1}$, respectively), indicating that the reaction of the peroxo species is rate limiting. The process in fact corresponds to one catecholase-like cycle and occurs with a stoichiometry of 2 catechol:1 O_2 (reaction 9 and Scheme 2). No dimers arising from C–C coupling, or C–O coupling reactions were detected in the oxidation of both substrates.

Indeed, the oxygenation of 4-carbomethoxy-phenolate by $[Cu_2(L66)]^{2+}$ and $[Cu_2(L55)]^{2+}$ was previously reported to occur also in acetonitrile at –40°C, where the reactions were normally carried out by exposing the Cu(I)-phenolate adduct to dioxygen (*136*). Since acetonitrile freezes above the temperature range of stability of the copper-peroxo complex, it could not be established whether the peroxo species binds and reacts directly with the phenolate in these systems. Moreover, when the phenol hydroxylation is carried out at room temperature, the main product obtained is the dimeric compound resulting from a formal Michael addition of the phenolate on 3-carbomethoxyquinone, that probably occurs on the bound catecholate (Scheme 10) (*134,136,137*). The reaction is not unusual for copper(I) mediated phenol oxidation and has been also reported for complexes with other ligands (*135*).

Treatment of $[Cu(L6Ph)]^{+}$, the mononuclear analogue of $[Cu_2(L66)]^{2+}$, with dioxygen at low temperature also produces a stable peroxo complex (*140*). Despite the oxygenation occurs to a partial extent, the reversibility of the process, the spectral features of the complex (LMCT bands at 356 nm, ε 20,000 $M^{-1}\,cm^{-1}$, and 560 nm, of very weak intensity), and further evidence

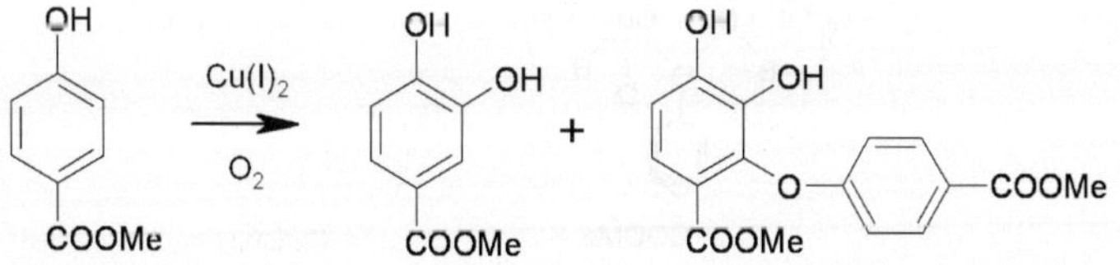

SCHEME 10. Products of the hydroxylation of 4-carbomethoxy-phenolate by $[Cu_2(L66)]^{2+}$ and dioxygen at room temperature.

indicates that peroxide binds in a side-on μ-η^2:η^2 fashion to a dimer of the complex. The spectral pattern also suggests that this Cu_2O_2 core must be highly planar, as it is in tyrosinase, or hemocyanin, with the absence of "butterfly" bending, or bis(μ-oxo)dicopper(III) isoforms (*6*). Upon reacting the dioxygen complex $[Cu_2(L6Ph)_2O_2]^{2+}$ with 4-carbomethoxy-phenolate in acetone at −85°C, the UV band at 356 nm rapidly disappears. This fast reaction is followed by a slower process, characterized by an increase in absorbance at 334 nm, due to the formation of the product copper(II)-catecholate complex. The 3-carbomethoxy-catechol product can be isolated in relatively yield (88 % based on the theoretical amount of $[Cu_2(L6Ph)_2O_2]^{2+}$), and isotope labeling experiment using $^{18}O_2$ confirmed the origin of the oxygen atom inserted into the phenol. The higher conversion of phenol into catechol obtained in the reaction with $[Cu_2(L6Ph)_2O_2]^{2+}$, with respect of the corresponding reaction by $[Cu_2(L66)O_2]^{2+}$, can be explained considering the higher accessibility of the phenolate to the Cu_2O_2 core of the L6Ph complex, where the ligand benzyl substituents are likely arranged in a trans disposition. The situation is different in the corresponding Cu_2O_2 complex derived from L66, where the tridentate amino-bis(benzimidazole) units are connected (with an obligatory *cis* arrangement) by the rigid xylyl linker. Kinetic measurements were carried out to characterize the two steps of the phenol hydroxylation promoted by $[Cu_2(L6Ph)_2O_2]^{2+}$ (Scheme 11). This is based on the electrophilic attack of the μ-η^2:η^2-dicopper(II) intermediate on the phenolate. The limited amount of $[Cu_2(L6Ph)_2O_2]^{2+}$ complex initially present is rapidly consumed by reaction with the phenolate and does not accumulate further in the solution. The high reactivity of the $[Cu_2(L6Ph)_2O_2]^{2+}$ complex ($k_2 = 359\,M^{-1}s^{-1}$) and the similarity between the value of the oxygenation rate ($k_1[O_2] = 1.1 \times 10^{-3}\,s^{-1}$) with the second step reaction, followed at 334 nm, ($k_r = 8.9 \times 10^{-4}\,s^{-1}$) suggest continuous formation and rapid reaction of the peroxodicopper(II) complex with

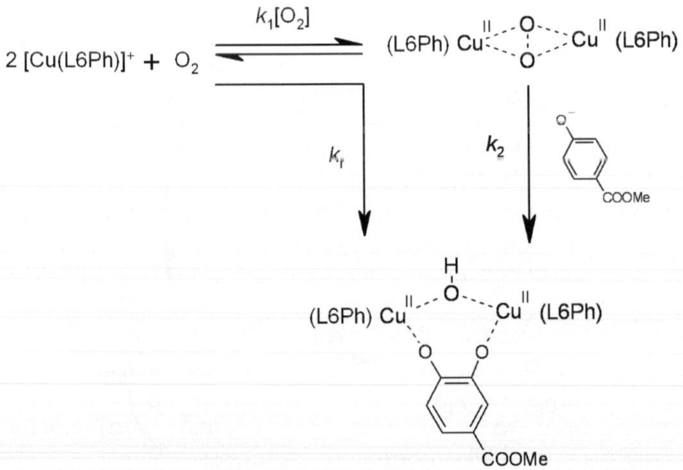

SCHEME 11. The two steps of the phenol hydroxylation promoted by the complex $[Cu(L6Ph)]^+$ in the presence of dioxygen at −85°C.

phenolate. This shifts the dioxygen binding equilibrium driving the formation of more peroxo intermediate (Scheme 11).

Extensive mechanistic information on the phenolate hydroxylation by a μ-η^2:η^2-peroxodicopper(II) complex has been provided by the work of Itoh and coworkers describing the conversion of a series of *para*-X-$C_6H_4O^-Li^+$ substrates to the corresponding catechols by reaction with the Cu_2O_2 adduct derived from the copper(I) complex with the ligand LPy2 (*138*). The dinuclear peroxodicopper(II) complex is relatively stable at low temperature and could be characterized. The Li salts of *para*-X-substituted phenols (X = *t*-Bu, Me, Cl, Br, CO_2Me) react at −94°C with the peroxo complex to give the corresponding catechols in good yield (99, 69, 90, 79, 60 %, respectively, based on the peroxo intermediate, using an excess of phenolate). Also in this case the reaction is highly selective and no C–C phenol coupling product was formed as a by-product. Isotope labeling experiment using $^{18}O_2$ confirmed the origin of the oxygen atom inserted into the substrates. Kinetic experiments carried out at low temperature showed a pseudo first-order behavior and, from the variation of the observed rates with the substrate concentration, Michaelis-Menten type saturation curves were obtained. This strongly indicates that a reversible complex between the phenolic substrate and the peroxo intermediate is formed prior to the oxygenation step. The proposed reaction mechanism as represented in Scheme 12. The formation constant of the active complexes K_f and the rate constant k_{ox} for the oxidation process vary from 465 to 940 M^{-1} and from 0.083 to 0.93 s^{-1}, respectively. Comparison of the association constants and hydroxylation rates for the series of substrates showed little effect on K_f, but revealed a clear influence of the phenol *para*-X substituent on k_{ox}, with electron donation increasing the hydroxylation rate constant. Interestingly, a Hammett plot of the *para*-X substituent effects gave a ρ value (−1.8) which

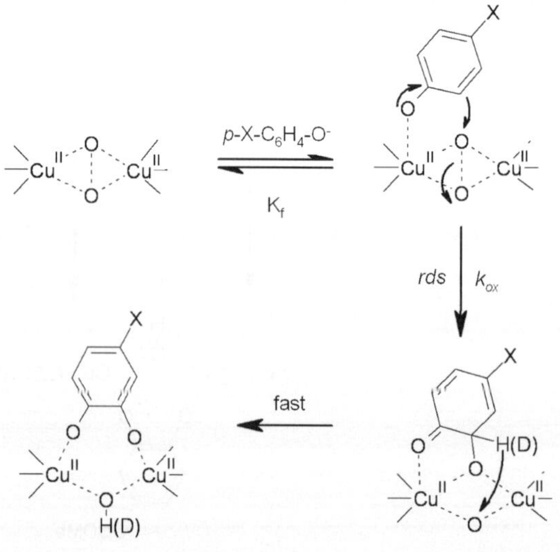

SCHEME 12. Mechanism of hydroxylation of a series of *para*-X-$C_6H_4O^-Li^+$ derivatives by the dioxygen complex $[Cu_2(LPy2)_2(O_2)]^{2+}$ at −94°C.

is very similar to the reaction constant recently determined by the same group on a similar series of phenolic substrates using tyrosinase (−2.4) (*36*) and fairly close to that of the ligand hydroxylation reaction exhibited by $[Cu_2(XYL)]^{2+}$ (−2.1) (*113*). The negative value of these constants for both the enzymatic and biomimetic reactions is consistent with the electrophilic aromatic substitution character for the phenol hydroxylation. Furthermore, no kinetic deuterium isotope effect was found, as in the case of the enzymatic reaction, when p-Cl-C_6D_4OLi was used as substrate. All the results are thus consistent with the mechanism reported in (Scheme 12), where the oxygenation of the phenolate occurs in the rate determining step (r.d.s.) and C–O bond formation occurs simultaneously with O–O bond cleavage in the intermediate. The proton transfer from the aromatic nucleus of the substrate to the Cu-bound oxygen atom follows in a fast step. Use of a phenolate instead of a phenol is essential in the biomimetic reaction, both for binding to the Cu_2O_2 species and for electronic activation. In fact, when a phenol is used in the reaction, instead of the phenolate salt, only the phenol C–C coupling dimer was obtained (see below).

The Cu(I) complex with the didentate secondary diamine ligand DBED has been shown to form a Cu_2O_2 species and activate it towards exogenous substrates in tyrosinase-like reactions (*139*). At low temperature $[Cu(DBED)]^+$ undergoes reversible oxygenation and the spectroscopic and analytical data (stoichiometry 2:1 Cu/O_2) strongly support the formation of the dioxygen complex $[Cu_2(DBED)_2O_2]^{2+}$ with μ-η^2:η^2-peroxo character (Scheme 13). This is confirmed by the characteristic LMCT bands at 350 nm, resonance Raman shift (721 cm^{-1}, shifted to 681 cm^{-1} with $^{18}O_2$) and Cu–Cu distance (3.45 Å) calculated from EXAFS data. The steric bulk by the *tert*-butyl substituent on the nitrogen donor bias the formation of μ-η^2:η^2-peroxodicopper(II) rather than bis(μ-oxo)dicopper(III) complex. The stability

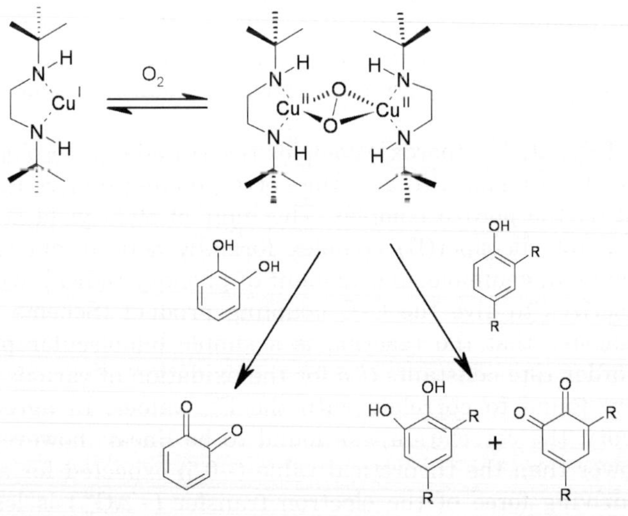

SCHEME 13. Oxygenation, phenol hydroxylation, and catechol oxidation by the complex $[Cu(DBED)]^+$ at −80°C.

of $[Cu_2(DBED)_2O_2]^{2+}$ allowed the reactivity towards exogenous substrates to be studied at low temperature. In particular, the peroxo complex reacts rapidly with sodium and lithium salts of 2,4-di-*tert*-butylphenol at 193 K producing a 1:1 mixture of 3,5-di-*tert*-1,2-benzoquinone and 3,5-di-*tert*-butylcatechol (80 % yield with respect to the oxidizing equivalents of the peroxo complex) (Scheme 13). The use of $^{18}O_2$ cleanly supports the direct oxygen-atom transfer to the phenolic substrate. In addition, $[Cu_2(DBED)_2O_2]^{2+}$ reacts with catechol giving the quinone in good yield (95 %) without any sign of radical oxidation. Not surprisingly, $[Cu_2(DBED)_2O_2]^{2+}$ is unreactive towards 2,4-di-*tert*-butylphenol, supporting the attenuated radical oxidation ability of peroxo complexes.

The reactions of neutral phenols with two distinct dicopper-dioxygen complexes, a μ-η^2:η^2-peroxodicopper(II) complex and a bis(μ-oxo)dicopper(III) complex, have been recently compared in a systematic study using the copper(I) complexes with the two related ligands LPy2 and LPy1 (*141*). The Cu(I) complex with LPy2 yields predominantly the μ-η^2:η^2-peroxo $[Cu_2(LPy2)_2O_2]^{2+}$ complex, while the Cu(I) complex with LPy1 yields the bis(μ-oxo) $[Cu_2(LPy1)_2O_2]^{2+}$ species. Since Tolman *et al.* discovered the existence of equilibria between peroxo and bis(μ-oxo) intermediates (*142*), the factors controlling which isomer predominates in the Cu_2O_2 complexes have been extensively studied and understood (*63,143,144*). Here, reducing the coordination capability of the LPy2 ligand, upon removal of one of the pyridine arms, the equilibrium between the isomeric Cu_2O_2 complexes is shifted from the μ-η^2:η^2-peroxo to the bis(μ-oxo) species (*132*). A quantitative comparison of the oxidizing ability of these two species is quite difficult, due to the fast equilibrium between them. In fact, both dioxygen complexes $[Cu_2(LPy1)_2(O)_2]^{2+}$ (oxo) and $[Cu_2(LPy2)_2O_2]^{2+}$ (peroxo) react with a series of neutral phenols under identical conditions yielding C–C coupling dimers as the major products. The formation of these dimers indicates that phenoxy radicals are produced as intermediates. Then, these experiments do not make clear which are the factors addressing the reaction pathway towards catechol formation (hydroxylation), or phenoxy radical formation and C–C coupling.

The $[Cu_2(LPy2)_2O_2]^{2+}$ (peroxo) complex reacts under anaerobic conditions with 4-substituted phenols to give the corresponding dimers in 30 to 50 % yield, based on the peroxo complex. The limit of 50 % yield suggests that the μ-η^2:η^2-peroxodicopper(II) complex formally acts as one-electron oxidant producing an equimolecular amount of phenoxy radical, which spontaneously dimerizes to give the C–C coupling product (Scheme 14). Kinetic analysis indicates that the reaction is a simple bimolecular process, and the second-order rate constants (k_2) for the oxidation of various substituted phenols were found to correlate with the E^0_{ox} values. In agreement with Marcus theory, the correlation was found to be linear; however, the slope (−0.72) is lower than the theoretical value (−0.5) expected for a process in which the driving force of the electron transfer ($-\Delta G^0_{et}$) is less than the reorganization energy. Therefore, it is conceivable that the reaction proceeds with a mechanism in which the electron transfer is coupled with a proton

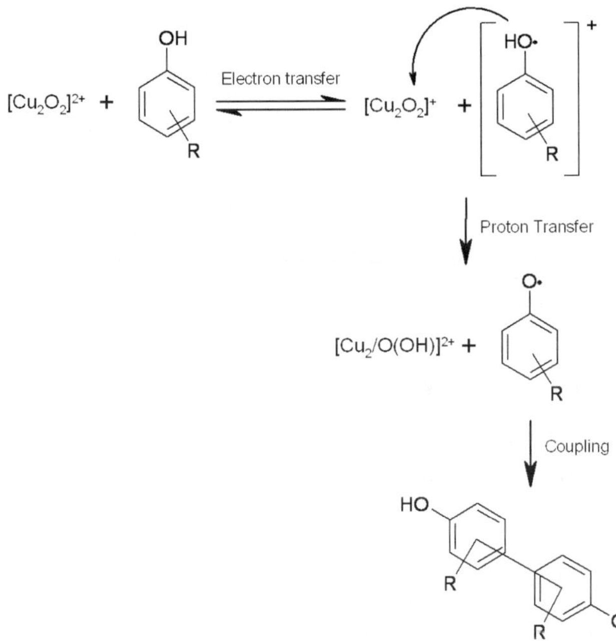

SCHEME 14. PCET mechanism of the phenol coupling reaction promoted by the Cu_2O_2 complexes $[Cu_2(LPy1)_2O_2]^{2+}$ (oxo) and $[Cu_2(LPy2)_2O_2]^{2+}$ (peroxo).

transfer (PCET), as illustrated in (Scheme 14). The concerted mechanism justifies the slope of the Marcus plot, which is between the values for a pure electron transfer r.d.s. (–0.5) and for a pure proton transfer r.d.s. (–1.0). Kinetic deuterium isotope effects measured for several substrates confirmed the involvement of the proton transfer in the r.d.s. and were comparable with the data obtained for other PCET reactions (*141*). The kinetic data obtained with the $[Cu_2(LPy1)_2O_2]^{2+}$ (oxo) complex and the deuterium isotope effects indicate that the electron transfer from the phenol to the dioxygen complex is also coupled with the subsequent proton transfer (Scheme 14), as in the previous case. The yields of C–C coupling products remain close to 50 %, confirming that the bis(μ-oxo)dicopper(III) complex also acts as one-electron oxidant of the phenol. The main difference in the behavior of $[Cu_2(LPy1)_2(O)_2]^{2+}$ (oxo) and $[Cu_2(LPy2)_2O_2]^{2+}$ (peroxo) complexes was found in the rate constants k_2. With the same substrate, the k_2 value for the $[Cu_2(LPy1)_2(O)_2]^{2+}$ (oxo) complex weas about two orders of magnitude larger than the corresponding value obtained with the $[Cu_2(LPy2)_2O_2]^{2+}$ (peroxo) complex.

The PCET mechanism for the two Cu_2O_2 complexes was further confirmed using cumylperoxyl radical, a well-known hydrogen atom transfer (HAT) reagent. The activity of this radical towards the same series of phenolic substrates in the reaction with *N,N*-dimethylaniline exhibits a completely different dependence from the oxidation potentials, in contrast with the one-electron oxidation of the phenols by the Cu_2O_2 complexes. Given that both the bis(oxo) and peroxo compounds oxidize neutral phenols following

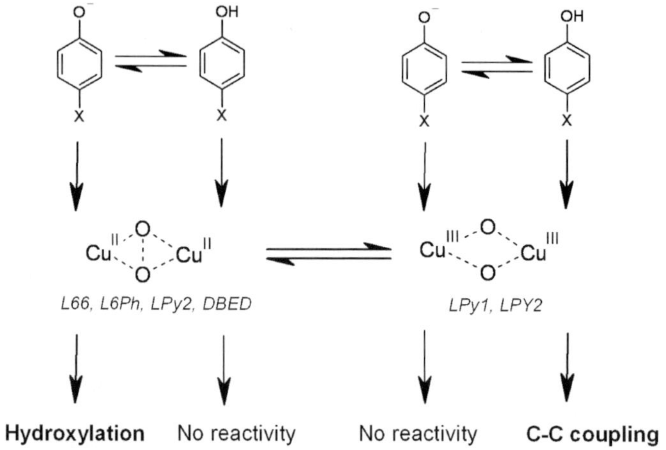

SCHEME 15. Pattern of reactivity of μ-η^2:η^2-peroxodicopper(II) and bis (μ-oxo)dicopper(III) complexes towards phenols and phenolate ions.

a pure PCET mechanism, the differences in the rate constants can be attributed to the difference in the inherent reactivity between the two different moieties. In such a case, the one-electron reduction potential (E^0_{red}) of the bis(oxo) complex should be ~0.1 V higher than that of the peroxo complex. Another possible explanation, however, is that the bis (μ-oxo)dicopper(III) complex is the real active species in both cases. Therefore, the difference in the rates can be attributed to the difference in the relative amount of the two species in the fast bis(oxo)/peroxo equilibria for the two systems. This is consistent with the general trend of reactivity that can be deduced from the examples reported in the literature so far (Scheme 15). Thus, the bis(μ-oxo)dicopper(III) complex coexists as a minor product when the μ-η^2:η^2-peroxodicopper(II) complex is prepared using $(PyCH_2CH_2)_2NR$ tridentate ligand (*63,130*). If the peroxo complex was the active species for the PCET oxidation of the phenol, the reaction of $[Cu_2(LPy2)_2O_2]^{2+}$ with lithium phenolate, having much more negative oxidation potential than the neutral form, would proceed *via* outer sphere electron transfer too. This is inconsistent with the previous data demonstrating that the oxygenation of phenolate by the μ-η^2:η^2-peroxodicopper(II) complex proceeds by an electrophilic aromatic substitution mechanism (Scheme 15). Thus, the bis(μ-oxo)dicopper(III) complex is like the active species in the phenol C–C coupling reaction. This appears to be the reason why the enzyme tyrosinase keeps the Cu... Cu distance at 3.5 Å, which can only accommodate the μ-η^2:η^2-peroxodicopper(II) form, and assists the deprotonation of the phenol in order to produce a highly efficient hydroxylation reaction.

IV. Catecholase Activity

The catecholase reaction 9 is energetically simpler than the phenolase reaction 8 and is therefore promoted by several copper(II) complexes in various conditions (*119,121,137,139,145–157*). Though, the catecholase mechanism is more complicated than the phenolase mechanism (compare Scheme 1 and 2) and, for this reason, in most cases very little insight is given on the reaction in support of a speculative mechanism. Genuine model systems for the enzymatic catecholase activity should form active Cu_2O_2 species and the reaction should conform to the stoichiometry of reaction 9, with O_2 reduced to water. The alternative would be the oxidation of one catechol molecule and reduction of O_2 to dihydrogen peroxide, as it has been established in some cases (*152,153,155*), while proof that the reaction occurs with complete reduction of O_2 to water has been obtained only in very few cases (*119,137*). Furthermore, the catecholase activity of some well-defined Cu_2O_2 adducts has been shown to proceed through a radical pathway, hence diminishing their relevance as models for the activity of tyrosinase and catechol oxidase (*128,158*).

Among the diphenols, 3,5-di-*tert*-butylcatechol ($DTBCH_2$) is widely used as a substrate in model reactions, due to its low redox potential and to the relative stability of the corresponding quinone (DTBQ). The most systematic and detailed mechanistic studies on the catalytic oxidation of $DTBCH_2$ have been reported by our group using the dicopper(II) complexes derived from the ligands L55, L66, and EBA (*119,121,137,154*). For these dinuclear copper complexes the oxidation rates were found to depend linearly on the catalyst concentration, showing that the reaction needs the cooperation of two copper centers, quite likely for the need to bind the substrate in a bridging mode. With the aim of obtaining insight into the mechanism of the catecholase reaction, we studied the catalytic reaction in various experimental conditions. When the catecholase reaction of $DTBCH_2$ was performed in a solvent mixture of methanol containing aqueous buffer at about pH 5, the reaction was slowed down sufficiently to allow the observation of a biphasic behavior in the curve of quinone produced vs. time. The reaction, in fact, occurs with an initial fast and stoichiometric step followed by a slower catalytic phase. The first step can be described as:

$$[Cu^{II}{-}OH - Cu^{II}]^{3+} + DTBCH_2 \rightleftarrows [Cu_2^{II}/DTBC]^{2+} + H_2O + H^+ \qquad (12)$$

$$[Cu_2^{II}/DTBC]^{2+} \longrightarrow [Cu_2^{I}]^{2+} + DTBQ \qquad (13)$$

and requires the binding of the catechol to the dicopper(II) form of the complex. Such adducts have been isolated and spectroscopically characterized for L55 and L66 complexes (*136,157*). The binding mode of the catechol residue was previously proposed to be $\eta^1{:}\eta^1$ (or μ-1,4), similar to that found in related dinuclear Cu(II) systems (*149,159*), but it is more likely $\eta^2{:}\eta^1$, as recently suggested by resonance Raman data (*38*) (Scheme 16). The latter binding mode resembles that found in the crystal structure of the catechol oxidase complex with phenylthiourea (*25,26*) (Fig. 2).

The initial step in the catechol oxidation is fast, basically for the high reactivity of the copper(II) complex toward $DTBCH_2$. Since quinone formation occurs after a pre-equilibrium binding of the substrate, the reaction

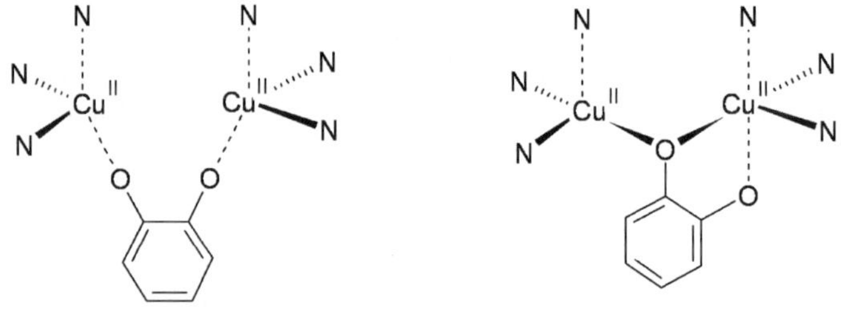

$\eta^1{:}\eta^1$ $\eta^2{:}\eta^1$

SCHEME 16. The $\eta^1{:}\eta^1$ (or μ-1,4) and $\eta^2{:}\eta^1$ binding modes of catecholate residues as bridging ligands at dicopper(II) centers.

shows $DTBCH_2$ saturation behavior, thus allowing the determination of the kinetic parameters K_M (substrate dissociation constant for reaction 12) and k_{cat} (oxidation rate of reaction 13). The data obtained for the various dinuclear complexes are collected in Table I. In the case of $[Cu_2(EBA)]^{4+}$, the plots show substrate inhibition at high catechol concentration. This behavior is explained considering that in the presence of a large excess of $DTBCH_2$ coordination of two of these molecules to the dicopper(II) complex occurs, each one chelating to a single metal center (Scheme 17). The two Cu(II) centers are forced to stay far from each other, preventing the possibility of the two-electron oxidation process to the bound catecholate ion, that is required to yield DTBQ in the first step.

The second phase of the reaction observed in the optical trace vs. time is the global catalytic cycle, which contains reactions 12 and 13 and the oxygenation of the dicopper(I) complex. The closure of the cycle can be achieved in different ways, which are outlined as path *a*, paths *b*+*d*, and path *c* in Scheme 17. What discriminates between them is the reactivity of the Cu_2O_2 species, since it can react with protons according to path *a* and *b*, or oxidize the catechol according to path *c*. For the reactions catalyzed by the Cu(II) cmplexes of L55, L66, and EBA no detectable amount of dihydrogen peroxide was found in solution; therefore, the only possible path is *c*. Thus, the catalytic cycle can be represented by the following reactions:

$$[Cu^{II}{-}OH - Cu^{II}]^{3+} + DTBCH_2 \rightleftarrows [Cu_2^{II}/DTBC]^{2+} + H_2O + H^+ \quad (12)$$

$$[Cu_2^{II}/DTBC]^{2+} \longrightarrow [Cu_2^{I}]^{2+} + DTBQ \quad (13)$$

$$[Cu_2^{I}]^{2+} + O_2 \rightleftharpoons [Cu_2/O_2]^{2+} \quad (14)$$

$$[Cu_2/O_2]^{2+} + DTBCH_2 \rightleftarrows [Cu_2/O_2/DTBC] + 2H^+ \quad (15)$$

$$[Cu_2/O_2/DTBC] \longrightarrow \xrightarrow{3H^+} [Cu^{II}{-}OH - Cu^{II}]^{3+} + DTBQ + H_2O \quad (16)$$

It is important to note that this mechanism conforms to the pathway of the enzymatic reaction (Scheme 2). The activity of the complex is controlled

TABLE I

KINETIC PARAMETERS FOR THE CATALYTIC OXIDATION OF $DTBCH_2$ BY COPPER(II) COMPLEXES

Conditions	Complex	First step			Second step		
		k_{cat} (s^{-1})	k_{cat}/K_M ($M^{-1}s^{-1}$)	K_M (mM)	k_{cat} (s^{-1})	k_{cat}/K_M ($M^{-1}s^{-1}$)	K_M (mM)
1	$[Cu_2(L55)]^{4+}$ [a]	1.4 ± 0.1	900 ± 90	1.5 ± 0.2	0.33 ± 0.01	140 ± 10	2.4 ± 0.2
1	$[Cu_2(EBA)]^{4+}$ [a]	0.7 ± 0.4	60 ± 5	13 ± 8	0.05 ± 0.01	7 ± 1	7 ± 2
1	$[Cu_2(L66)]^{4+}$ [a]	[i]	9300 ± 800				
2	$[Cu_2(EthOPy4)]^{3+}$ [b]				0.032	11.2	2.92
2	$[Cu_2(Sab4)_2]$ [c]				1.05		8.7
2	$[Cu_2(SabI)_2]$ [c]				0.36		10.4
3	$[Cu_2(L_{OCH3})]^{3+}$ [d]				0.0088		0.25
2	$[Cu_2(L^1)]^{3+}$ [e]				0.058		0.24
3	$[Cu_2(BPMP)(OH)]^{2+}$ [f]				0.0044		1.49
2	$[Cu(pzmhe)N_3]^{+}$ [g]				0.0016		
2	$[Cu(tpyea)]^{2+}$ [h]				0.83		

[1] Methanol-buffer pH 5.1.
[2] Methanol.
[3] Acetonitrile, or acetonitrile-buffer pH 7.0
[a] Ref. *137.*
[b] Ref. *155.*
[c] Ref. *162.*
[d] Ref. *156.*
[e] Ref. *163.*
[f] Ref. *148.*
[g] Ref. *164.*
[h] Ref. *146.*
[i] Too fast to be measured accurately.

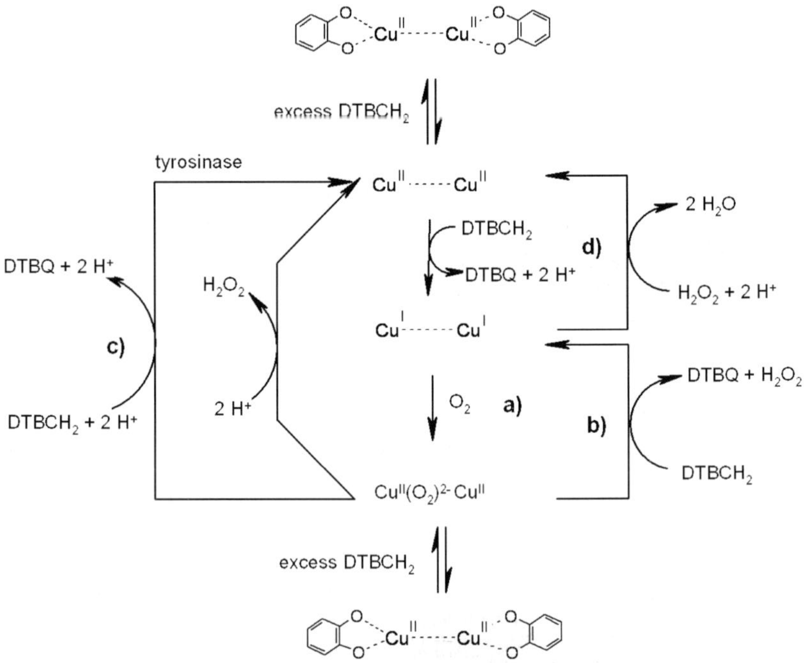

SCHEME 17. Pathways of the catalytic oxidations of catechols at dicopper(II) centers. Inhibition at high substrate concentration is also shown.

by the r.d.s. of the catalytic cycle. In the methanol/buffer solution the rates of the catalytic reactions depend on dioxygen only for $[Cu_2(L66)]^{4+}$, whereas for L-55 and EBA complexes the activities are unaffected by O_2 concentration. Therefore, for the catalytic activity of the $[Cu_2(L66)]^{4+}$ complex the r.d.s. is the formation of the dioxygen adduct (reaction 14), whereas the rate vs. $[DTBCH_2]$ profile observed for $[Cu_2(L55)]^{4+}$ and $[Cu_2(EBA)]^{4+}$ indicates that the substrate reacts in the r.d.s. of the reaction, represented as reactions 15 and 16. Treatment of the rate data allows determination of the kinetic parameters K_M (substrate dissociation constant for reaction 15) and k_{cat} (oxidation rate of reaction 16) for the second phase (Table I). For $[Cu_2(EBA)]^{4+}$ substrate inhibition is observed also in the second phase. For the complexes $[Cu_2(L66)]^{4+}$ and $[Cu_2(L55)]^{4+}$ the inhibition process does not occur, because the rigid *m*-xylyl linker keeps a folded structure of the complexes.

The addition of dihydrogen peroxide modifies the pathway of $DTBCH_2$ oxidation. In particular, for $[Cu_2(L66)]^{4+}$ the rate is strongly and linearly increased by peroxide. This behavior can be associated with the slow oxygenation of the dicopper(I) intermediate (reaction 14), which instead of reacting with dioxygen prefers to react with dihydrogen peroxide, following path *d*. For $[Cu_2(L55)]^{4+}$ and $[Cu_2(EBA)]^{4+}$ the reaction rate is slightly reduced by the addition of dihydrogen peroxide. This could be associated with the partial transformation of the dicopper(II) complex to the less reactive peroxo complex, according to the reverse reaction of path *a*.

Comparing the activity of the catechol oxidase reaction, in terms of the rate constants for the two phases of $DTBCH_2$ oxidation (Table I), the following trends emerge. In the first step the reactivity order is: $[Cu_2(L66)]^{4+} \gg [Cu_2(L55)]^{4+} > [Cu_2(EBA)]^{4+}$, and in the second step: $[Cu_2(L55)]^{4+} > [Cu_2(EBA)]^{4+} \gg [Cu_2(L66)]^{4+}$. The first step involves the electron transfer form the bridging catecholate to the Cu(II) pair. The process is clearly dominated by the reduction potential of the Cu(II)/Cu(I) couple and is always faster than the second step. The latter step involves oxygenation of the dicopper(I) species, binding of the catechol to the Cu_2O_2 intermediate, and the final electron transfer from the catecholate anion to the peroxo species. For $[Cu_2(L66)]^{4+}$, the oxygenation of the Cu(I) species is rate limiting, and this restrains the efficiency of the catalytic process. On the other hand, since oxygenation of $[Cu_2(L55)]^{2+}$ and $[Cu_2(EBA)]^{2+}$ is fast in methanol/aqueous buffer, the step limiting the overall catalytic cycle of these complexes is the reaction of the catechol with the Cu_2O_2 species. The small size of the chelate rings formed upon coordination of L55 and EBA to the copper ions, with a weak apical coordination of the tertiary nitrogen atom (*107,119*), makes the peroxo species particularly reactive. From the structural coordinates of the bis(μ-hydroxo)dicopper(II) derivatives of L55 and EBA, it can be proposed that the incoming substrate can easily displace the weakly coordinated axial nitrogen donor in the putative μ-η^2:η^2-peroxodicopper(II) complex.

Even though supported by spectral evidence only in a few cases, the active Cu_2O_2 intermediate in the catechol oxidation seems to be identifiable with a μ-η^2:η^2-peroxodicopper(II) species rather than a bis(μ-oxo)dicopper(III) species (Scheme 18). As discussed in the previous section, the complex $[Cu_2(L66)]^{2+}$ yields a Cu_2O_2 adduct of side-on μ-η^2:η^2-peroxodicopper(II)

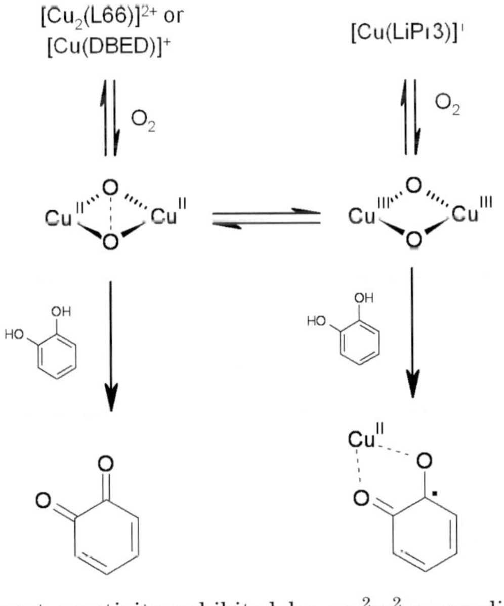

SCHEME 18. Different reactivity exhibited by μ-η^2:η^2-peroxodicopper(II) and bis (μ-oxo)dicopper(III) complexes towards catechols.

type (*121*). At low temperature, this intermediate was shown to quantitatively convert $DTBCH_2$ into DTBQ, and also selectively hydroxylate a phenol without forming radical C–C, or C–O coupling products. Also the mononuclear complex $[Cu(DBED)]^+$ reversibly reacts with dioxygen at low temperature to form a dioxygen complex of μ-η^2:η^2-peroxo type (see Section III) (*139*). This species reacts with catechol giving the quinone in very good yield (95 %), without any sign of radical oxidation (Scheme 18). On the other hand, in systems where the μ-η^2:η^2-peroxodicopper(II) complex exhibits a certain propensity to isomerize to the bis(μ-oxo)dicopper(III) isomer, as in the case of the complexes derived from LiPr3, the reaction of this species with catechol affords a mononuclear semiquinonato-copper(II) complex (Scheme 18) (*128,158*).

Further validation of the mechanism proposed for the catecholase activity of the dicopper complexes $[Cu_2(L66)]^{4+}$, $[Cu_2(L55)]^{4+}$, and $[Cu_2(EBA)]^{4+}$ (Scheme 17) has been obtained investigating the inhibitory effect of kojic acid [5-hydroxy-2-(hydroxymethyl)-γ-pyrone] (*154*). This fungal metabolite is one of the most efficient inhibitors of mushroom tyrosinase and other polyphenol oxidases (*160,161*). When the catalytic oxidation of $DTBCH_2$ was studied in the presence of kojic acid, strong competitive inhibition was observed in the steps exhibiting substrate concentration dependence, i.e., step 1 in all cases and step 2 for $[Cu_2(L55)]^{4+}$ and $[Cu_2(EBA)]^{4+}$. As expected, for all the complexes the first step is more influenced by the inhibition than the second one and, in addition, the inhibition constants are similar to the values found for kojic acid inhibition of tyrosinase (*160,161*). The effect of kojic acid on the $DTBCH_2$ oxidation catalyzed by the dinuclear Cu(II) complexes of L66, L55, and EBA is reported in Scheme 19. The lack of inhibition in the second step of the reaction catalyzed by $[Cu_2(L66)]^{4+}$ shows that kojic acid does not compete with dioxygen for binding to the dicopper(I) species. However, it interacts with the μ-η^2:η^2-peroxodicopper(II) complex. From the multicomponent pattern of the LMCT bands in the UV-VIS spectra and the NMR hyperfine shifts observed for the kojic acid adducts with the dinuclear Cu(II) complexes, it has been proposed that the inhibitor binds to the metal ions in a bridging mode (*154*); the situation is different when kojic acid binds to mononuclear Cu(II) complexes, as shown in Scheme 20.

Recently, the activation and thermodynamic parameters for the catalytic $DTBCH_2$ oxidation promoted by $[Cu_2(L55)]^{4+}$ in a mixed aqueous/organic, cryogenic solvent have been obtained (*157*). The catecholase reaction proceeds also in this case with a biphasic rate of DTBQ development, and the rates of the two steps still exhibit hyperbolic dependence on $DTBCH_2$ concentration. However, the second step under these conditions also depends linearly on O_2 concentration. The spectroscopic and kinetic analysis of the system allow the formulation of the following mechanism:

$$[Cu^{II}-OH-Cu^{II}]^{3+} + DTBCH_2 \rightleftarrows [Cu_2^{II}/DTBC]^{2+} + H_2O + H^+ \qquad (12)$$

$$[Cu_2^{II}/DTBC]^{2+} \longrightarrow [Cu_2^{I}]^{2+} + DTBQ \qquad (13)$$

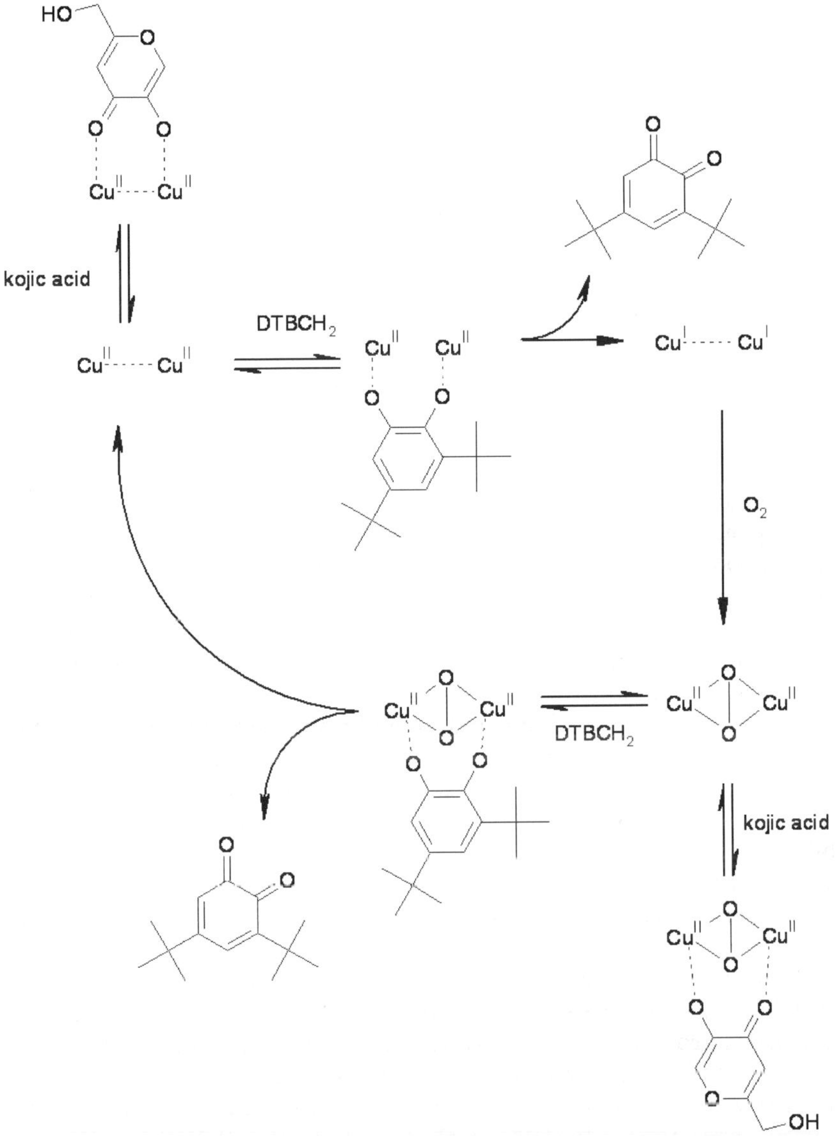

SCHEME 19. Inhibition of the catalytic oxidation of catechols at dicopper(II) centers by kojic acid.

$$[Cu_2^1]^{2+} + DTBCH_2 \rightleftarrows [Cu_2^1/DTBC] + 2H^+ \tag{17}$$

$$[Cu_2^1/DTBC] + O_2 \rightleftarrows [Cu_2/O_2/DTBC] \tag{18}$$

$$[Cu_2/O_2/DTBC] \longrightarrow \xrightarrow{3H^+} [Cu^{II}{-}OH - Cu^{II}]^{3+} + DTBQ + H_2O \tag{19}$$

where the initial step consisting of reactions 12 and 13 is the same as before, but DTBQ formation according to reaction 19 requires the sequential

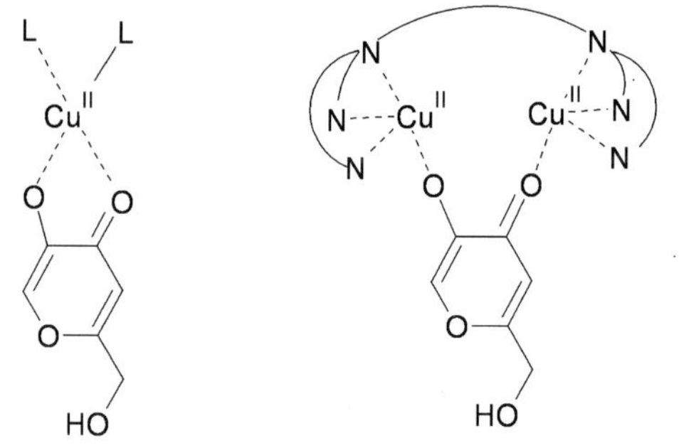

SCHEME 20. Binding of kojic acid at mononuclear and dinuclear copper(II) centers.

binding of the catechol and dioxygen to the dicopper(I) species (compare with reactions 14–16).

The spectroscopic features of the DTBC complex with $[Cu_2(L55)]^{4+}$, which has been detected at low temperature, suggests that catecholate binds as bridging ligand to the dicopper(II) complex in a η^2:η^1 fashion, as proposed for nitrocatecholate-copper(II) complexes (*38*). In the complex formulated as $[Cu^I_2/DTBC]$ in reaction 17 the catechol is supposed to bind to the dicopper(I) complex in a symmetric η^1:η^1 bridging mode (*157*). A kinetic treatment of the two steps of the catalytic reaction yielded the parameters k_1, K_{D1}, k_2, and K_{D2}. The activation and thermodynamic parameters were obtained from the temperature dependence of k_1, K_{D1}, k_2, and K_{D2}. The constants K_{D1} and K_{D2} represent the dissociation constants of the catechol adducts of the Cu^{II}_2 and Cu^I_2 forms of the complex, i.e., the reverse of reactions 12 and 17, respectively, while k_1 is the rate constant for reaction 12 and k_2 is the product of the rate constant for reaction 19, the binding constant of dioxygen to the Cu^I_2 form (18), and $[O_2]$. The ΔH° of $DTBCH_2$ binding to $[Cu_2(L55)]^{4+}$, −29.3(9) kJ mol^{-1}, is surprisingly less favorable than for $[Cu_2(L55)]^{2+}$, −63(8) kJ mol^{-1}, because in the first step the displacement of the hydroxo group from the $[Cu_2(L55)OH]^{3+}$ moiety is involved. On the other hand, the OH group keeps the Cu(II) centers in close proximity and this favors, from an entropy point of view, coordination of the catechol in a bridging mode to $[Cu_2(L55)]^{4+}$ with respect to $[Cu_2(L55)]^{2+}$ (ΔS° amounts to −41(3) J mol^{-1} K^{-1} in the first step and to −156(26) J mol^{-1} K^{-1} in the second step). The activation enthalpy for the first step, $\Delta H^{\neq} = 51(2)$ kJ mol^{-1}, is associated with the difference in redox potential of the two partners and the overlap between the orbitals involved, while the activation entropy, $\Delta S^{\neq} = -73(6)$ J mol^{-1} K^{-1} shows a negative value. The analysis of the activation parameters for the second step, $\Delta H^{\neq} = 82(4)$ kJ mol^{-1} and $\Delta S^{\neq} = 86(12)$ J mol^{-1} K^{-1}, is complicated by the fact that k_2 is a composite constant. The large positive $\Delta H^{\neq}$ value is probably determined by the O–O bond cleavage in the transition state of reaction 19, in agreement with theoretical calculations (*42*). In spite of a negative contribution from the dioxygen binding of reaction 18, $\Delta S^{\neq}$ has a large positive value, possibly because cleavage of the O–O and

Cu–O(catechol) bonds in the transition state relieves the steric crowding of the ternary complex.

Table I contains also rate constants for the catalytic $DTBCH_2$ oxidation by copper(II) complexes reported by different groups (*148,150,155,162–164*). The rate data cannot be directly compared one to another, because of the different experimental conditions employed. In general, the experiments were carried out using unbuffered solutions in organic solvents. Furthermore, in most cases only global turnover rates are reported for these reactions and therefore we assigned the rate constants to the second, slow step in our mechanism, i.e., the turnover step. Several parameters clearly affect the activity of the complexes, including the nuclearity of the complex, the reduction potential of the Cu(II)/Cu(I) couple, the geometry of the copper coordination sphere (*146*), and the Cu–Cu distance (*148,150*). Malachowski *et al.* found that for a series of five-coordinate copper(II) complexes with polyamine ligands the optimum reduction potential range for effective catalysis was around 0.0 V, but steric effects were also important and, in general, the problem of the nuclearity of the complex in the catechol oxidation reaction was not addressed (*146*). Structural and steric features controlling the catecholase reactions have been addressed by studies on a series of dinuclear complexes derived from μ-phenoxo bridging, polyamine ligands (*147,148,150*). The μ-phenoxo bridge can keep the two Cu ions at a distance of 3 Å, or below, which is similar to that present in the met form of catechol oxidase (*25*), and this distance is found to optimize the efficiency in the catechol oxidation reaction within homologous series of complexes.

A detailed study of the catecholase activity of dinuclear complexes with ligands carrying ethylpyridine chelating arms has been performed by the Speier group (*155*). The dicopper(II) complexes of EthPy4 and EthOPy4 have been structurally and spectroscopically characterized. $DTBCH_2$ is catalytically oxidized by those systems with the parallel formation of H_2O_2 as reduction product of dioxygen. On the basis of the rate dependence on dioxygen and $DTBCH_2$, the following mechanism has been proposed:

$$[Cu^{II}---Cu^{II}] + DTBCH_2 \longrightarrow [Cu^{I}---Cu^{I}] + DTBQ + 2H^+ \qquad (20)$$

$$[Cu^{I}---Cu^{I}] + O_2 \rightleftarrows [Cu^{II}(O_2^{2-})Cu^{II}] \qquad (21)$$

$$[Cu^{II}(O_2^{2-})Cu^{II}] + DTBCH_2 \longrightarrow [Cu^{I}---Cu^{I}] + DTBQ + H_2O_2 \qquad (22)$$

The catalysis is initiated by reaction 20, which is too fast to be detected; the catalytic cycle is composed by reactions 21 and 22. The efficiency of the reaction turnover is lower than for benzimidazole containing systems described above. The treatment of the rate data for the dicopper(II) complex of EthPy4, on the basis of the proposed model in which the oxygenation of the Cu^{I}_2 moiety is taken as r.d.s., provided the kinetic and thermodynamic parameters of the process. The parameters related to the oxygenation step ($k = 7.6\ M^{-1}s^{-1}$; $\Delta H^{\neq} = 47.6$ kJ mol^{-1} and $\Delta S^{\neq} = -76$ J K^{-1} mol^{-1}) are not congruent with the data already reported for the homologue pyridine dicopper systems (*62*). The different ligand chain length and the different conditions could be responsible for the consistent disagreement. For the reaction of the copper(II) complex of EthOPy4, the activation parameters have been determined ($\Delta H^{\neq} = 57.0\,kJ\,mol^{-1}$, $\Delta S^{\neq} = -6.9\,J\,K^{-1}\,mol^{-1}$). Both the enthalpy and entropy contributions have a reduced absolute value compared with those found for the reaction catalyzed by $[Cu_2(L55)]^{4+}$ (see above), probably due to the fact that in this case the reaction does not involve an O–O bond cleavage.

Some mononuclear Cu(II) complexes have been also reported to act as catalysts in the catechol oxidation (*146,151,165*). The efficiency of these systems is very low and in general it is always possible that the actual species responsible for catechol oxidation is a dimeric species resulting from aggregation of two mononuclear units, as it has been recently shown for an amino-carbohydrate-copper(II) system (*156*).

V. Oxidations by Trinuclear Complexes

Multicopper oxidases are typically active in the catalytic one-electron oxidation of a variety of diphenolic, polyphenolic, enediolic, and aminophenolic substrates (*1,53,166,167*). The mechanism of these reactions is complex and, as discussed in Section I, it involves a sequence of four one-electron oxidations of substrate molecules. The radical products of these reactions undergo dismutation, as shown in Scheme 21 for the oxidation of ascorbate to semidehydroascorbate radical (*168,169*). The substrate binds to the enzymes close to type 1 Cu, whereas the trinuclear cluster is only accessible to dioxygen, or other small molecules. This situation is clearly difficult to reproduce in a model system and for this reason the type of model oxidation reactions that have been studied so far using synthetic trinuclear copper complexes is more related to the activity of type 3 Cu enzymes than multicopper oxidases. Nevertheless, such trinuclear complexes open new perspectives in stereoselective catalysis, because one of the metal centers

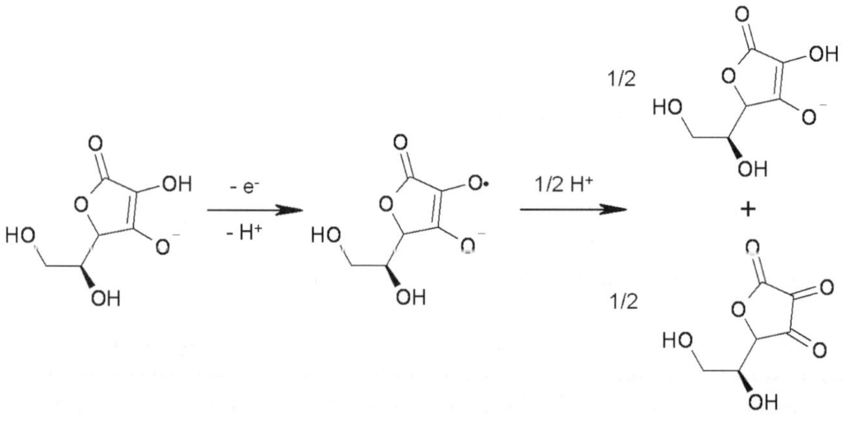

SCHEME 21. Enzymatic one-electron oxidation of ascorbate and dismutation of the semidehydroascorbate radical.

can provide a site for substrate binding close to the dinuclear, dioxygen binding unit.

As an extension of our work on synthetic models for type 3 Cu proteins, we recently started an extensive investigation on trinuclear copper complexes that reproduce the coordination features of the trinuclear clusters of multicopper oxidases. Such biomimetic chemistry is, in fact, surprisingly little developed compared to the large amount of research effort dedicated to the study of dinuclear Cu complexes. A mixed valence trinuclear complex containing the bis(μ_3-oxo) core (Scheme 4) was obtained upon oxygenation at low temperature of the Cu(I) complex with tetramethyl-cyclohexane-diamine (*170,171*). Similar trigonally symmetric $[Cu_3(\mu_3\text{-}O)]^{3+}$ complexes were obtained from Cu(I) complexes with strong σ-donor, didentate ligands that usually promote the formation of bis(μ-oxo)dicopper(III) species (*63,172*). The spectral signatures of these trinuclear cores are appreciably different from those of Cu_2O_2 species; for instance, the optical LMCT bands occur at 355 (ε 15,000 $M^{-1}cm^{-1}$), 480 (1400), and 620 nm (800) in one case (*170*), and 342 (ε 12,000 $M^{-1}cm^{-1}$), 515 (1000), and 685 nm (800) in the other case (*172*). The preferential formation of one, or the other type of adduct depends on the conditions and subtle steric effects on the ligand, with the more compact bis(μ_3-oxo)tricopper(II,II,III) species favored by less sterically demanding ligands.

Trinuclear complexes with multidentate ligands that could serve as biomimetic models for the type 2/type 3 Cu sites were scarce until a few years ago (*173–175*). This certainly reflects at least in part the synthetic difficulty to develop suitable polydentate ligands capable to generate complexes with structural features approximately similar to those of biological trinuclear clusters. Trinuclear copper(II) complexes with simple nitrogen donor ligands but containing tris(μ-hydroxo), or μ_3-hydroxo (or oxo) bridges, encompassing structural motives that are proposed to be present as intermediates in the catalytic cycle of multicopper oxidases (*57,58*), have recently attracted increasing interest for their intriguing magnetic exchange and spin-frustration properties (*176–180*). Genuine biomimetic models of the biological Cu_3N_8 cores, with a correct distribution of donor atoms among the three Cu ions, have been developed in the last few years by our group (*181–186*). The structures of the ligands employed in these investigations are collected in Fig. 6. Except for PBM, we were able to introduce chiral residues into the ligands. In PHI these chiral centers are derived from L-histidine residues (*182*), whereas (*R*)-DBN (*184*) and (*S*)-DBN (*186*) are enantiomeric ligands derived from (*R*)- and (*S*)-1,1′-binaphthyl-2,2′-diamine, respectively.

In the resting, oxidized type 2/type 3 cluster of multicopper oxidases the three Cu(II) ions are at distances in the range of 3.6–3.9 Å; the pair of Cu ions representing type 3 Cu are bridged by an hydroxo (or oxo) ligand (Fig. 3). As a result of this bridging interaction, the two Cu atoms are strongly magnetically coupled and EPR silent (*1,6*). A total of eight histidine ligands is distributed among the Cu ions: three histidines bind to each Cu ion of the type 3 Cu pair, and two to type 2 Cu. The design of the octadentate ligands reported in Fig. 6, aimed at reproducing the coordination set present

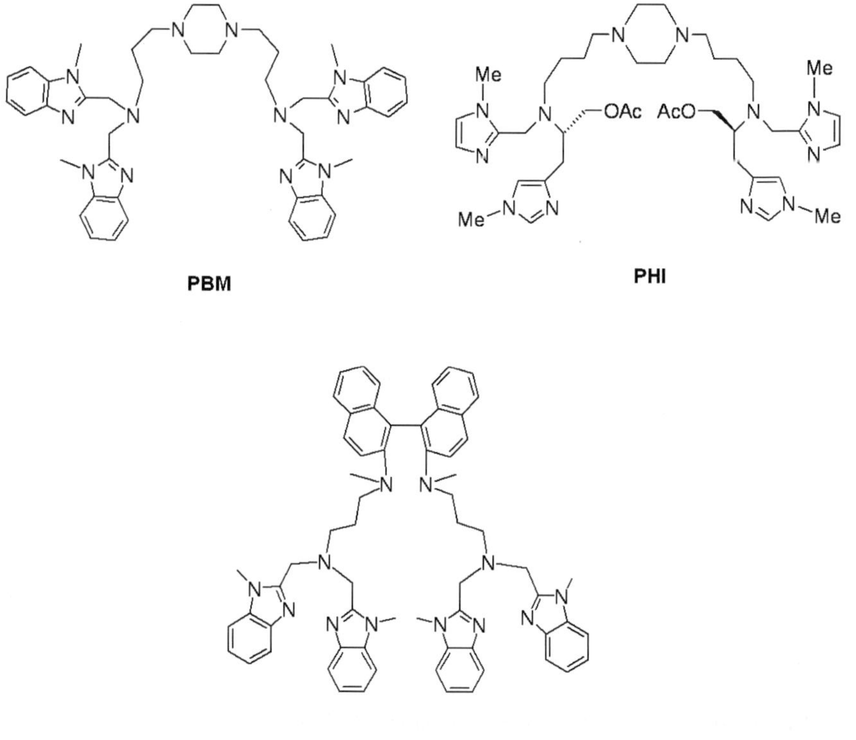

FIG. 6. Structure of the octadentate ligands used for the synthesis of biomimetic trinuclear copper complexes.

in the enzymes, is based on the covalent attachment of two lateral arms containing tridentate N_3 binding units to the two amine N donors present in the central residue. The topology of the resulting metal binding sites is shown in Scheme 22 (structure I), where the two metal ions at sites A are coordinated by the lateral tridentate units and the third metal ion at site B is bonded to the central didentate residue. As shown by EPR studies, two of the Cu(II) ions, i.e., one CuA and CuB, are strongly coupled and EPR non-detectable (*181,182,186*), and the electronic and CD spectral properties are consistent with the establishment of one, or more probably two hydroxo bridges between these Cu centers in solution. Therefore, the structure of the trinuclear complexes is better represented by structure II in Scheme 22. The collective spectral properties of the complexes indicate that the two CuA centers are five coordinated and CuB is four-coordinated (*186*). Unlike multicopper oxidases, the design of the present trinuclear complexes does not allow to establish the coupling between the two three-coordinated CuA sites through bridging ligands. However, we note that, at least in laccase, recent EPR investigations carried out on various derivatives of the enzyme show that the EPR active copper in the cluster is one of the type 3 Cu centers and, therefore, the magnetic coupling must occur between the remaining type 3 Cu and type 2 Cu (*187*). Then the coupling pattern within the

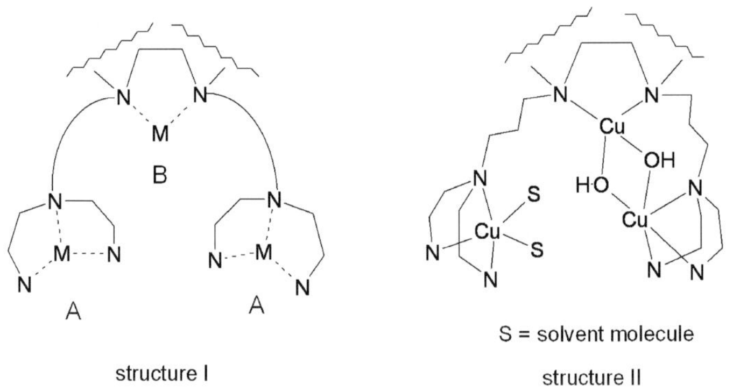

SCHEME 22. Schematic representation of the structural arrangement of the metal ions in the biomimetic models for biological trinuclear copper clusters (structure I), and diagram showing the coupling interaction through hydroxo bridges between one CuA and CuB (structure II).

trinuclear cluster may be more dynamic than shown by the crystal structure of the enzymes.

The trinuclear Cu(II) complexes derived from the ligands PBM, PHI, (*R*)-DBN, and (*S*)-DBN are all active in the catalytic oxidations of catechols. Experiments carried out with 3,5-di-*tert*-butylcatechol as a substrate showed in most cases the same type of biphasic behavior described previously for the dinuclear model systems in Section IV. Therefore, a similar two-step mechanism can be assumed, with two catechol molecules oxidized per catalytic cycle and reduction of dioxygen to water. The catalytic activity of the trinuclear complexes against this substrate decreases in the order: $[Cu_3(PBM)]^{6+} > [Cu_3(PHI)]^{6+} > [Cu_3\{(R)\text{-DBN}\}]^{6+}$ (*181,183,185*), but the differences in the turnover rates are restricted to within one order of magnitude. In general, the rate constants are comparable with, but slightly lower than those found for the dinuclear Cu(II) complexes with benzimidazole ligands described in Section IV, except for $[Cu_2(L55)]^{4+}$, which is indeed very active. Anaerobic experiments carried out by reacting the trinuclear complexes and the catechol in a 1:1 ratio invariably show reduction of one CuA and the CuB ion, indicating that this Cu pair represents the site of substrate binding. The hydroxo bridges likely favor this type of interaction, presumably allowing the catechol to bind in anionic form and as a bridging ligand between the two coppers. The most interesting aspect of the reactivity of the trinuclear complexes $[Cu_3(PHI)]^{6+}$, $[Cu_3\{(R)\text{-DBN}\}]^{6+}$, and $[Cu_3\{(S)\text{-DBN}\}]^{6+}$ is the possibility to exploit their chirality to perform stereoselective oxidation processes. We have so far investigated the discriminating ability of these complexes in the catalytic oxidation of L-/D-Dopa and L-/D-Dopa methyl esters (DopaOMe). The complex $[Cu_3(PHI)]^{6+}$ exhibits a remarkable enantio differentiation in the oxidation of L- and D-Dopa, with a strong preference for the L enantiomer, but very poor discrimination between L- and D-DopaOMe (*183*). This behavior can be accounted for by the substrate anchoring effect provided by the chiral CuA center which does not participate in the catalytic reaction (Scheme 23). In fact, while the

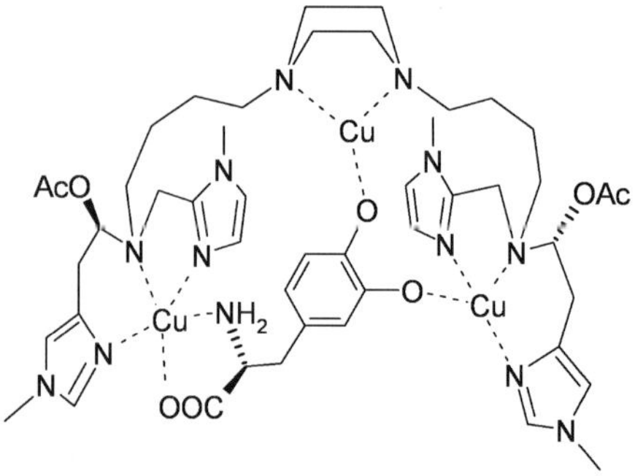

SCHEME 23. Schematic representation of the interaction between L-Dopa and the trinuclear complex $[Cu_3(PHI)]^{6+}$ showing the substrate anchoring effect and the origin of chiral recognition by the CuA center.

catechol residue of the substrate binds to the CuA-CuB pair, the chiral portion of L- and D-Dopa is allowed to interact strongly with the latter CuA center and thereby generate the discriminating effect. With the methyl ester derivatives of Dopa the interaction of the amino ester portion of the substrate with the CuA anchor is obviously much weaker and chiral recognition is poor. The discriminating ability can be quantitatively expressed by the absolute value of the enantioselectivity index $R = [(k_{cat}/K_M)_L - (k_{cat}/K_M)_D]/[(k_{cat}/K_M)_L + (k_{cat}/K_M)_D]$ (*186*). This parameter amounts to 61 % for the catalytic oxidation of L-/D-Dopa, but drops to less than 5 % in the case of L-/D-DopaOMe (*183*). In the case of $[Cu_3\{(R)\text{-DBN}\}]^{6+}$ and $[Cu_3\{(S)\text{-DBN}\}]^{6+}$ the chiral unit is in the non-polar binaphthyl residue; this is close to CuB, which is involved in the oxidation of the catechol residue but cannot give strong interaction with the chiral amino acidic portion of the substrates. The enantioselectivity is therefore low (R between 0 and 14 %) for both types of substrates (*185,186*). It should be added that for the dinuclear complexes derived from the DBN ligands $[Cu_2\{(R)\text{-DBN}\}]^{4+}$ and $[Cu_2\{(S)\text{-DBN}\}]^{4+}$, the enantioselectivity in the oxidation of both L-/D-Dopa and L-/D-DopaOMe couples is larger than that of the trinuclear complexes (R between 18 and 70 %) (*185,186*), but here the origin of the discriminating effect is the more classical "ligand induced". We plan to extend the scope of the "metal induced" discrimination by chiral trinuclear complexes to the oxidation reactions of other bifunctional substrates. In this respect, note that it is possible to prepare mixed-metal trinuclear complexes, containing different metal ions, using the chiral octadentate ligands in Fig. 6 by stepwise addition of the metal ions, as we previously did with PBM (*181*). Studies aimed at characterizing the reduced forms of the trinuclear complexes and their reaction with dioxygen are currently on the way in our laboratory.

VI. Conclusions

In this review we have given an overview on the biological and biomimetic oxidations promoted by dinuclear and trinuclear copper centers, with special emphasis to our contributions in the field. Many biomimetic systems aimed at reproducing characteristic structural and functional features of copper enzymes containing mononuclear, or non-coupled dinuclear active sites, i.e., superoxide dismutase, galactose oxidase, dopamine β-hydroxylase, and peptidylglycine α-hydroxylating monooxygenase, have been reported, but their chemistry could not be described here for space limitations. For the same reason, we did not give appropriate emphasis to structural and magnetic studies carried out by several other groups on complexes that can be considered as enzyme mimics. The properties and reactivity of the key Cu_nO_2 complexes involved in biomimetic oxidations have been covered by recent comprehensive reviews by Lewis and Tolman (*62*) and by Stack and coworkers (*63*). In the last decade, the progress in the understanding of the reactivity of these Cu_nO_2 systems using biomimetic compounds has been enormous. The main conclusion from these studies is that although many types of copper-dioxygen complexes can be formed, a prominent role is played by Cu_2O_2 complexes of μ-η^2:η^2-peroxo type, which seems the most suitable to carry out smooth monooxygenase and oxidase reactions by type 3 Cu enzymes. The current frontier in the synthetic biomimetic chemistry of copper complexes is the characterization of the ternary complexes formed by the key Cu_nO_2 species and the substrate undergoing the oxidation, or oxygenation reaction.

From a practical point of view, the application of biomimetic models to processes of synthetic, or industrial interest depends strongly on the transformation of the current stoichiometric reactions into catalytic reactions (e.g., in the phenol oxygenation reactions), the stability of the catalysts, and the introduction of stereoselectivity characteristics in the reactions, by using structurally more sophisticated and possibly chiral ligands. In addition, the biomimetic chemistry of the more complex multinuclear Cu sites, such as the trinuclear Cu sites of blue oxidases, is still in its infancy, and important contributions are expected in the next few years from the study of new and suitable model compounds.

Acknowledgments

We thank all our students and the collaborators from other laboratories, whose names are listed in the references to our papers. They gave excellent contributions to the research carried out by our group and their efforts are greatly appreciated. Financial support to our work was given by the Italian MIUR, through PRIN projects, the C.I.R.C.M.S.B. and the University of Pavia.

References

1. Solomon, E. I.; Sundaram, U. M.; Manchonkin, T. E. *Chem. Rev.* **1996**, *96*, 2563–2605.
2. Klinman, J. P. *Chem. Rev.* **1996**, *96*, 2541–2562.
3. Kaim, W.; Rall, J. *Angew. Chem. Int. Ed.* **1996**, *35*, 43–60.
4. Malmström, B. G.; Leckner, J. *Curr. Opinin. Chem. Biol.* **1998**, *2*, 286–292.

5. Abolmali, B.; Taylor, H. V.; Weser, U. *Struct. Bonding* **1998**, *91*, 92–190.
6. Solomon, E. I.; Chen, P.; Metz, M.; Lee, S.-K.; Palmer, A. E. *Angew. Chem. Int. Ed.* **2001**, *40*, 4570–4590.
7. Halcrow, M. A. *Angew. Chem. Int. Ed.* **2001**, *40*, 346–349.
8. Whittaker, J. W. *Chem. Rev.* **2003**, *103*, 2347–2364.
9. Wilmot, C. M.; Hadju, J.; McPherson, M. J.; Knowles, P. F.; Phillips, S. E. *Science* **1999**, *286*, 1724–1728.
10. Evans, J. P.; Ahn, K.; Klinman, J. P. *J. Biol. Chem.* **2003**, *278*, 49691–49698.
11. Chen, P.; Bell, J.; Eipper, B. A.; Solomon, E. I. *Biochemistry* **2004**, *43*, 5735–5747.
12. Abramson, J.; Svensson-Ek, M.; Byrne, B.; Iwata, S. *Biochim. Biophys. Acta* **2001**, *1544*, 1–9.
13. Steiner, R. A.; Kalk, K. H.; Dijkstra, B. W. *Proc. Natl. Acad. Sci. USA* **2002**, *99*, 16625–16630.
14. van Holde, K. E.; Miller, K. I. *Adv. Protein Chem.* **1995**, *47*, 1–81.
15. Volbeda, A.; Hol, W. G. J. *J. Mol. Biol.* **1989**, *209*, 249–279.
16. Magnus, K. A.; Ton-That, H.; Carpenter, J. E. *Chem. Rev.* **1994**, *94*, 727–735.
17. Cuff, M. E.; Miller, C.; van Holde, K. E.; Hendrickson, W. A. *Nat. Struct. Biol.* **2000**, *278*, 855–870.
18. Ghiretti, F. *Arch. Biochem. Biophys.* **1956**, *63*, 165–167.
19. Zlateva, T.; Santagostini, L.; Bubacco, L.; Casella, L.; Salvato, B.; Beltramini, M. *Biochemistry* **1998**, *72*, 211–215.
20. Salvato, B.; Santamaria, M.; Beltramini, M.; Alzuet, G.; Casella, L. *Biochemistry* **1998**, *40*, 14065–14077.
21. Beltramini, M.; Bubacco, L.; Casella, L.; Alzulet, G.; Gullotti, M.; Salvato, B. *Eur. J. Biochem.* **1995**, *232*, 98–105.
22. Decker, H.; Rimke, T. J. *J. Biol. Chem.* **1998**, *273*, 25889–25892.
23. Decker, H.; Dillinger, R.; Tuczek, F. *Angew. Chem. Int. Ed.* **2000**, *39*, 1591–1595.
24. Cary, J. W.; Lax, A. R.; Flurkey, W. H. *Plant. Mol. Biol.* **1992**, *20*, 245–253.
25. Klabunde, T.; Eicken, C.; Sacchettini, J. C.; Krebs, B. *Nat. Struct. Biol.* **1998**, *5*, 1084–1090.
26. Gerdemann, C.; Eicken, C.; Krebs, B. *Acc. Chem. Res.* **2002**, *35*, 183–191.
27. Gielens, C., et al. *Eur. J. Biochem.* **1997**, *248*, 879–888.
28. van Gelder, C. W.; Flurkey, W. H.; Wichers, H. J. *Phytochemistry* **1997**, *45*, 1309–1323.
29. Eicken, C.; Zippel, F.; Buldt-Karentzopoulos, K.; Krebs, B. *FEBS Lett.* **1998**, *436*, 293–299.
30. Rompel, A.; Fischer, H.; Meiwes, D.; Buldt-Karentzopoulos, K.; Dillinger, R.; Tuczek, F.; Witzel, H.; Krebs, B. *J. Biol. Inorg. Chem.* **1999**, *4*, 56–63.
31. Decker, H.; Tuczek, F. *Trends Biochem. Sci.* **2000**, *25*, 392–397.
32. Prota, G. "*Melanins and Melanogenesis*"; Academic Press: San Diego, CA, **1992**.
33. Land, E. J.; Ramsden, C. A.; Riley, M. J. *Acc. Chem. Res.* **2003**, *36*, 300–308.
34. Battaini, G.; Monzani, E.; Casella, L.; Lonardi, E.; Tepper, A. W.; Canters, G. W.; Bubacco, L. *J. Biol. Chem.* **2002**, *277*, 44606–44612.
35. Rodriguez-Lopez, J. N.; Fenoll, L. G.; Penalver, M. J.; Garcia-Ruiz, P. A.; Varon, R.; Martinez-Ortiz, F.; Garcia-Canovas, F.; Tudela, J. *Biochim. Biophys. Acta* **2001**, *1548*, 238–256.
36. Yamazaki, S.-I.; Itoh, S. *J. Am. Chem. Soc.* **2003**, *125*, 13034–13035.
37. Bubacco, L.; van Gastel, M.; Groenen, E. J.; Vijgenboom, E.; Canters, G. W. *J. Biol. Chem.* **2003**, *278*, 7381–7389.
38. Plenge, T.; Dillinger, R.; Santagostini, L.; Casella, L.; Tuczek, F. *Z. Anorg. Allg. Chem.* **2003**, *629*, 2258–2265.
39. Jolley Jr., R. L.; Evans, L. H.; Makino, N.; Mason, H. S. *J. Biol. Chem.* **1974**, *249*, 335–345.
40. Yamazaki, M.; Morioka, C.; Itoh, S. *Biochemistry* **2003**, *43*, 11546–11553.
41. Jaenicke, E.; Decker, H. *ChemBioChem.* **2004**, *5*, 163–169.
42. Siegbahn, P. E. *J. Biol. Inorg. Chem.* **2003**, *8*, 567–576; **2004**, *9*, 577–590.
43. Ferrari, R. P.; Laurenti, E.; Ghibaudi, E. M.; Casella, L. *J. Inorg. Biochem.* **1997**, *68*, 61–69.
44. Granata, A.; Monzani, E.; Bubacco, L.; Casella, L. *Chem. Eur. J.* **2005**, in the press.
45. Messerschmidt, A.; Ladenstein, R.; Huber, R.; Bolognesi, M.; Avigliano, L.; Petruzzelli, M.; Rossi, A.; Finazzi-Agro, A. *J. Mol. Biol.* **1992**, *224*, 179–205.

46. Messerschmidt, A. In: "*Multicopper Oxidase*"; World Scientific: Singapore, **1997**, pp. 23–79.
47. Bertrand, T.; Jolivat, C.; Briozzo, P.; Caminade, E.; Joly, N.; Mazdak, C.; Mougin, C. *Biochemistry* **2002,** *41*, 7325–7333.
48. Hakulinem, N.; Kiiskinen, L.; Kruus, K.; Saloheimo, M.; Paananem, A.; Koivula, A.; Rouvinem, J. *Nat. Struct. Biol.* **2002,** *9*, 601–605.
49. Lindley, P. F.; Card, G.; Zaitseva, I.; Zaitsev, V.; Reinhammar, B.; Selin-Lindgren, E.; Yoshida, K. *J. Biol. Inorg. Chem.* **1997,** *4*, 579–587.
50. Zaitsev, V. N.; Zaitseva, I.; Papiz, M.; Lindley, P. F. *J. Biol. Inorg. Chem.* **1999,** *4*, 579–587.
51. Quintanar, L.; Gebhard, M.; Wang, T.-P.; Kosman, D. J.; Solomon, E. I. *J. Am. Chem. Soc.* **2004,** *126*, 6579–6589.
52. Roberts, S. A.; Weichestel, A.; Grass, G.; Thakali, K.; Hazzard, J. T.; Tollin, G.; Rensing, C.; Montford, W. R. *Proc. Natl. Acad. Sci. USA* **2002,** *99*, 2766–2771.
53. Reinhammar, B.; Malmström, B. G. In: "*Copper Proteins*"; Ed. Spiro, T. G.; Wiley: New York, **1981**, pp. 109–150.
54. Adman, E. T. *Adv. Protein Chem.* **1991,** *42*, 145–197.
55. Nakamura, K.; Kawabata, T.; Yura, K.; Go, N. *FEBS Lett.* **2003,** *553*, 239–244.
56. Shin, W.; Sundaram, U. M.; Cole, J. L.; Zhang, H. H.; Hedman, B.; Hodgson, K. O.; Solomon, E. I. *J. Am. Chem. Soc.* **1996,** *118*, 3202–3215.
57. Palmer, A. E.; Lee, S. K.; Solomon, E. I. *J. Am. Chem. Soc.* **2001,** *123*, 6591–6599.
58. Lee, S. K.; DeBeer, G. S.; Antholine, W. E.; Hedman, B.; Hodgson, K. O.; Solomon, E. I. *J. Am. Chem. Soc.* **2002,** *124*, 6180–6193.
59. Gaspard, S.; Monzani, E.; Casella, L.; Gullotti, M.; Maritano, S.; Marchesini, A. *Biochemistry* **1997,** *36*, 4852–4859.
60. Santagostini, L.; Gullotti, M.; De Gioia, L.; Fantucci, P.; Franzini, E.; Marchesini, A.; Monzani, E.; Casella, L. *Int. J. Biochem. Cell. Biol.* **2004,** *36*, 881–892.
61. Karlin, K. D.; Zuberbuhler, A. D. In: "*Bioinorganic Catalysis*"; 2nd edn. Revised and Expanded; Eds. Reedijk, J.; Bouwman, E.; Dekker: New York, **1999**, pp. 469–534.
62. Lewis, E.; Tolman, W. B. *Chem. Rev.* **2004,** *104*, 1047–1076.
63. Mirica, L. M.; Ottenwaelder, X.; Stack, T. D. P. *Chem. Rev.* **2004,** *104*, 1013–1045.
64. Hatcher, L. Q.; Karlin, K. D. *J. Biol. Inorg. Chem.* **2004,** *9*, 669–683.
65. Schindler, S. *Eur. J. Inorg. Chem.* **2000,** 2311–2326.
66. Gamez, P.; Aubel, P. G.; Driessen, W. L.; Reedijk, J. *Chem. Soc. Rev.* **2001,** *30*, 376–385.
67. Itoh, S.; Fukuzumi, S. *Bull. Chem. Soc. Jpn.* **2002,** *75*, 2081 2095.
68. Hatcher, L. Q.; Karin, M. *Adv. Inorg. Chem.* **2005,** *58*, Chapter 4.
69. Arends, I. W. C. E.; Gamez, P.; Sheldon, R. A. *Adv. Inorg. Chem.* current issue, Chapter 6.
70. Aboelella, N. W.; Lewis, E.; Reynolds, A. M.; Brennessel, W. W.; Cramer, C. J.; Tolman, W. B. *J. Am. Chem. Soc.* **2002,** *124*, 10660–10661.
71. Kitajima, N.; Fujisawa, K.; Moro-oka, Y.; Toriumi, K. J. *J. Am. Chem. Soc.* **1989,** *111*, 8975–8976.
72. Bol, J. E.; Driessen, W. L.; Ho, R. Y. N.; Maase, B.; Que Jr., L.; Reedijk, J. *Angew. Chem. Int. Ed.* **1997,** *36*, 998–1000.
73. Kodera, M.; Katayama, K.; Tachi, Y.; Kano, K.; Hirota, S.; Fujinami, S.; Suzuki, M. *J. Am. Chem. Soc.* **1999,** *121*, 11006–11007.
74. Cvetkovivic, M.; Batten, S. R.; Moubaraki, B.; Murray, K. S.; Spiccia, L. *Inorg. Chim. Acta* **2001,** *324*, 131–140.
75. Hu, Z.; Williams, R. D.; Tran, D.; Spiro, T. G.; Gorun, S. M. *J. Am. Chem. Soc.* **2000,** *122*, 3556–3557.
76. Hu, Z.; George, G. N.; Gorun, S. M. *J. Am. Chem. Soc.* **2001,** *123*, 4812–4813.
77. Blackman, A. G.; Tolman, W. B. *Struct. Bonding (Berlin)* **2000,** *97*, 179–211.
78. Kitajima, N.; Moro-oka, Y. *Chem. Rev.* **1994,** *94*, 737–757.
79. Karlin, K. D.; Hayes, J. C.; Gultneh, Y.; Cruse, R. W.; McKown, T. W.; Hutchinson, J. P.; Zubieta, J. *J. Am. Chem. Soc.* **1984,** *106*, 2121–2128.
80. Gelling, O. J.; van Bolhuis, F.; Meetsma, A.; Feringa, B. L.; *J. Chem. Soc. Chem. Commun.* **1988**, 552–554.
81. Casella, L.; Gullotti, M.; Pallanza, G.; Rigoni, R. *J. Am. Chem. Soc.* **1988,** *110*, 4221–4227.

82. Menif, R.; Martell, A.E. *Chem. Commun.* **1989**, 1521–1523.
83. Drew, M. G. B.; Trocha-Grimshaw, J.; McKillop, K. P. *Polyhedron* **1989,** *8*, 2313–2315.
84. Menif, R.; Martell, A. E.; Squattrito, P. J.; Clearfield, A. *Inorg. Chem.* **1990,** *29*, 4723–4729.
85. Casella, L.; Gullotti, M.; Bartosek, G.; Pallanza, G.; Laurenti, E. *Chem. Commun.* **1991**, 1235–1237.
86. Becker, M.; Schindler, S.; van Eldik, R. *Inorg. Chem.* **1994,** *33*, 5370–5371.
87. Ryan, S.; Adams, H.; Fenton, D. E.; Becker, M.; Schindler, S. *Inorg. Chem.* **1998,** *37*, 2134–2140.
88. Ma, H.; Allmendinger, M.; Thewalt, U.; Lentz, A.; Klinga, M.; Rieger, B. *Eur. J. Inorg. Chem.* **2002**, 2857–2867.
89. Utz, D.; Heinemann, F. W.; Hampel, F.; Richens, D. T.; Schindler, S. *Inorg. Chem.* **2003,** *42*, 1430–1436.
90. Battaini, G.; Monzani, E.; Perotti, A.; Para, C.; Casella, L.; Santagostini, L.; Gullotti, M.; Dillinger, R.; Nather, C.; Tuczek, F. *J. Am. Chem. Soc.* **2003,** *125*, 4185–4198.
91. Foxon, P.S.; Shindler, S.; Thaler, F.; Weitzer, M.; Liehr, G.; Mukherjee, J.; Balamurugan, V.; Ghosh, D.; Mukherjee, R. *J. Chem. Soc. Dalton Trans.* **2004**, 2321–2328.
92. Itoh, S.; Nakao, H.; Berreau, L.; Kondo, T.; Komatsu, M.; Fukuzumi, S. *J. Am. Chem. Soc.* **1998,** *120*, 2890–2899.
93. Itoh, S.; Taki, M.; Nakao, H.; Holland, P. L.; Tolman, W. B.; Que Jr., L.; Fukuzumi, S. *Angew. Chem. Int. Ed.* **2000,** *39*, 398–400.
94. Blain, I.; Bruno, P.; Giorgi, M.; Lojou, M.; Lexa, D.; Reglier, M. *Eur. J. Inorg. Chem.* **1998,** 1297–1304.
95. Xie, D.; Gutsche, C. D. *J. Org. Chem.* **1998,** *63*, 9270–9278.
96. Allen, W. E.; Sorrel, T. N. *Inorg. Chem.* **1997,** *36*, 1732–1734.
97. Mahapatra, S.; Halfen, J. A.; Tolman, W. B. *J. Am. Chem. Soc.* **1996,** *118*, 11575–11586.
98. Mahadevan, V.; Hou, Z.; Cole, A. P.; Root, D. E.; Solomon, E. I.; Stack, T. D. P. *J. Am. Chem. Soc.* **1997,** *119*, 11996–11997.
99. Mahapatra, S.; Halfen, J. A.; Wilkinson, E. C.; Pan, G.; Wang, X.; Young Jr., V. G.; Cramer, C. J.; Que Jr., L.; Tolman, W. B. *J. Am. Chem. Soc.* **1996,** *118*, 11555–11574.
100. Mahapatra, S.; Young Jr., V. G.; Tolman, W. B. *Angew. Chem. Int. Ed.* **1997,** *36*, 130–133.
101. Mahapatra, S.; Kaderli, S.; Llobet, A.; Neuhold, Y.-M.; Palanche, T.; Halfen, J. A.; Young Jr., V. G.; Kaden, T. A.; Que, L.; Zuberbuehler, A. D.; Tolman, W. B. *Inorg. Chem.* **1997,** *36*, 6343–6356.
102. Enomoto, M.; Aida, T. *J. Am. Chem. Soc.* **1999,** *121*, 874–875.
103. Taki, M.; Teramae, S.; Nagatomo, S.; Tachi, Y.; Kitagawa, T.; Itoh, S.; Fukuzumi, S. *J. Am. Chem. Soc.* **2002,** *124*, 9332–9333.
104. Zhang, C. X.; Liang, H.-C.; Kim, E.-i.; Gan, Q.-F.; Tyeklar, Z.; Karlin, K. D.; Lam, K.-C.; Rheingold, A. L.; Kaderli, S.; Zuberbuehler, A. D. *Chem. Commun.* **2001**, 631–632.
105. Klein Gebbink, R. J. M.; Martens, C. F.; Kenis, P. J. A.; Jansen, R. J.; Nolting, H.-F.; Sole, V. A.; Feiters, M. C.; Karlin, K. D.; Nolte, R. J. M. *J. Am. Chem. Soc.* **1999,** *121*, 5755–5768.
106. Hayaishi, H.; Fujinami, S.; Nagatomo, S.; Ogo, S.; Suzuki, E.; Uehara, A.; Watanabe, Y.; Kitagawa, T. *J. Am. Chem. Soc.* **2000,** *122*, 2124–2125.
107. Battaini, G.; Perotti, A.; Casella, L.; Monzani, E.; Gullotti, M.; Santagostini, L.; Nardin, G.; Randaccio, L.; Schindler, S. *Eur. J. Inorg. Chem.* **2003**, 1197–1205.
108. Karlin, K. D.; Shi, J.; Hayes, J. C.; McKown, J. W.; Hutchinson, J. P.; Zubieta, J. *Inorg. Chim. Acta* **1984,** *91*, L3–L7.
109. Karlin, K. D., Nasir, M. S., Cohen, B. I., Cruse, R. W., Kaderli, S.; Zuberbühler, A. D. *J. Am. Chem. Soc.* **1994,** *116*, 1324–1336.
110. Becker, M.; Schindler, S.; Karlin, K. D.; Kaden, T. A.; Kaderli, S.; Palanche, T.; Zuberbuehler, A. D. *Inorg. Chem.* **1999,** *38*, 1989–1995.
111. Pidcock, E.; Obias, H. V.; Zhang, C. X.; Karlin, K. D.; Solomon, E. I. *J. Am. Chem. Soc.* **1998,** *120*, 7841–7847.
112. Karlin, K. D.; Cohen, B. I.; Jacobson, R. R.; Zubieta, J. *J. Am. Chem. Soc.* **1987,** *109*, 6194–6196.
113. Nasir, M. S.; Cohen, B. I.; Karlin, K. D. *J. Am. Chem. Soc.* **1992,** *114*, 2482–2494.

114. Chen, P.; Solomon, E. I. *J. Inorg. Biochem.* **2002,** *88,* 368–374.
115. Sorrell, T. N.; Garrity, M. L. *Inorg. Chem.* **1991,** *30,* 210–215.
116. Sorrell, T. N.; Vankai, V. A.; Garrity, M. L. *Inorg. Chem.* **1991,** *30,* 207–210.
117. Mahapatra, S.; Kaderli, S.; Llobet, A.; Neuhold, Y.-M.; Palanche, T.; Halfen, I.; Young, A. T.; Kaden, T. A.; Que Jr., L.; Zuberbuehler, A. D.; Tolman, W. B. *Inorg. Chem.* **1997,** *36,* 6343–6356.
118. Sorrell, T. N. *Tetrahedron* **1989,** *45,* 3–68.
119. Monzani, E.; Battaini, G.; Perotti, A.; Casella, L.; Gullotti, M.; Santagostini, L.; Nardin, G.; Randaccio, L.; Geremia, S.; Zanello, P.; Opromolla, G. *Inorg. Chem.* **1999,** *38,* 5359–5369.
120. Casella, L.; Carugo, O.; Gullotti, M.; Garofani, S.; Zanello, P. *Inorg. Chem.* **1993,** *32,* 2056–2067.
121. Santagostini, L.; Gullotti, M.; Monzani, E.; Casella, L.; Dillinger, R.; Tuczek, F. *Chem. Eur. J.* **2000,** *6,* 519–522.
122. Cruse, R. W.; Kaderli, S.; Meyer, C. J.; Zuberbuehler, A. D.; Karlin, K. D. *J. Am. Chem. Soc.* **1988,** *110,* 5020–5024.
123. Karlin, K. D.; Kaderli, S.; Zueberbuehler, A. D. *Acc. Chem. Res.* **1997,** *30,* 139–147.
124. Zhang, C. X.; Liang, H.-C.; Kim, E.-i.; Shearer, J.; Helton, M. E.; Kim, E.; Kaderli, S.; Incarvito, C. D.; Rheingold, A. L.; Karlin, K. D. *J. Am. Chem. Soc.* **2003,** *125,* 634–635.
125. Paul, P. P.; Tyeklar, Z.; Jacobson, R. R.; Karlin, K. D. *J. Am. Chem. Soc.* **1991,** *113,* 5322–5332.
126. Karlin, K. D.; Ghosh, P.; Cruse, R. W.; Farooq, A.; Gultneh, Y.; Jacobson, R. R.; Blackburn, N. J.; Strange, R. W.; Zubieta, J. *J. Am. Chem. Soc.* **1988,** *110,* 6769–6780.
127. Ghosh, P.; Tyeklar, Z.; Karlin, K. D.; Jacobson, R. R.; Zubieta, J. *J. Am. Chem. Soc.* **1987,** *109,* 6889–6891.
128. Kitajima, N.; Koda, T.; Iwata, Y.; Moro-oka, Y. *J. Am. Chem. Soc.* **1990,** *112,* 8833–8839.
129. Obias, H. V.; Lin, Y.; Murthy, N. N.; Pidcock, E.; Solomon, E. I.; Ralle, M.; Blackburn, N. J.; Neuhold, Y.-M.; Zuberbüehler, A. D.; Karlin, K. D. *J. Am. Chem. Soc.* **1998,** *120,* 12960–12961.
130. Pidcock, E.; DeBeer, S.; Obias, H. V.; Hedman, B.; Hodgson, K. O.; Karlin, K. D.; Solomon, E. I. *J. Am. Chem. Soc.* **1999,** *121,* 1870–1878.
131. Taki, M.; Itoh, S.; Fukuzumi, S. *J. Am. Chem. Soc.* **2001,** *123,* 6203–6204.
132. Mahadevan, V.; DuBois, J. L.; Hedman, B.; Hodgson, K. O.; Stack, T. D. P. *J. Am. Chem. Soc.* **1999,** *121,* 5583–5584.
133. Henson, M.; Vance, M.; Zhang, C. X.; Liang, H.-C.; Karlin, K. D.; Solomon, E. I. *J. Am. Chem. Soc.* **2003,** *125,* 5186–5192.
134. Sayre, L. M.; Nadkarni, D. V. *J. Am. Chem. Soc.* **1994,** *116,* 3157–3158.
135. Mandal, S.; Macikenas, D.; Protasiewicz, J. D.; Sayre, L. M. *J. Org. Chem.* **2000,** *65,* 4804–4809.
136. Casella, L.; Monzani, E.; Gullotti, M.; Cavagnino, D.; Cerina, G.; Santagostini, L.; Ugo, R. *Inorg. Chem.* **1996,** *35,* 7516–7525.
137. Monzani, E.; Quinti, L.; Perotti, A.; Casella, L.; Gullotti, M.; Randaccio, L.; Geremia, S.; Nardin, G.; Faleschini, P.; Tabbi, G. *Inorg. Chem.* **1998,** *37,* 553–562.
138. Itoh, S.; Kumei, H.; Taki, M.; Nagatomo, S.; Kitagawa, T.; Fukuzumi, S. *J. Am. Chem. Soc.* **2001,** *123,* 6708–6709.
139. Mirica, L. M.; Vance, M.; Rudd, M.; Hedman, B.; Hodgson, K. O.; Solomon, E. I.; Stack, T. D. P. *J. Am. Chem. Soc.* **2002,** *124,* 9332–9333.
140. Battaini, G.; De Carolis, M.; Monzani, E.; Casella, L. *Chem. Commun.* **2003,** 726–727.
141. Osako, T.; Ohkubo, K.; Masayasu, T.; Tachi, Y.; Fukuzumi, S.; Itoh, S. *J. Am. Chem. Soc.* **2003,** *125,* 11027–11033.
142. Halfen, J. A.; Mahapatra, S.; Wilkinson, E. C.; Kaderli, S.; Young Jr., V. G.; Que Jr., L.; Zuberbuehler, A. D.; Tolman, W. B. *Science* **1996,** *271,* 1397–1400.
143. Que Jr., L.; Tolman, W. B. *Angew. Chem. Int. Ed.* **2002,** *41,* 1114–1137.
144. Tolman, W. B. *Acc. Chem. Res.* **1997,** *30,* 227–237.
145. Oishi, N.; Nishida, Y.; Ida, K.; Kida, S. *Bull. Chem. Soc. Jpn.* **1980,** *53,* 2847–2850.

146. Malachowski, M. R.; Huynh, B. H.; Tomlinson, L. J.; Kelly, R. S.; Furbee, J. W. *J. Chem. Soc. Dalton trans* **1995**, 31–36.
147. Than, R.; Feldman, A. A.; Krebs, B. *Coord. Chem. Rev.* **1999**, *182*, 211–241.
148. Torelli, S.; Belle, C.; Gautier-Luneau, I.; Pierre, J. L.; Saint-Aman, E.; Latour, J. M.; Le Pape, L.; Luneau, D. *Inorg. Chem.* **2000**, *39*, 3526–3536.
149. Borzol, H.; Comba, P.; Pritzkow, H. *Chem. Commun.* **2001**, 97–98.
150. Belle, C.; Beguin, C.; Gautier-Luneau, I.; Hamman, S.; Philouze, C.; Pierre, J. L.; Thomas, F.; Torelli, S.; Saint-Aman, E.; Bonin, M. *Inorg. Chem.* **2002**, *41*, 479–491.
151. Speier, G. *J. Mol. Catal. A* **1986**, *37*, 259–267.
152. Chyn, J. P.; Urbach, F. L. *Inorg. Chim. Acta* **1991**, *189*, 157–163.
153. Balla, J.; Kiss, T.; Jameson, R. F. *Inorg. Chem.* **1992**, *31*, 58–62.
154. Battaini, G.; Monzani, E.; Casella, L.; Santagostini, L.; Pagliarin, R. *J. Biol. Inorg. Chem.* **2000**, *5*, 262–268.
155. Selmeczi, K.; Reglier, M.; Giorgi, M.; Speier, G. *Coord. Chem. Rev.* **2003**, *245*, 191–201.
156. Wegner, R.; Gottshaldt, M.; Poppitz, W.; Jager, E.-G.; Klemm, D. *J. Mol. Catal. A* **2003**, *201*, 93–118.
157. Granata, A.; Monzani, E.; Casella, L. *J. Biol. Inorg. Chem.* **2004**, *9*, 903–913.
158. Berreau, L. M.; Mahapatra, S.; Halfen, J. A.; Houser, R. P.; Young Jr., V. G.; Tolman, W. B. *Angew. Chem. Int. Ed.* **1998**, *38*, 207–210.
159. Karlin, K. D.; Gultneh, Y.; Nicholson, T.; Zubieta, J. *Inorg. Chem.* **1985**, *24*, 3725–3727.
160. Chen, J. S.; Rolle, R. S.; Otwell, W. S.; Balan, M. O.; Marshall, M. R. *J. Agric. Food Chem.* **1991**, *39*, 1396–1401.
161. Chen, J. S.; Wei, C.; Marshall, M. R. *J. Agric. Food Chem.* **1991**, *39*, 1897–1901.
162. Yang, C.-T.; Vetrichelvan, M.; Yang, X.; Moubaraby, B.; Murray, K. S.; Vittal, J. J. *Dalton Trans.* **2004**, 113–121.
163. Reim, J.; Krebs, B. *J. Chem. Soc. Dalton Trans.* **1997**, 3793–3804.
164. Malachowski, M. R.; Dorsey, B.; Sackett, J. G.; Kelly, R. S.; Ferko, A. L.; Hardin, R. N. *Inorg. Chim. Acta* **1996**, *249*, 85–92.
165. Casella, L.; Gullotti, M.; Pessina, C.; Pintar, A. *Gazz. Chim. Ital.* **1986**, *116*, 41–49.
166. Messerschmidt, A. (Ed.), *"Multicopper Oxidases"*; World Scientific: Singapore, **1997**.
167. Wimalasena, K.; Dharmasena, S. *Biochem. Biophys. Res. Commun.* **1994**, *203*, 1471–1476.
168. Kroneck, P. M. H.; Armstrong, F. A.; Merkle, H.; Marchesini, A. *Adv. Chem. Ser.* **1982**, *200*, 223–248.
169. Casella, L.; Monzani, E.; Santagostini, L.; De Gioia, L.; Gullotti, M.; Fantucci, P.; Beringhelli, T.; Marchesini, A. *Coord. Chem. Rev.* **1999**, *185/186*, 619–628.
170. Cole, A. P.; Root, D. E.; Mukherjee, P.; Solomon, E. I.; Stack, T. D. P. *Science* **1996**, *273*, 1848–1850.
171. Root, D. E.; Henson, M. J.; Machonkin, T.; Mukherjee, P.; Stack, T. D. P.; Solomon, E. I. *J. Am. Chem. Soc.* **1998**, *120*, 4982–4990.
172. Taki, M.; Teramae, S.; Nagatomo, S.; Tachi, Y.; Kitagawa, T.; Itoh, S.; Fukuzumi, S. *J. Am. Chem. Soc.* **2002**, *124*, 6367–6377.
173. Karlin, K. D.; Gan, Q.-F.; Farooq, A.; Liu, S.; Zubieta, J. *Inorg. Chem.* **1990**, *29*, 2549–2551.
174. Adams, H.; Bailey, N. A.; Dwyer, M. J. S.; Fenton, D. E.; Hellier, P. C.; Hempstead, P. D.; Latour, J. M. *J. Chem. Soc. Dalton Trans.* **1993**, 1207–1216.
175. Singh, K.; Long, J. R.; Stavropulos, P. *Inorg. Chem.* **1998**, *37*, 1073–1079.
176. Ferrer, S.; Lloret, F.; Bertomeu, I.; Alzuet, G.; Borràs, J.; Garcia-randa, S.; Liu-Gonzàlez, M.; Haasnoot, J. G. *Inorg. Chem.* **2002**, *41*, 5821–5830.
177. Angaridis, P. A.; Baran, P.; Boca, R.; Cervantes-Lee, F.; Haase, W.; Mezei, G.; Raptis, R. G.; Werner, R. *Inorg. Chem.* **2002**, *41*, 2219–2228.
178. Cage, B.; Cotton, F. A.; Dalal, N. S.; Hillard, E. A.; Rakvin, B.; Ramsey, C. M. *J. Amer. Chem. Soc.* **2003**, *125*, 5270–5271.
179. Yoon, J.; Mirica, L. M.; Stack, T. D. P.; Solomon, E. I. *J. Amer. Chem. Soc.* **2004**, *126*, 12586–12595.
180. Mirica, L. M.; Stack, T. D. P. *Inorg. Chem.* **2005**, *44*, 2131–2133.
181. Monzani, E.; Casella, L.; Zoppellaro, G.; Gullotti, M.; Pagliarin, R.; Bonomo, R. P.; Tabbì, G.; Nardin, G.; Randaccio, L. *Inorg. Chim. Acta* **1998**, *282*, 180–192.

182. Santagostini, L.; Gullotti, M.; Pagliarin, R.; Bianchi, E.; Casella, L.; Monzani, E. *Tetrahedron: Asymmetry* **1999**, *10*, 281–295.
183. Santagostini, L.; Gullotti, M.; Pagliarin, R.; Monzani, E.; Casella, L. *Chem. Commun.* **2003**, 2186–2187.
184. Mimmi, M.C.; Gullotti, M.; Santagostini, L.; Pagliarin, R.; De Gioia, L.; Monzani, E.; Casella, L. *Eur. J. Inorg. Chem.* **2003**, 3934–3944.
185. Mimmi, M. C.; Gullotti, M.; Santagostini, L.; Saladino, A.; Casella, L.; Monzani, E.; Pagliarin, R. *J. Mol. Catal. A: Chem.* **2003**, *204/205*, 381–389.
186. Mimmi, M. C.; Gullotti, M.; Santagostini, L.; Battaini, G.; Monzani, E.; Pagliarin, R.; Zoppellaro, G.; Casella, L. *Dalton Trans.* **2004**, 2192–2201.
187. Fraterrigo, T. L.; Miller, C.; Reinhammar, B.; McMillin, D. R. *J. Biol. Inorg. Chem.* **1999**, *4*, 183–187.

GREEN OXIDATION OF ALCOHOLS USING BIOMIMETIC Cu COMPLEXES AND Cu ENZYMES AS CATALYSTS

ISABEL W.C.E ARENDS[a,*], PATRICK GAMEZ[b] and ROGER A. SHELDON[a]

[a]Laboratory for Biocatalysis and Organic Chemistry, Department of Biotechnology, Delft University of Technology, Julianalaan 136, 2628 BL Delft

[b]Leiden Institute of Chemistry, Gorlaeus Laboratories, Leiden University, P.O. Box, 2300 RA Leiden, The Netherlands

I. Introduction

The oxidation of alcohols to carbonyl compounds is one of the most fundamental and important processes in the fine chemical industry. The classical methodology is based on the stoichiometric use of heavy metals, notably Cr and Mn (*1,2*). Alternatively metal-free oxidation, such as the Swern and Pfitzner-Moffat protocols, is based on e.g., dimethylsulfoxide as oxidant in the presence of an activating reagent such as N,N′-dicyclohexylcarbodiimide, an acid anhydride or acid halide (*3*). Although the latter methods avoid the use of heavy metals, they usually involve moisture-sensitive oxidants and environmentally undesirable reaction media, such as chlorinated solvents. The desired oxidation of alcohols only requires the formal transfer of two hydrogen atoms, and therefore the atom economy of these methods is extremely disadvantageous. The current state of the art in alcohol oxidations

ADVANCES IN INORGANIC CHEMISTRY
VOLUME 58 ISSN 0898-8838 / DOI: 10.1016/S0898-8838(05)58006-1

relies on the use of chemocatalysts employing molecular oxygen or dihydrogen peroxide as the final oxidant. In this way water is the only byproduct formed. Among the best methods reported Pd (*4*), Pt (*5,6*), Ru (*6*) and W-based catalysts (*7*) have demonstrated large potential. In the search for new catalysts, Cu-based methods would constitute an attractive alternative (*8,9*). Copper is a cheap, non-toxic metal which is predominant in many enzymes which are involved in the oxidation of alcohols and phenolic structures (as e.g., in lignin). These copper-containing oxidases provide a wealth of inspiration to the design of efficient copper-catalysts. In this chapter we will deal with examples of biomimetic and biocatalytic copper-based redox-catalysts that could be used for the green and selective oxidation of alcohols.

Copper-containing oxidases can be divided into enzymes with mononuclear copper-sites, dinuclear copper-sites, and enzymes with multinuclear copper sites (*10*). Many of these are employed by nature for the oxidative dehydrogenation of natural alcohols employing dioxygen as the ultimate oxidant. Besides alcohols, aliphatic and aromatic amines are also substrates for copper-oxidases. Major advances have been made toward understanding the structure/function relationship of the active sites in copper-containing oxidases. The publication of the crystal structures of galactose oxidase (*11*), catechol oxidase (*12*), and laccase (*13*) are important recent milestones. They also represent notable examples of mononuclear, dinuclear, and multinuclear active sites of copper oxidases respectively. The most appealing example for biocatalytic alcohol oxidation is no doubt formed by galactose oxidase (GOase). Galactose oxidase [E.C. 1.1.3] is secreted by certain fungi into the extracellular environment, where it functions as a broad-spectrum catalyst for oxidation of primary alcohols and a source of dihydrogen peroxide (*14*). The active site of GOase involves a mononuclear copper(II) species with a distorted square pyramidal geometry. It combines two distinct one-electron acceptors, a Cu(II) metal center and a stable protein free radical, into a metalloradical complex that functions as a two-electron redox unit during catalytic turnover. The enzyme itself shows – up till now – limited practical use, due to its low reactivity (see below), although some interesting examples have been reported. However, galactose oxidase has been the major source of inspiration for scientists over the last two decades in designing an efficient copper catalyst for the aerobic oxidation of alcohols, and excellent examples have been reported in the literature. A large part of this chapter will be devoted to biomimetic models of galactose oxidase and an overview will be given.

A second copper-enzyme which has demonstrated large potential in the catalytic oxidation of alcohols is laccase (*15*). This blue-copper, one-electron transfer, enzyme requires mediators to extend its catalytic action. Significant progress in its use has been made in the last five years and an overview thereof will be given as well. Finally, the class of dinuclear-copper oxidases, catechol oxidases and tyrosinases, and mimics thereof, have shown to be of limited use in the oxidation of aliphatic and aromatic alcohols. Their potential lies in the oxidation of *ortho*-diphenols to the corresponding quinones. For an overview of these dinuclear copper oxidases, the reader is referred to the contributions of L. Casella and K.D. Karlin elsewhere in this Volume.

II. Enzymatic Systems Using Cu-oxidases

A. Galactose Oxidase

Galactose oxidase (EC 1.1.3) in nature catalyzes the oxidation of D-galactose to D-galactohexodialdose (see Fig. 1, reaction 1). A detailed

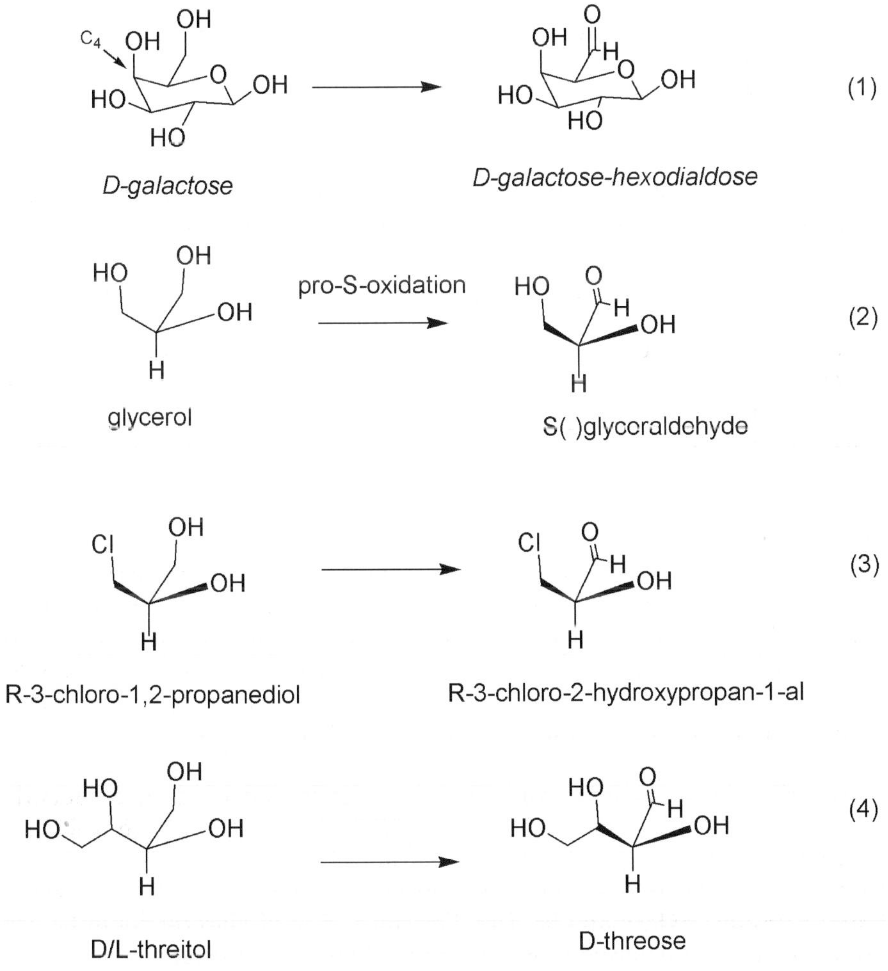

Fig. 1. Stereospecific oxidation of substrates by galactose oxidase (adapted from Klibanov *et al.* (*29*) and Drueckhammer *et al.* (*30*).

discussion on its structure and mechanism is given below in Section III. This fungal GOase can now be produced using recombinant techniques (*16,17*) making it possible to economically produce large quantities of the enzyme. GOase is highly selective for D-galactose in biological samples containing unsubstituted hexoses. Investigations reveal that the substituents at C_4 in carbohydrates must be axial and not significantly larger than –OH, in order for the GOase to be active. Thus, in the native enzyme glucose is not a substrate. Substitution at C_1, C_2, or C_3 however is without significant effect (*18–23*). It must be noted that recently by directed evolution, a galactose oxidase was obtained that could oxidize D-glucose (*13,24*). Simpler molecules such as dihydroxyacetone (*25*), glycerol, 1,3-propanediol, salicyl alcohol (*26*), hydroxypyruvate, 2-methylene-1,3-propanediol, and glycolaldehyde (*27*) are also oxidized at reasonable rates.

The primary alcohol group at C-6 of galactose is oxidized stereospecifically with removal of the pro-S hydrogen (*28*). Klibanov showed that D-galactose, glycerol, and R-3-chloro-1,2-propanediol are oxidized by GOase with the same stereospecificity leading to S (−) glyceraldehyde, and R-3-chloro-2-hydroxypropan-1-al, respectively (*29*) (Fig. 1). It thus appears that the stereospecificity of galactose oxidase catalyzed oxidations of unnatural substrates can be predicted on the basis of the conformation of the natural substrate of the enzyme, D-galactose. An application thereof is the kinetic resolution of D/L-threitol, to yield D-threose, as exemplified in reaction 4 (*30*).

Despite its high selectivity in the oxidation of hexoses, GOase displays a rather broad substrate range for nearly all primary alcohols. In Table I the relative activities of alcohol substrates are given. As shown in the table, α-carbonyl alcohols as a class are the best substrates. Unsaturated or aromatic substrates constitute another reasonable good group of substrates. The inherent chemical reactivity of the alcohol is not a determinant in itself since substrates with comparable electronic properties differ markedly as substrates. For pyridine carbinols, the *m*- and *p*-derivatives are more reactive than the o-derivative. Remarkably *m*-substituted benzyl alcohols display larger reactivities than *p*-substituted benzyl alcohols.

Whittaker performed quantitative structure-activity correlation studies and showed that within a series of *m*-substituted alcohols the rates correlate with σ_m. Thus the greater reactivity of *m*-methoxybenzyl alcohol relative to chloro- and nitro-benzyl alcohol is consistent with the electronic properties of the benzylic bond. This would correspond to hydrogen atom transfer as a key mechanistic step (see below).

Examples of preparative scale conversions with galactose oxidase are scarce (*31*). The oxidations proceed rather slowly and yields are generally low. An interesting application was published recently, where galactose oxidase was employed for the transformation of glycerol to L-glyceraldehyde, which in turn was used in a one-pot multienzyme synthesis of L-fructose (*32*). Using 75 U ml^{-1} GOase and 93 U ml^{-1} catalase, 6 % of glycerol could be converted to L-glyceraldehyde in 48 h (see Fig. 2). In this case galactose oxidase has a pH optimum at 6.7 but is active over a wide range of 5.5 to 8.0. Catalase was added to the system to destroy dihydrogen peroxide and thus prevent

TABLE I

RELATIVE REACTIVITIES OF ALCOHOL SUBSTRATES TO BE OXIDIZED BY GALACTOSE OXIDASE

galactose oxidase

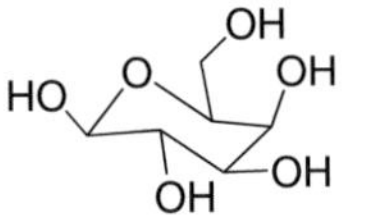

100

Aliphatic alcohols		α,β-Unsaturated alcohols		Aromatic alcohols	
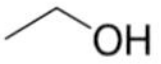	0.003	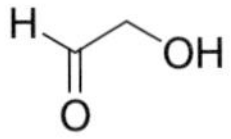	75	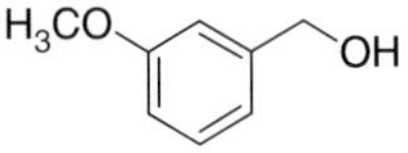	32
	0.002	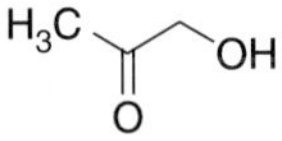	4.8	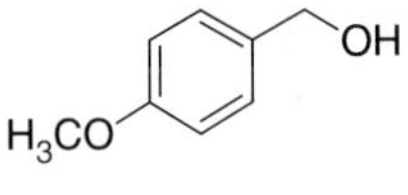	3.4[b]
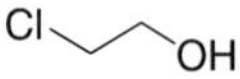	0.06	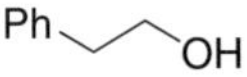	4.8	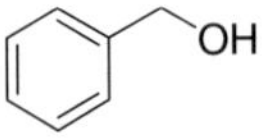	3.3[b]
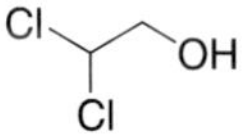	0	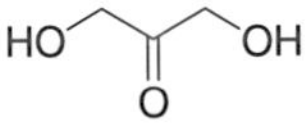	377	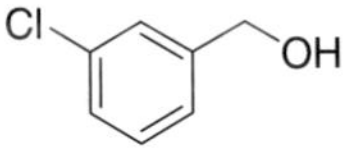	17[b]
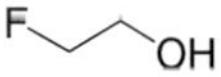	0	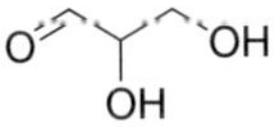	301	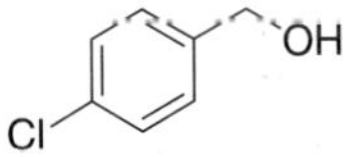	3.5
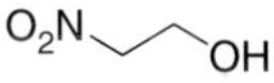	1.6		3.5	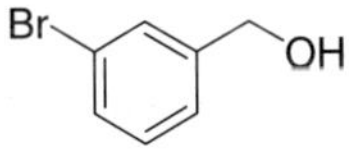	16[b]
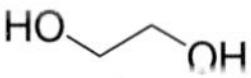	0.1	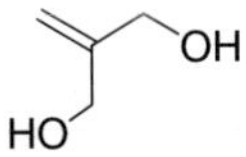	112	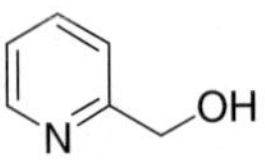	3.2
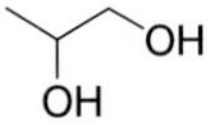	0.85	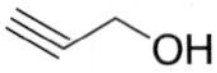	60	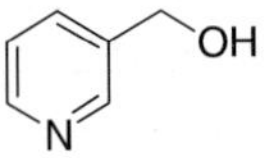	14
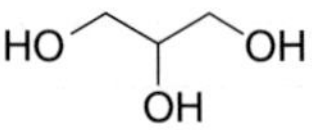	3.4[a]			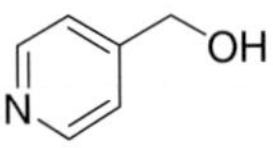	14

Adapted from Kosman (*108*).
[a]From Wong (*32*).
[b]From Whittaker (*109*); Data were calculated using the value for *m*-methoxy benzyl alcohol as a reference.

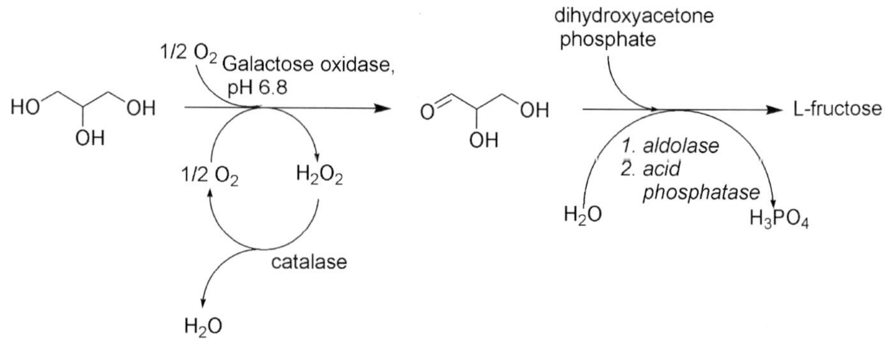

FIG. 2. One pot synthesis of L-Fructose, starting from glycerol and dihydroxyacetone phosphate using GOase and three other enzymes.

inactivation of the enzyme. The K_m for glycerol was determined as 1.15 M (vs. 0.055 M for the natural substrate galactose) and k_{cat} amounts to 7.7 min^{-1} (vs. 11 min^{-1} for galactose).

B. Laccase

Recently several laccases have been crystallized and their three-dimensional crystal structures are now available (*13,33–35*). Laccase belongs to the class of multicopper-oxidases and contains four copper centers per protein molecule: type 1 (T1) or blue copper, type 2 (T2) or normal copper, and type 3 (T3) or coupled dinuclear copper centers (*36*). It catalyzes the oxidation of electron rich aromatic substrates, usually phenols or aromatic amines, *via* four single electron oxidation steps concomitant with the four electron reduction of O_2 to H_2O (*15,37*). Electron transfer takes place at the T1 site, which is followed by electron transfer to the T2/T3 assembly. This is where the reduction of dioxygen takes place.

Laccases play an important role in the delignification of lignocellulose: They are secreted by white rot fungi as extracellular, glycosylated enzymes (*38,39*). Its good thermal stability coupled with its lack of substrate inhibition and high oxidation rates (10–100 fold higher than those of lignin peroxidase or manganese peroxidase) make laccase an ideal candidate for the development of enzymatic oxidation processes (*40,41*).

In lignin degradation, as well as in most applications, laccase alone is ineffective, since it is too large a molecule to penetrate the cell walls of wood and react with the lignin. Consequently, so-called mediators, low-molecular weight electron transfer agents, are employed to shuttle electrons from the lignin to the enzyme. For example, 3-hydroxy anthranilic acid is produced by the white rot fungus *Pycnoporos cinnabarinus* and is believed to play the role of an electron mediator (*42*). 2,2′-Azinobis-(3-ethylbenzothiazoline-6-sulfonic acid (ABTS) was the first compound found to be capable of mediating the laccase-catalyzed oxidation of non-phenolic model compounds, such as the oxidation of veratryl alcohol to the corresponding aldehyde (*38*). Subsequently, 1-hydroxybenzotriazole (*43,44*) and other N-hydroxy compounds

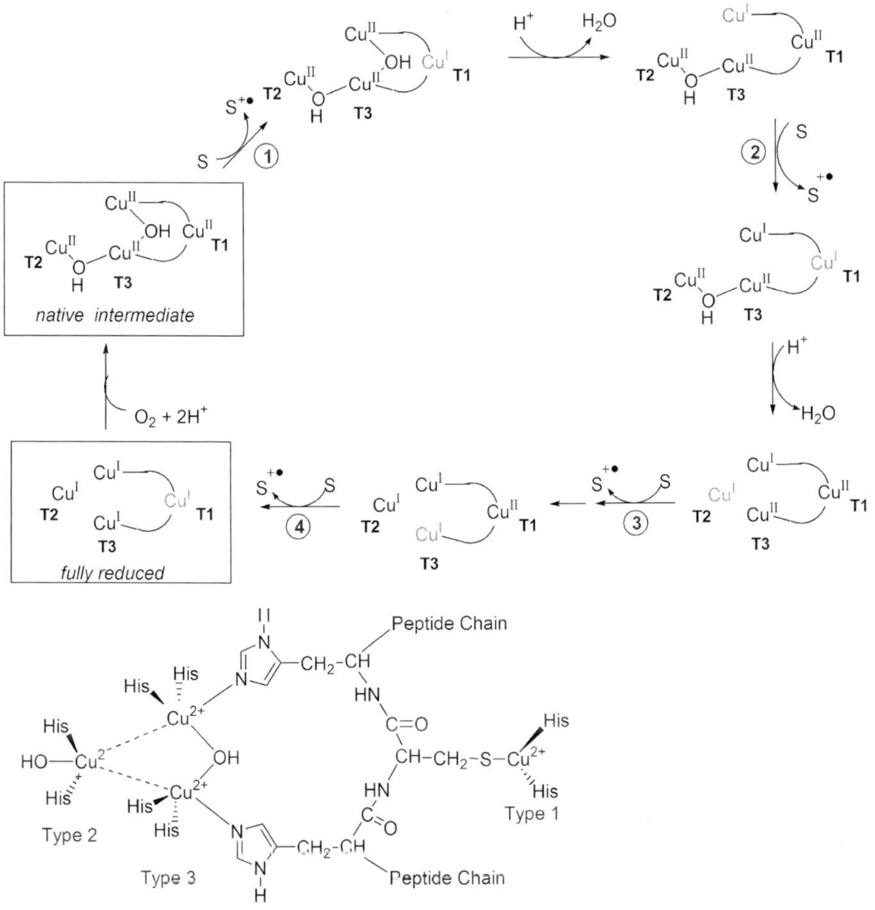

FIG. 3. Catalytic cycle of blue copper proteins, adapted from Solomon *et al.* (*36*).

such as N-hydroxyacetanilide, violuric acid, and N-hydroxyphthalimide were shown to act as mediators (*45*). A common feature of these mediators appears to be the propensity to form N-oxy (nitroxyl) radicals (see Fig. 4).

In 1996 it was shown that by using ABTS as a mediator, laccase from *Trametes versicolor* was able to catalyze the aerobic oxidation of a series of benzylic alcohols to the corresponding benzaldehydes (*46*). Subsequently, Galli, and co-workers found that the stable N-oxy radical, 2,2,6,6-tetramethyl-piperidinyl-1-oxy (TEMPO, Fig. 4) in combination with laccase from *Trametes villosa*, catalyzes the aerobic oxidation of primary benzylic alcohols (*47*). The selective oxidation of the primary alcohol moiety in carbohydrates had been previously reported in two patents (*48,49*).

The laccase/mediator system can be schematically represented as in Fig. 5.

In a subsequent comparison of various mediators, in the laccase-catalyzed aerobic oxidation of benzylic alcohols (Table II), TEMPO proved to be the most effective (*47,50*).

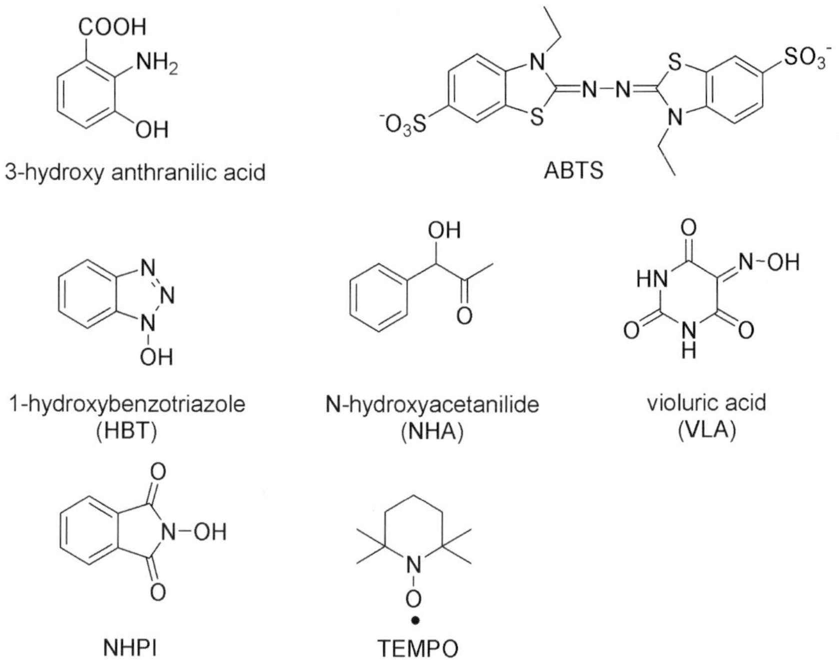

FIG. 4. Laccase mediators.

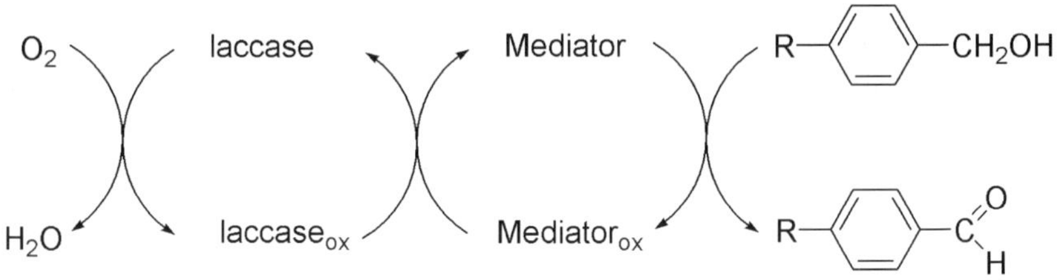

FIG. 5. Laccase/mediator catalyzed oxidation of alcohols.

TABLE II

LACCASE/MEDIATOR CATALYZED OXIDATION OF BENZYLIC ALCOHOLS (*47,50*)

Mediator	Yield of aldehyde (%)	
	Benzyl alcohol	Veratryl alcohol
ABTS	2	37
VLA	42	61
NHPI	54	74
HBT	30	92
TEMPO	92	99

[alc] 20 mM; [med] 30 mol %, 24 h, pH 5, laccase from *Trametes villosa*: 150 U/mmol.

The mechanistic details of these laccase/mediator catalyzed aerobic oxidations are still a matter of conjecture (*51–54*). However, experiments with a probe alcohol point towards one-electron oxidation of the mediator by the oxidized (cupric) form of the laccase followed by reaction of the oxidized mediator with the substrate, either *via* electron transfer (ET), e.g., with ABTS, or *via* hydrogen atom transfer (HAT), e.g., with N-hydroxy compounds which form N-oxy radicals (*55*). TEMPO and its derivatives form a unique case: one-electron oxidation of TEMPO affords the oxoammonium cation which oxidizes the alcohol *via* a heterolytic pathway (Fig. 6), giving the carbonyl product and the hydroxylamine. The T1 copper center in fungal laccases has a redox potential of *ca.* 0.8 V vs. NHE. Consequently, fungal laccases can easily oxidize TEMPO to the corresponding oxoammonium cation, since the oxidation potential of the latter, which was first measured by Golubev and co-workers (*56,57*), is 0.75 V. This was confirmed by EPR measurements, which showed that laccase is reduced in the presence of TEMPO: One equivalent of laccase could oxidize at least three equivalents of TEMPO within a few minutes under anaerobic conditions (*58*).

There are various alternatives for reoxidizing the hydroxylamine back to TEMPO to complete the catalytic cycle. It can be oxidized by dioxygen, laccase or the oxoammonium cation. The active oxidant is the same as that in the TEMPO catalyzed oxidations of alcohols with hypochlorite (or other single oxygen donors), a method which is widely used in the oxidation of a broad range of alcohols using low catalyst loadings (1 mol % or less) (*59*).

The laccase/TEMPO system was the subject of further investigations by Galli and Gentili and co-workers (*47,51*) and Sheldon and co-workers (*58*). The substrate scope is large (see Table III), although the reactivity of

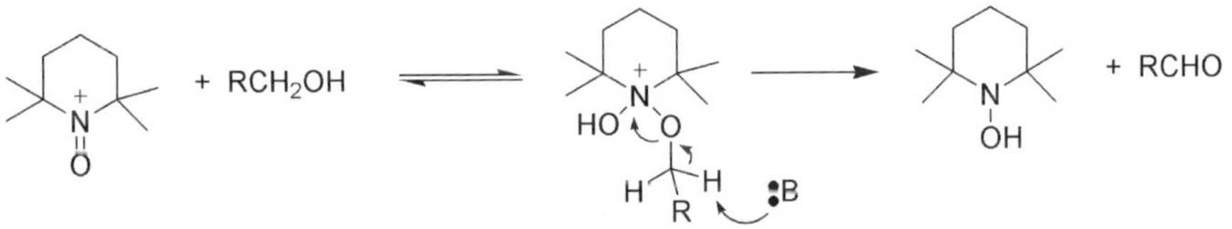

FIG. 6. Pathway of oxoammonium-catalyzed oxidation of alcohols.

TABLE III

OXIDATIONS WITH DIOXYGEN USING A LACCASE/TEMPO SYSTEM (*47,58*)

Substrate	Laccase (U/mmol)	Time	TEMPO	Yield
Benzylalcohol	*Trametes villosa* (150)	24 h	15 mol%	92 %
Benzylalcohol	*Trametes versicolor* (47)	24 h	10 mol%	100 %
1-Phenylethanol	*Trametes versicolor* (47)	24 h	10 mol%	100 %
Benzylalcohol/ 1-Phenylethanol	*Trametes versicolor* (47)	5.5 h	15 mol%	77 %/35 %
Cyclohexanol	*Trametes villosa* (150)	24 h	15 mol%	35 %
1-(*p*-Methoxy-phenyl) ethanol	*Trametes villosa* (150)	24 h	15 mol%	94 %
1-Decanol	*Trametes villosa* (150)	24 h	15 mol%	15 %

aliphatic alcohols is still rather poor. Also secondary alcohols react slower than primary ones. In a competition experiment the conversion reached after 5 h for benzyl alcohol was about twice that of 1-phenylethanol.

The recyclability of the laccase/TEMPO system was investigated as well: When an aqueous buffer solution from a benzyl alcohol oxidation was reused with fresh substrate, the conversion observed after 24 h decreased from 100 to 26 % (*58*). It turned out that severe deactivation was taking place when the laccase solution was incubated with TEMPO in the absence of substrate (oberved conversions after 4 h, were 60, 27 and 18 % when the preincubation times were 0, 19 and 37 h, respectively). In contrast, when the laccase/TEMPO system was subjected to preincubation under an atmosphere of nitrogen, negligible deactivation was taking place. Hence, it was concluded that laccase is not stable in the presence of TEMPO and dioxygen. It seems likely that this involves a suicide deactivation by the enzyme whereby the oxoammonium cation oxidizes amino acid residues in the protein that are essential for activity, and/or oxidizes glycosyl moieties, resulting in loss of activity. In the presence of alcohol substrate the oxoammonium cation preferentially oxidizes the alcohol and, hence, deactivation is to be expected only at high conversion/low concentration of (alcohol) substrate.

III. Biomimetic Systems for Alcohol Oxidation

Galactose oxidase (GOase; EC 1.1.3) is a single polypeptide of 68,000 Dalton molecular mass with a type-2 copper site (*60*). It was first isolated in 1959 (*18*) and its crystal structure was reported in 1991 (*11,61*). This fungal enzyme selectively catalyzes the efficient oxidation of primary alcohols to the corresponding aldehydes with concomitant reduction of molecular oxygen to dihydrogen peroxide (*62*). The active site of GOase involves a mononuclear copper(II) species with a distorted square pyramidal geometry (Fig. 7). As the oxidation of the alcoholic substrate requires 2 e^- and copper is a 1 e^- redox metal ion, one of the phenol ligands is most likely involved in the oxidative transformation. Indeed, in the active form of the biocatalyst, the equatorial plane is formed by two nitrogen atoms from two histidine imidazoles (His496 and His581), one oxygen atom of an *ortho*-S-modified tyrosyl radical (Tyr272) and one oxygen atom of a water molecule. The axial position is occupied by a second tyrosinate ligand (Tyr495).

The catalytic mechanism of GOase has been extensively studied (Fig. 8) (*63,64*). The primary alcohol first coordinates to the active species **A**, leading to the formation of the metal-phenoxyl radical complex **B**. This species undergoes proton abstraction from the substrate by the axial tyrosinate (Tyr495), followed by a rapid intramolecular electron transfer from the intermediate ketyl radical anion with reduction of Cu^{II} to Cu^{I}. The copper(I) species **C** reacts with dioxygen to form the hydroperoxo copper(II) complex **D** with the liberation of the aldehyde. Finally, dihydrogen peroxide is released to give back the active form of the enzyme.

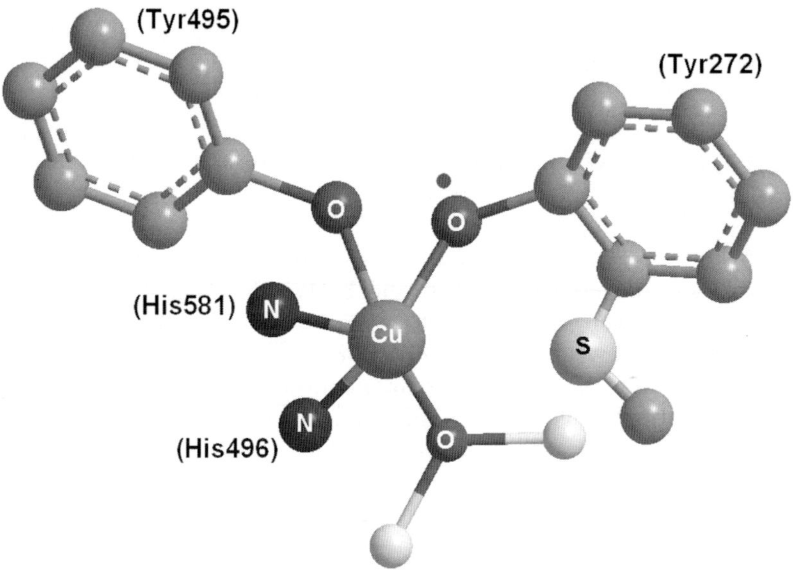

FIG. 7. Active site of galactose oxidase (GOase) (*11*).

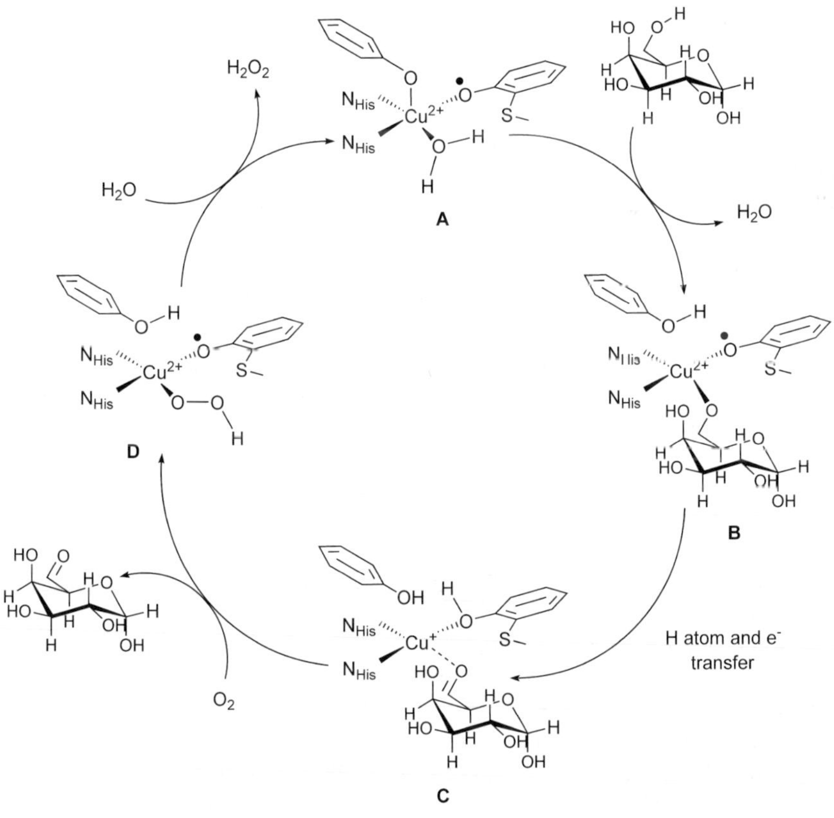

FIG. 8. Proposed catalytic cycle of GOase (*63,64*).

Since the past twenty years, numerous structural models have been reported in the literature (*65–68*), but it appeared to be more difficult to mimic the catalytic activity of GOase (*69–72*).

A. Bio-inspired Catalysts Using Dioxygen as Oxidant: [Copper(II)-Schiff Base] Catalyst (*73*)

The first efficient synthetic biomimetic copper catalysts were published in 1998 by Stack and co-workers (*73*) and by Wieghardt and co-workers (*74*). Stack and Wang were one of the first to describe model Cu(II) complexes that reproduce many of the particular properties of GOase, including the catalytic conversion of an alcohol to an aldehyde in the presence of 20 equivalents of tris(4-bromophenyl)aminium hexachloroantimonate $(TPA^+)[SbCl6^-]$ as oxidant, and 20 equivalents of a strong base, i.e., *n*-BuLi (*75*). The ligands used for this study are salen-type compounds with a binaphthyl backbone and thioether functions (Fig. 9), which provide a single copper ion with an N_2O_2 coordination environment.

Thus, the crystal structure of the copper compound obtained from $Cu(OAc)_2 \cdot H_2O$ and the ligand BSI (Fig. 9, $R_1 = SPr^i$ and $R_2 = Bu^t$) in methanol confirms the formation of a monomeric, four-coordinate, distorted square planar Cu(II) complex (Fig. 10). It should be noted that a complex with the very same (chiral) ligand framework was already published almost fifty years ago by Lions and Martin (*75*).

This Cu^{II} complex has spectroscopic characteristics similar to those of GOase, such as the EPR spectrum, but it is only able to perform the stoichiometric oxidation of benzyl alcohol to benzaldehyde (*76*). Two years later, Stack and co-workers reported a significant improvement of the activities: High turnover numbers (TON) were reached for reactions carried out at 1 atm of O_2 and room temperature (*73*). Two N_2O_2 donors, namely BSP (Fig. 9, $R_1 = SPh$ and $R_2 = Bu^t$) and BDB (Fig. 9, $R_1 = Bu^t$ and $R_2 = Bu^t$), were used as copper ligands to generate catalysts that can efficiently convert neat benzylic and allylic alcohols to their respective aldehydes or ketones in the presence of a basic co-catalyst, i.e., lithium or sodium methoxide (Table IV) (*73*).

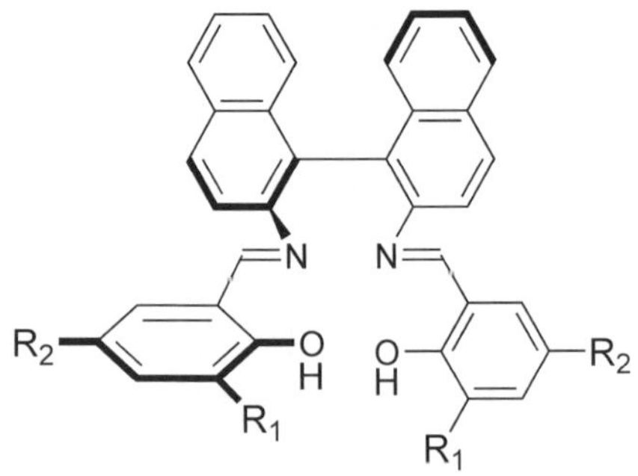

Fig. 9. Structure of the N_2O_2 ligand designed to mimic the active site of GOase upon coordination with copper(II) ion (*75*).

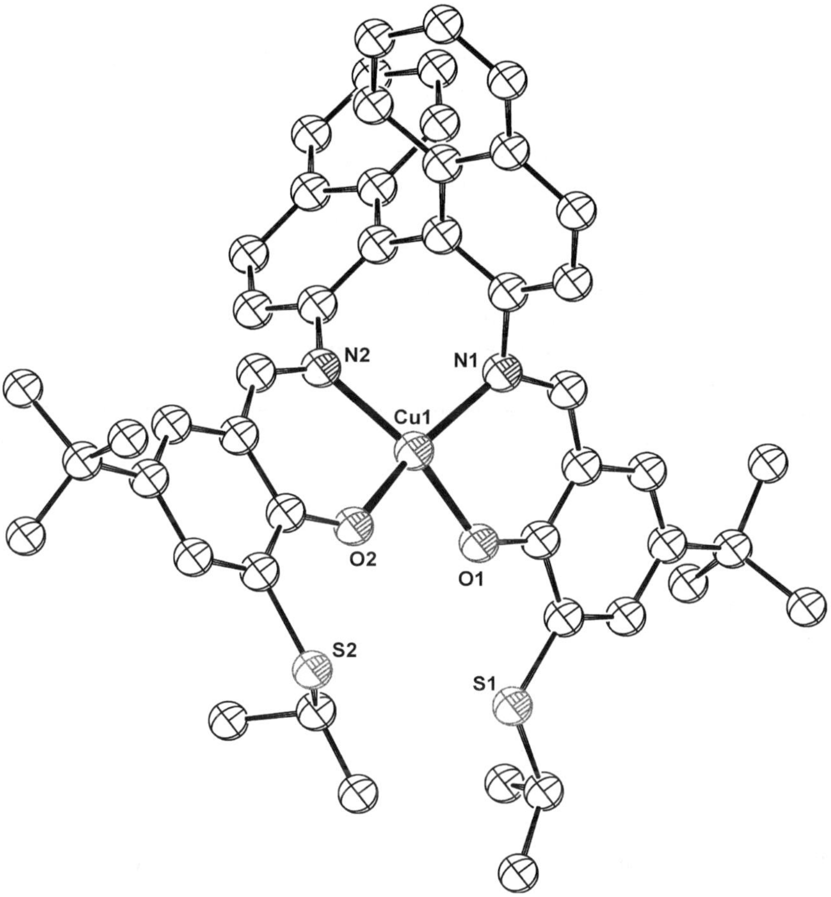

FIG. 10. ORTEP representation (50 % thermal ellipsoids) of the X-ray crystal structure of [Cu(II)BSI]. Selected distances (Å) are as follows: Cu1–O1, 1.900; Cu1–O2, 1.894; Cu1–N1, 1.959; Cu1–N2, 1.982 (*75*).

TABLE IV

TURNOVERS ACHIEVED IN THE [Cu^{II}(BSP)]- AND [Cu^{II}(BDB)]-CATALYZED OXIDATION OF ALCOHOLS AT ROOM TEMPERATURE (*73*)

Catalyst[a]	TON/Solvent-free		TON/1.5 M solution in acetonitrile		
	Benzyl alcohol[b]	1-Phenylethanol	Benzyl alcohol[c]	1-Phenylethanol	Cinnamyl alcohol
[Cu^{II}BSP(BF_4)]$^{\bullet}$	1300	–	–	–	–
[Cu^{II}BSP]	1300	60	400	30	40
[Cu^{I}BSP]$^-$	900	–	–	–	–
[Cu^{II}BDB]]	200	30	40	30	30

[a]Typical conditions: 0.05 mol% Cu-complex, 0.2–0.5 mol% base, 1 atm O_2, RT, 20 h

[b]0.01 % catalyst and 2 % base.

[c]0.005 mol% catalyst and 0.2 mol% base.

Thus, a TON of more than 1000 was achieved using 0.01 to 0.06 % of the [Cu^{II}BSP] coordination compound as catalyst and no further oxidation of the aldehyde products was observed. The oxidation reactions can also be performed in acetonitrile solutions, although the catalyst is much less efficient in this solvent (Table IV). The lower activity of [Cu^{II}BDB] may find two explanations: (1) steric hindrance generated by the four bulky *tert*-butyl groups which hinders an easy coordination of the alcoholic substrate to the copper ion; (2) absence of a thioether bond which may play a role during the catalytic cycle of GOase.

A mechanism for this reaction has been proposed (Fig. 11) from investigations with the most effective catalyst and substrate, [Cu^{II}BSP], and benzyl alcohol, respectively. The catalytic reaction may be started with the catalyst in three different forms (Table IV and Fig. 11): the aerobically isolated form

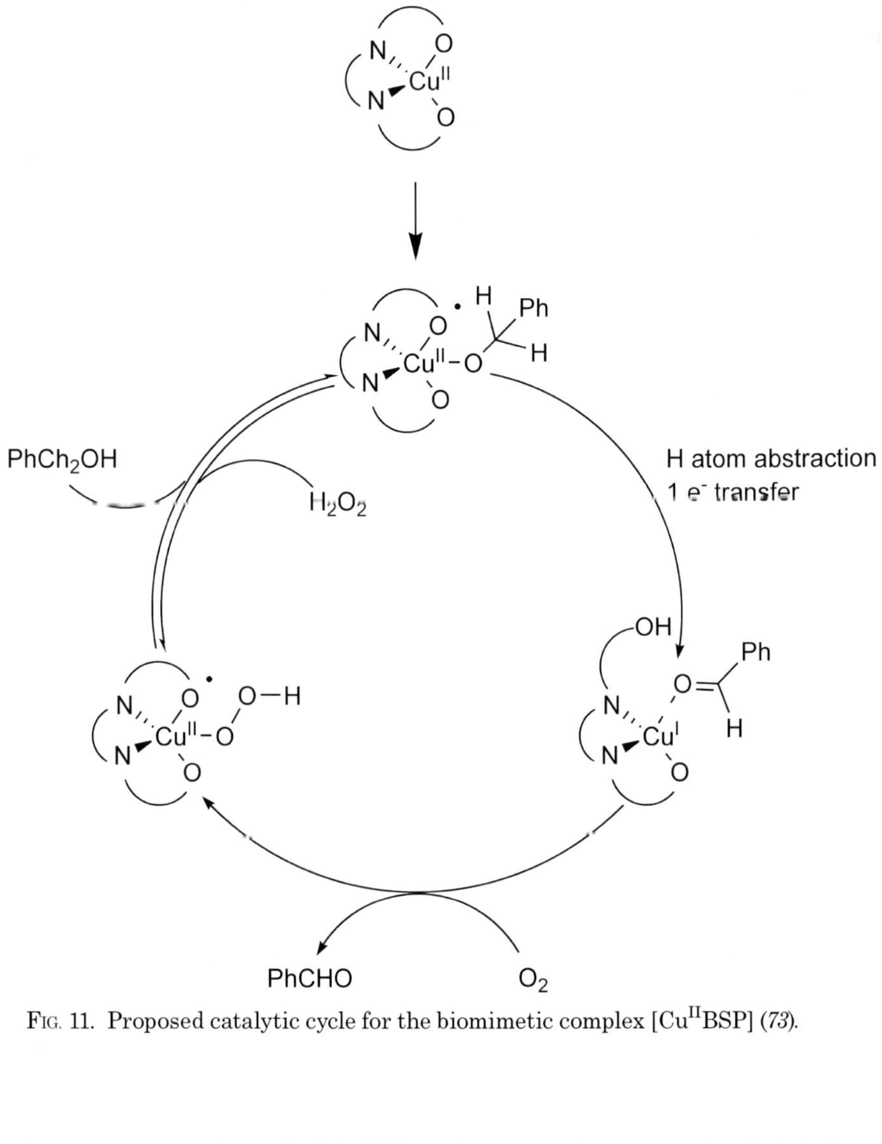

FIG. 11. Proposed catalytic cycle for the biomimetic complex [Cu^{II}BSP] (*73*).

$[Cu^{II}BSP]$, the preoxidized form $[Cu^{II}BSP(BF_4)]^{\bullet}$, and and the pre-reduced form $[Cu^{I}BSP]^{-}$. The alcohol first coordinates to the catalyst precursor $[Cu^{II}BSP]$, leading to the formation of a metal-phenoxyl radical complex. This species undergoes the substrate C^{α}-H proton abstraction by the radical, followed by a rapid intramolecular electron transfer with reduction of Cu^{II} to Cu^{I}. After this rate-determining step, the copper(I) species reacts with dioxygen to form an hydroperoxo copper(II) with the release of the carbonyl product. Finally, dihydrogen peroxide is replaced by a new alcohol molecule to give back the active species (Fig. 11).

More recently, Stack and co-workers reported new model compounds based on the investigations described above (*70,77*). A monophenolate-copper(II) complex (Fig. 12b) (*70*) was obtained from 2,4-di-*tert*-butyl-6-(2-*tert*-butyl-1,10-phenanthrolin-9-yl)phenol (Fig. 12a, HL^{Phen}), $[Cu^{I}(CH_3CN)_4](CF_3SO_3)$, and triethylamine in 1:1 dichloromethane/methanol followed by exposure to dioxygen. The crystal structure of $[L^{Phen}Cu^{II}(CF_3SO_3)]$ is depicted in Fig. 12b. The highly distorted copper(II) complex becomes EPR-silent upon oxidation with the strong one-electron oxidant tris(4-bromophenyl)aminium hexachloroantimonate, reflecting the formation of a radical species analogous to the antiferromagnetically coupled Cu^{II}-phenoxyl radical in the oxidized form of GOase (*78*).

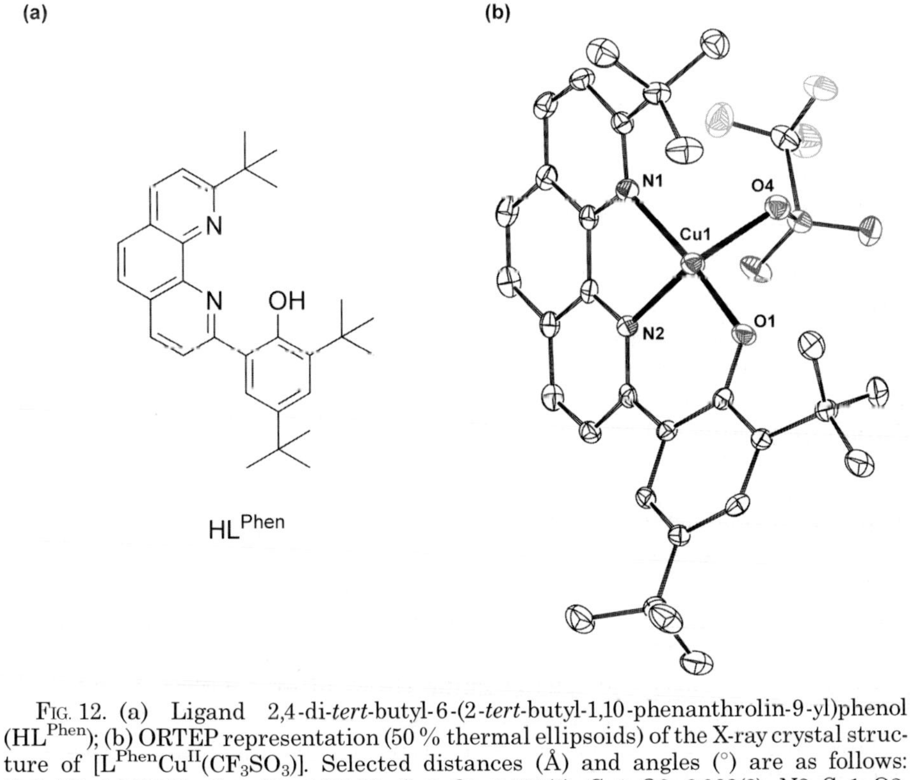

FIG. 12. (a) Ligand 2,4-di-*tert*-butyl-6-(2-*tert*-butyl-1,10-phenanthrolin-9-yl)phenol (HL^{Phen}); (b) ORTEP representation (50 % thermal ellipsoids) of the X-ray crystal structure of $[L^{Phen}Cu^{II}(CF_3SO_3)]$. Selected distances (Å) and angles (°) are as follows: Cu1–N1, 1.9802(2); Cu1–N2, 1.921(2); Cu1–O1, 1.838(2); Cu1–O2, 2.082(2); N2–Cu1–O2, 137.8(1) (*70*).

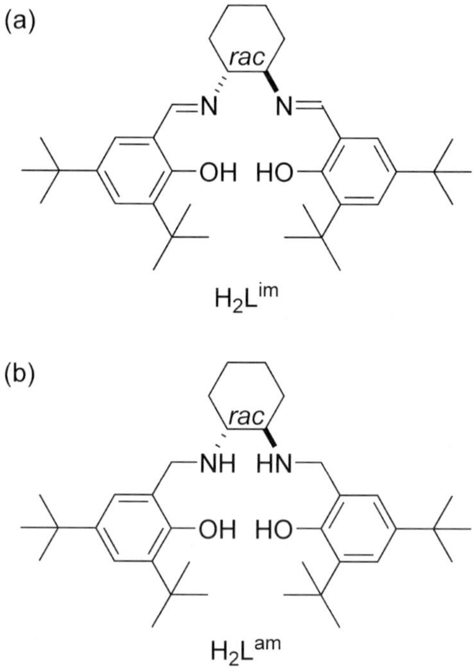

FIG. 13. (a) (±)-*N,N'*-bis(3,5-di-*tert*-butylsalicylidene)-*trans*-cyclohexane-1,2-diamine (H_2L^{im}); (b) (±)-*N,N'*-bis(3,5-di-*tert*-butylsalicyl)-*trans*-cyclohexane-1,2-diamine (H_2L^{am}) (*76*).

GOase model compounds were lately prepared by Stack and Pratt from two Schiff base derivatives, namely (±)-*N,N'*-bis(3,5-di-*tert*-butylsalicylidene)-*trans*-cyclohexane-1,2-diamine and (±)-*N,N'*-bis(3,5-di-*tert*-butylsalicyl) *trans*-cyclohexane-1,2-diamine (H_2L^{am}) as shown in Fig. 13 (*77*). Reaction of $Cu(OAc)_2 \cdot H_2O$ and sodium hydroxide with H_2L^{im} (Fig. 13a) or H_2L^{am} (Fig. 13b) in methanol yielded two monocopper complexes, which showed spectroscopic features similar to those of GOase upon oxidation with silver hexafluoroantimonate. In addition, these two radical compounds could perform the (stoichiometric) oxidation of benzyl alcohol to benzaldehyde, thereby mimicking the oxidizing half-reaction of GOase.

B. [Cu^{II}(2,2′-thiobis(2,4-di-*tert*-butylphenol))] CATALYST (*74*)

Wieghardt and co-workers similarly reported the catalytic oxidation of primary and secondary alcohols by a dinuclear Cu(II)–phenoxyl complex at 20°C under air (*74*). The reaction of $Cu^{I}Cl$ with 2,2′-thiobis(2,4-di-*tert*-butylphenol) (Fig. 14, LH_2) and triethylamine (1:1:2) in dry methanol under air leads to a neutral mononuclear blue [$Cu^{II}L(NEt_3)$] complex (Fig. 14).

This coordination compound was prepared to model the active site of GOase. Its crystal structure shows a non-expected complex where the sulfur atom of the tridentate dianion L^{2-} is coordinated to the uncommon

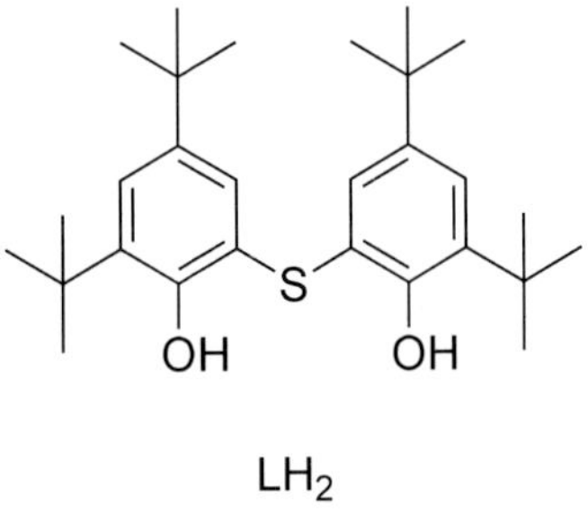

FIG. 14. 2-(3,5-di-*tert*-butyl-2-hydroxyphenylthio)-4,6-di-*tert*-butylphenol (LH_2) (*74*).

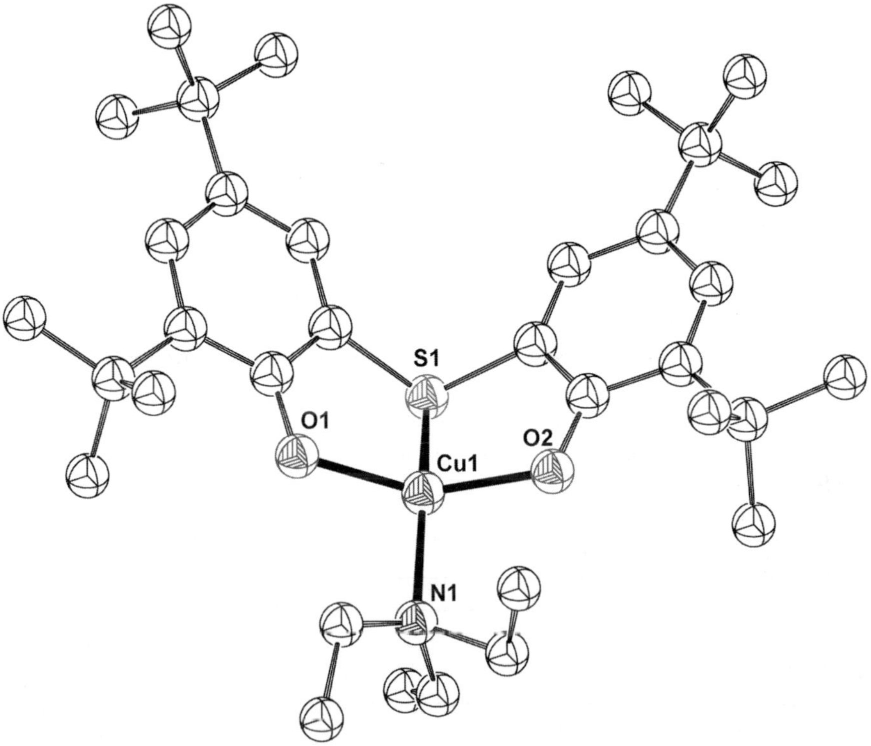

FIG. 15. ORTEP representation (50 % thermal ellipsoids) of the X-ray crystal structure of $[Cu^{II}L(NEt_3)]$. Selected distances (Å) and angles (°) are as follows: Cu1–O1, 1.889(3); Cu1–O2, 1.895(3); Cu1–N1, 2.050(4); Cu1–S1, 2.290(2); O1–Cu1–O2, 152.2(2); O1–Cu1–N1, 97.9(2); O2–Cu1–N1, 98.0(2); O1–Cu1–S1, 87.6(1); O2–Cu1–S1, 88.0(1); N1–Cu1–S1, 154.0(1) (*74*).

tetracoordinated Cu^{II} ion, resulting in a coordination geometry intermediate between square planar and tetrahedral (Fig. 15).

Only one triethylamine is coordinated to the metal center while two are required to mimic the N donors of the histidine residues (see Fig. 7), and no methanol molecule is present as substrate model. In THF solution, a diamagnetic copper compound was formed which exhibited spectral features

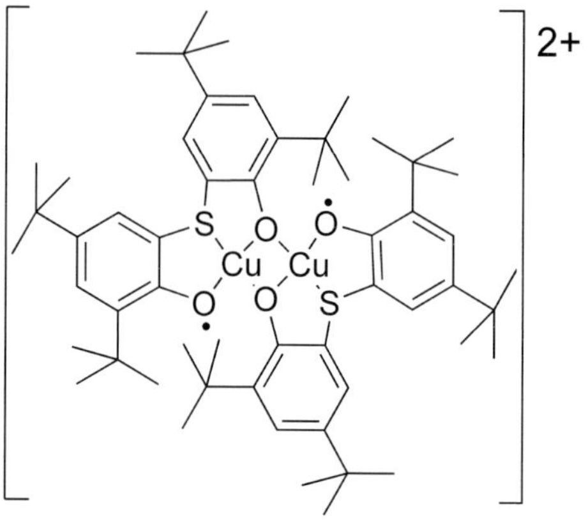

FIG. 16. Proposed dinuclear species present in solution (*74*).

characteristic for coordinated phenoxyl radicals ($\lambda_{max} \approx 404$ nm, $\varepsilon = 8.0 \times 10^3$ L mol^{-1} cm^{-1}; Raman: $\lambda_{exc} = 458$ nm, υ(C–O$^\bullet$) = 1451 cm^{-1}). This complex is most likely dinuclear in solution with two bridging phenolate groups and two phenoxyl ligands as depicted in Fig. 16 (*74*).

For catalysis, 0.2 mol % of the [Cu^{I}Cl(2,2′-thiobis(2,4-di-*tert*-butylphenol)) $(NEt_3)_2$] catalyst prepared *in situ* from copper(I) chloride, LH_2, and triethylamine at 50°C in THF, was used to oxidize several alcohols at room temperature in air. The most important results are summarized in Table V.

Thus, ethanol was converted to acetaldehyde with a yield of 63 % (TON = 315). However the reaction is not chemoselective and several side-products resulting from C–C coupling were detected, i.e., butane-2,3-diol, 3-hydroxybutan-2-one, and butane-2,3-dione (Table V). Similar results were observed for the oxidation of benzyl alcohol with the formation of 60 % (TON = 300) benzaldehyde, small amounts of 1,2-glycol and α-hydroxyketone, and 3–4 % of α-diketone (Table V). Contrary to GOase, this copper catalyst is also able to oxidize secondary alcohols. Thus, 2-propanol and diphenylmethanol could be used as substrates but, in both cases, only the 1,2-diol derivatives, namely 2,3-dimethylbutane-2,3-diol and 1,1,2,2-tetraphenylethane-1,2-diol, resulting from an oxidative C–C coupling were formed, with a yield of 61 and 68 %, respectively (Table V). In contrast, the aerobic oxidation of 2-butanol exclusively led to the formation of butan-2-one (yield = 39 %; TON = 195), suggesting two different reaction pathways for the formation of the carbonyl compounds and for the formation of the 1,2-glycols (Fig. 17, A and B) (*74*).

Ethanol, benzyl alcohol or 2-butanol coordinate to the proposed dinuclear copper complex (Fig. 16) to produce the active species **I** (Fig. 17, A). After H abstraction by one of the two phenoxyl groups, a radical is created which is oxidized to the corresponding carbonyl derivative by the second phenoxyl radical. In the case of 2-propanol and diphenylmethanol, two alcohol molecules coordinate to the dinuclear moiety resulting in species **II** (Fig. 17, B). The H abstractions by the two phenoxyl ligands generate two

TABLE V

[Cu^{II}(2,2′-Thiobis(2,4-di-*tert*-butylphenol))$_2Cl_2$]-Catalyzed Aerobic Oxidation of Primary and Secondary Alcohols[a] (*74*)

Substrate	Products (**Yield** [%])			
Ethanol	Ethanal (**63**)	Butane-2,3-diol (**1.5–2.0**)	3-Hydroxybutan-2-one (**1.0–2.0**)	Butane-2,3-dione (**3–5**)
Benzyl alcohol	Benzaldehyde (**60**)	1,2-Diphenylethane-1,2-diol (**1**)	2-Hydroxy-1,2-diphenylethanone (~**1**)	Benzil (**3–4**)
Propan-2-ol	Propan-2-one (< **2**)	2,3-Dimethylbutane-2,3-diol (**61**)	–	–
Diphenylmethanol	–	1,1,2,2-Tetraphenylethane-1,2-diol (**68**)	–	–
Butan-2-ol	Butan-2-one (**39**)	–	–	–

[a]Typical conditions: 0.125 M alcohol, 0.2 mol % Cu-complex, 0.4 mol % NEt_3, 20°C, air.

secondary radicals which can undergo C–C coupling to give the 1,2-glycol products and the release of dihydrogen peroxide (*74*).

The group of Wieghardt continued to design related phenol-based ligands to prepare synthetic GOase models and an interesting successor compound was described one year later (*79*). The trifluoroacetate salt of the cation *N,N*-bis(2-hydroxy-3,5-di-*tert*-butylphenyl)ammonium $[H_4(L^3)](CF_3CO_2)$ (Fig. 18a)

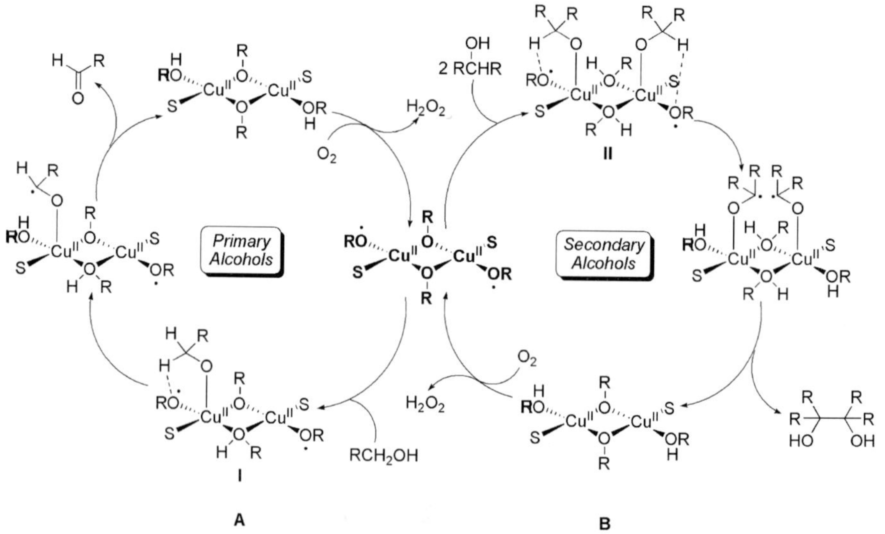

FIG. 17. Proposed catalytic cycles for primary (A) and secondary alcohols (B) (*74*).

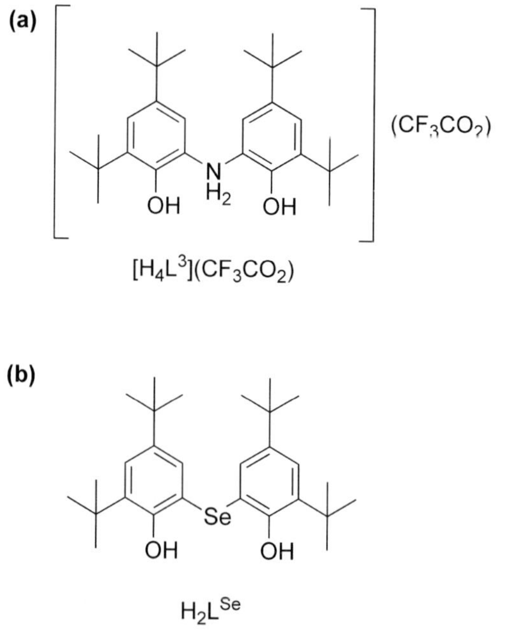

FIG. 18. Bis-phenol ligands developed by Wieghardt and co-workers (*79,82*).

(*80*) reacts with $[Cu^{I}(CH_3CN)_4]ClO_4$ in CH_3CN in the presence of excess NEt_3 to yield a yellow solution, which upon exposure to air at room temperature provides the neutral complex $[Cu^{II}(L^2)(NEt_3)]$ (Fig. 19). The Copper(II) ion is in a slightly distorted square-planar N_2O_2 environment formed by the tridentate $(L^2)^{2-}$ ligand (Fig. 20) (*81*) along with a triethylamine molecule.

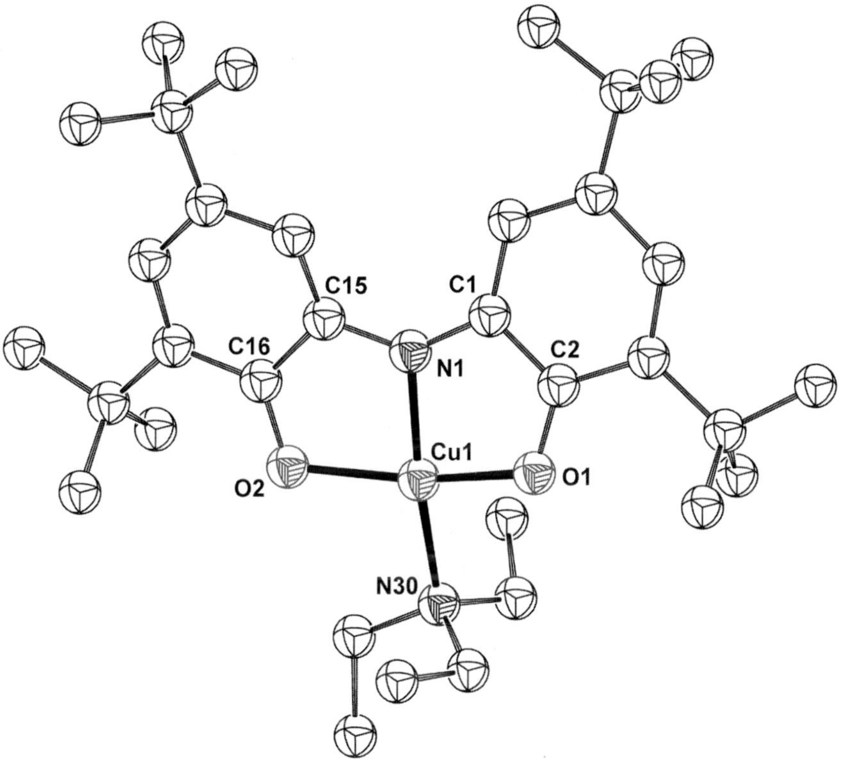

FIG. 19. ORTEP representation (50 % thermal ellipsoids) of the x-ray crystal structure of $[Cu^{II}(L^2)(NEt_3)]$. Selected distances (Å) and angles (°) are as follows: Cu(1)–N(1), 1.906(2); Cu(1)–O(1), 1.926(1); Cu(1)–O(2), 1.959(1); Cu(1)–N(30), 2.010(2); O(1)–C(2), 1.322(2); O(2)–C(16), 1.310(2); N(1)–C(1), 1.384(2); N(1)–C(15), 1.358(2); N(1)–Cu(1)–O(1), 83.83(6); N(1)–Cu(1)–O(2), 83.34(6); O(1)–Cu(1)–O(2), 159.73(6); N(1)–Cu–N(30), 168.13(7); O(1)–Cu(1)–N(30), 96.08(6), O(2)–Cu(1)–N(30), 99.68(6) (*79*).

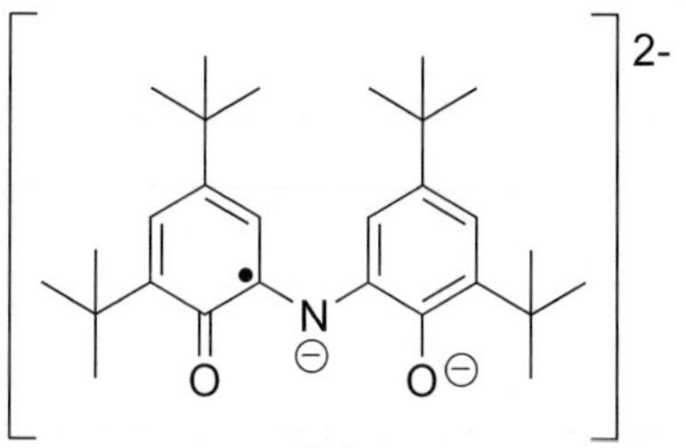

FIG. 20. Tridentate ligand $(L^2)^{2-}$ with a spin $S = 1/2$ (*106*).

The formation of ligand $(L^2)^{2-}$ was proven by variable-temperature susceptibility measurements on solid samples of $[Cu^{II}(L^2)(NEt_3)]$. Indeed, this biomimetic complex exhibits a diamagnetic ground state ($S = 0$) arising from the intramolecular antiferromagnetic coupling between the spins of the radical ligand ($S_L{}^2 = 1/2$) and that of the Cu^{II} ion ($S_{Cu} = 1/2$). The same phenomenon is observed with the active form of the natural enzyme GOase where the copper(II) ion is coupled with a tyrosyl radical.

A solution of $[Cu^{II}(L^2)(NEt_3)]$ in THF can stoichiometrically convert benzyl and ethyl alcohol to benzaldehyde and ethanal, respectively, under anaerobic conditions at room temperature. These oxidation reactions can be performed in open air with catalytic amounts of $[Cu^{II}(L^2)(NEt_3)]$. Thus, 2.65×10^{-5} M catalyst in THF could oxidize a 0.125 M ethanol or benzyl alcohol solution (55 % yield, 2594 TON) within 20 hours reaction time, at room temperature under air. The catalytic reactions could be carried out without solvent, and no over-oxidation products or oxidative C–C coupling compounds could be detected by gas chromatography, which is a significant improvement of the catalytic system described above (see Table V).

The latest variation reported by the group of Wieghardt involves a selenium derivative (Fig. 18b) of the ligands previously described (*82*). Reaction of H_2L^{Se} with Cu^ICl in dry methanol in the presence of triethylamine leads to $[CuL^{Se}(Et_3N)]$ upon exposure to air. $[CuL^{Se}(Et_3N)]$ is isostructural with $[Cu^{II}L(NEt_3)]$ (see Fig. 15).

This mononuclear Cu(II) complex in combination with a base, i.e., tetrabutylammonium methoxide, in acetonitrile catalytically oxidized primary alcohols to the corresponding aldehydes with up to 95 TON in the case of benzyl alcohol. This oxidation could also be performed under solvent-free conditions (e.g., pure benzyl alcohol) with a strong base like butyl lithium. Once again, the only organic product observed is the aldehyde.

C. $[Cu_2^{II}$(2-(Methylthio)-4-*tert*-butyl-6-[[bis[2-(2-pyridyl)ethyl]amino]-methyl]-phenol$)_2](PF_6)_2$ CATALYST (*83–85*)

S-containing phenolic ligands related to LH_2 (Fig. 14) were developed by Itoh *et al.* (Fig. 21) (*83–85*).

Reaction of 1 equivalent $Cu^{II}Cl_2$ with 1 equivalent MeL-H in methanol in the presence of 1 equivalent triethylamine leads to a coordination compound which is further reacted with $NaPF_6$ (*83*). Single crystals suitable for X-ray diffraction were obtained by recrystallization from water–acetonitrile. The dinuclear species, whose molecular structure is depicted in Fig. 22, can be converted into monomeric species by the addition of an external ligand such as pyridine (Py), which exhibit structural and spectroscopic features similar to those of GOase (*84*).

The activated form of the monomeric cation, namely the phenoxyl radical $[Cu^{II}(tBuL^{\bullet})(NO_3)]^+$, can be obtained by treating the dinuclear complex $[Cu_2{}^{II}(tBuL)_2](PF_6)_2$ with ceric ammonium nitrate, i.e., $(NH_4)_2[Ce^{IV}(NO_3)_6]$ (*84*). This $[Cu^{II}(tBuL^{\bullet})(NO_3)]^+$ model complex of the active form of GOase

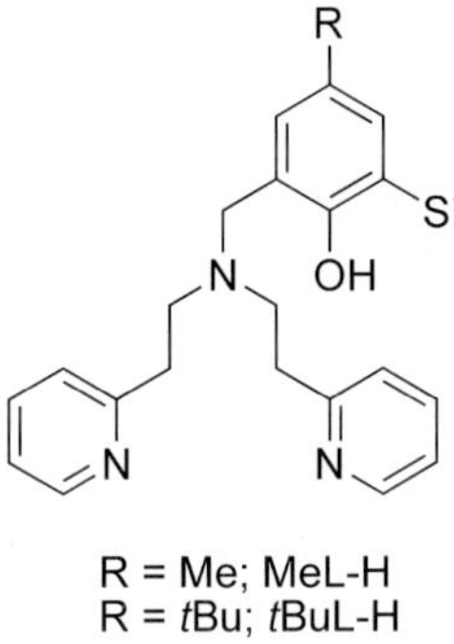

FIG. 21. 2-(Methylthio)-4-methyl-6-[[bis[2-(2-pyridyl)ethyl]amino]methyl]-phenol (MeL-H) and 2-(Methylthio)-4-*tert*-butyl-6-[[bis[2-(2-pyridyl)ethyl]amino]methyl]-phenol (*t*BuL-H) (*83,84*).

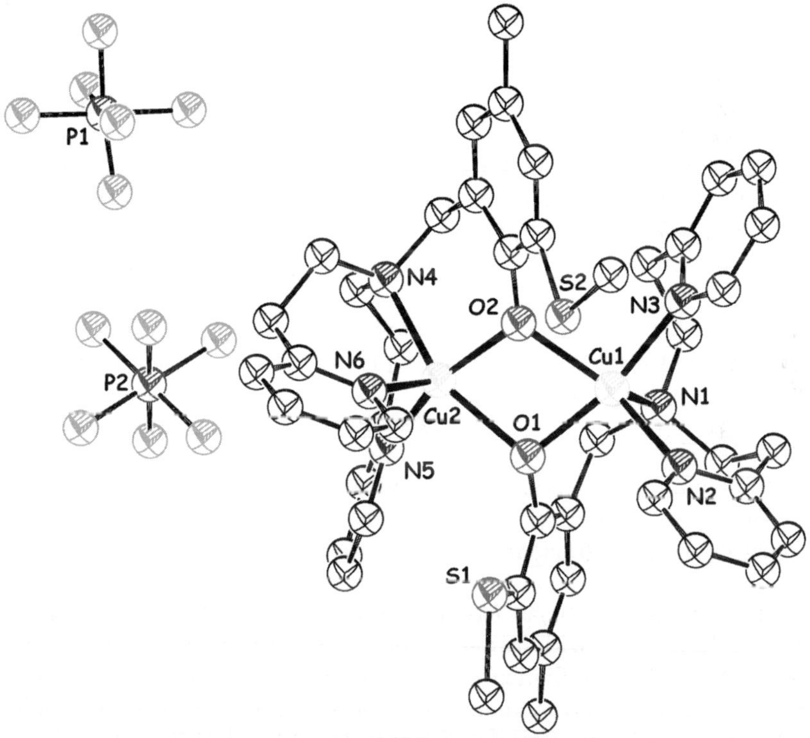

FIG. 22. ORTEP representation (50% thermal ellipsoids) of the X-ray crystal structure of $[Cu^{II}_2(MeL^-)_2](PF_6)_2$. Selected distances (Å) and angles (°) are as follows: Cu(1)–O(1), 1.955(7); Cu(1)–O(2), 2.025(7); Cu(1)–N(1), 2.270(10); Cu(1)–N(2), 2.023(10); Cu(1)–N(3), 1.990(10); Cu(2)–O(1), 2.096(7); Cu(2)–O(2), 1.913(7); Cu(2)–N(4), 2.122(9); Cu(2)–N(5), 1.97(1); Cu(2)–N(6), 2.167(10); O(1)–Cu(1)–O(2), 74.4(3); O(1)–Cu(1)–N(1), 90.9(3); O(1)–Cu(1)–N(2), 97.1(3); O(1)–Cu(1)–N(3), 167.3(4); O(2)–Cu(1)–N(1), 117.3(3); O(2)–Cu(1)–N(2), 146.4(4); O(2)–Cu(1)–N(3), 94.2(3); N(1)–Cu(1)–N(2), 94.9(4); N(1)–Cu(1)–N(3), 89.5(4); N(2)–Cu(1)–N(3), 95.5(4); O(1)–Cu(2)–O(2), 73.6(3); O(1)–Cu(2)–N(4), 155.7(3); O(1)–Cu(2)–N(5), 93.0(3); O(1)–Cu(2)–N(6), 105.8(3); O(2)–Cu(2)–N(4), 92.3(3); O(2)–Cu(2)–N(5), 159.1(4); O(2)–Cu(2)–N(6), 97.4(3); N(4)–Cu(2)–N(5), 94.3(4); N(4)–Cu(2)–N(6), 95.3(4); N(5)–Cu(2)–N(6), 101.7(4); Cu(1)–O(1)–Cu(2), 102.3(3), Cu(1)–O(2)–Cu(2), 106.5(3) (*83,85*).

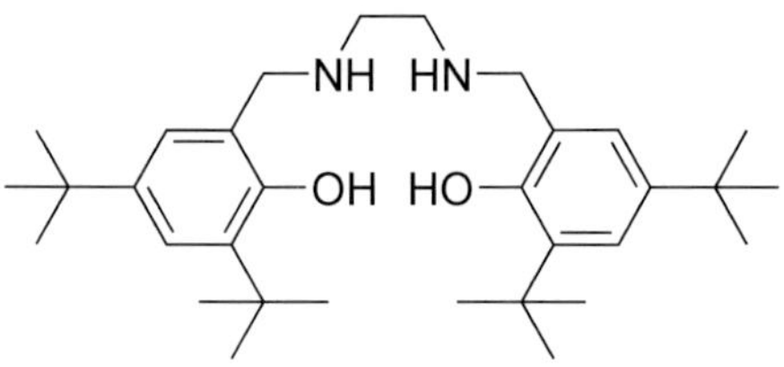

FIG. 23. *N, N'*-bis(2-hydroxy-3,5-di-*tert*-butylbenzyl)-1,2-ethylenediamine (H_2L^{diam}) (*87*).

is able to stoichiometrically oxidize benzyl alcohol to benzaldehyde with concomitant formation of the reduced species $[Cu^I(t\text{BuL–H})]^+$.

D. $[Cu^{II}$(*N,N'*-bis(2-hydroxy-3,5-di-*tert*-butylbenzyl)-1,2-ethylenediamine)] CATALYST (*86*)

In 1998, a third efficient functional model of galactose oxidase was published by Saint-Aman *et al.* (*86*) who performed the electrochemical catalytic oxidation of primary alcohols to the corresponding aldehydes. $[Cu^{II}(L^{diam})]$ was obtained from one equivalent Cu^IClO_4, 2 equivalents of triethylamine, and one equivalent of the ligand *N,N'*-bis (2-hydroxy-3,5-di-*tert*-butylbenzyl)-1,2-ethylenediamine (H_2L^{diam}) as shown in Fig. 23 in dichloromethane in air (*87*).

This dark green $[Cu^{II}(L^{diam})]$ complex was then used for the chemoselective electrocatalytic oxidation of primary alcohols (methanol, ethanol, propan-1-ol, and butan-1-ol), which led to the formation of the corresponding aldehydes with a Faradaic yield close to 90%. The number of turnovers achieved by the copper complex during the electrolysis was close to 32. Only the primary alcohols could be oxidized by this catalytic system (*86*).

E. Cu^ICl(phen)–$DBADH_2$ (*88*)

Markó *et al.* developed an efficient copper-based catalytic system for the aerobic oxidation of a wide range of alcohols to the corresponding carbonyl products (Fig. 24) (*88*). Typically, 0.40 mol of the alcohol was reacted with pure dioxygen in 800 mL toluene between 80 and 90°C. The aerobic oxidation is catalyzed by 5 mol% [copper(I) chloride(1,10-phenanthroline)], 5 mol% di-*tert*-butylhydrazinedicarboxylate ($DBADH_2$) and two equivalents of potassium carbonate.

A number of primary and secondary aliphatic, allylic, and benzylic alcohols could be selectively converted to the corresponding carbonyl compounds within 2 hours reaction time, and with excellent isolated yields ranging from 71 to 97% as listed in Table VI (*88*).

A mechanism has been proposed for this reaction, which is depicted in Fig. 25 (*88–90*) In the presence of $DBADH_2$ under basic conditions (K_2CO_3), the chloride anion of the initial $[Cu^ICl(Phen)]_2$ complex **1** is displaced by the

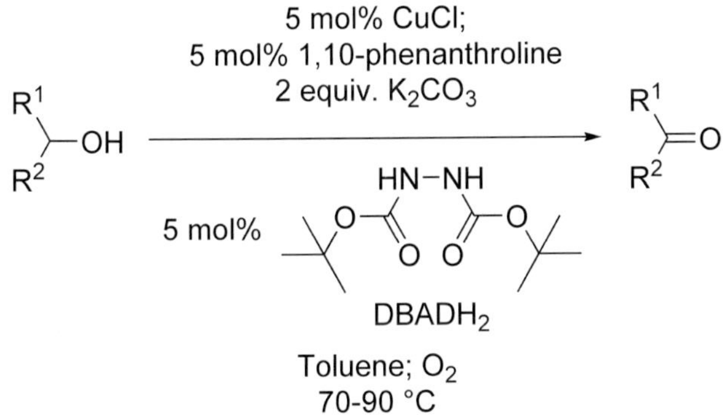

FIG. 24. [Copper-$DEADH_2$]-catalyzed oxidation of alcohols to aldehydes and ketones (*88*).

hydrazide nucleophile to form the hydrazino copper(I) complex **2**. Reaction with dioxygen leads to the peroxo derivative **3** where the Cu^{I} ions have been oxidized to Cu^{II}. Upon heating, compound **3** is then homolytically cleaved to give the oxy-copper radical **4**. This radical undergoes intramolecular hydrogen abstraction from the hydrazine moiety to afford the captodatively stabilized copper-hydroxy nitrogen-centered radical **5**. This species is in equilibrium with its mesomeric azo substituted copper(I) hydroxyl form **6**. The active species **6** exchanges its hydroxyl group with a substrate molecule, resulting in copper complex **7** with release of a water molecule. An intramolecular hydride transfer takes place to produce the carbonyl product and the catalytic cycle is completed through the formation of copper(I) complex **2** (*90*).

Two years later, Markó *et al.* reported an improved catalytic system which only required 0.25 equivalent of potassium carbonate instead of 2 equivalents (*89*). The oxidation reaction described above is dramatically influenced by the nature of the solvent. Thus, if the reaction was performed in fluorobenzene, total conversion of undecan-2-ol to undecan-2-one could be reached with 0.25 equivalent K_2CO_3, whereas 2 equivalents of base were necessary in toluene to convert 90 % of this secondary aliphatic alcohol (Table VI). These optimized conditions were applied to a variety of functionalized alcohols and the results are reported in Table VII. The catalyst tolerates both sulphur and nitrogen substituents on the substrate. Indeed, (thiophen-2-yl)methanol, N-protected (*S*)-valinol or (*S*)-prolinol could be oxidized to the corresponding aldehydes with very good yields. In addition, no racemization was detected for the two β-amino alcohols as well as for (2*S*,5*R*)-2-isopropyl-5-methylcyclohexanol. The hindered *endo*- and *exo*-borneol are both converted to camphor with similar reaction rates, despite their distinctly different steric properties.

The oxidations could also be performed using diethylhydrazinedicarboxylate ($DEADH_2$) instead of $DBADH_2$ and optimum conversions were achieved in toluene using 25 mol% of $DEADH_2$ (*90*). 1-Phenylethanol, 2,2,2-trifluoro-1-phenylethanol, and 2-hydroxy-1,2-diphenylethanone are excellent substrates for this catalytic system, affording the respective ketones in high yields (Table VIII) (*90*).

TABLE VI

[CuCl-DBADH$_2$]-Catalyzed Aerobic Oxidation of Alcohols with Two Equivalents K_2CO_3[a] (88)

Alcohol	Product	Yield % (Conv.)	Time (h)	Alcohol	Product	Yield % (Conv.)	Time (h)
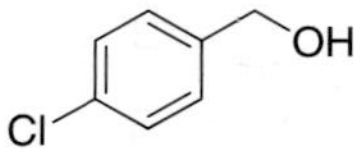	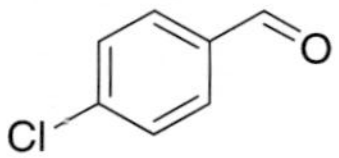	83(100)	1.5	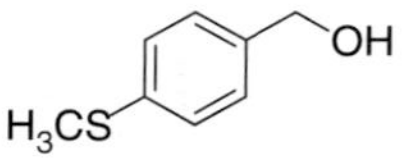	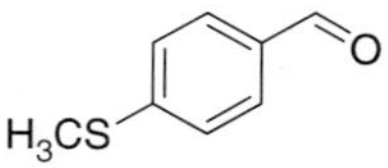	92(95)	1
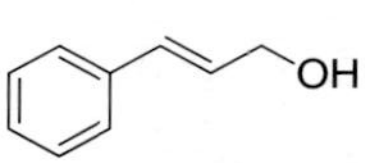	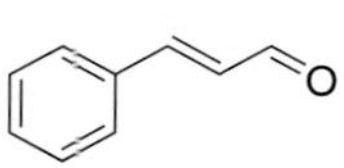	89(100)	1	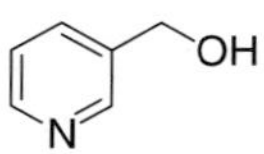	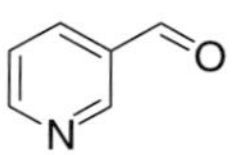	81(92)	1
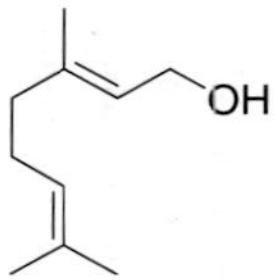	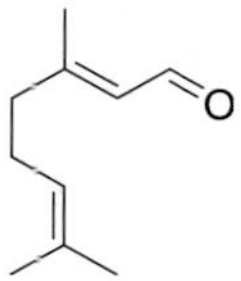	71(75)	1[b]	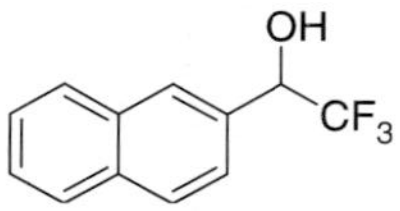	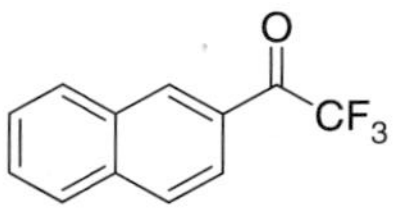	97(100)	1[e]
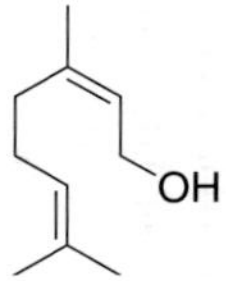	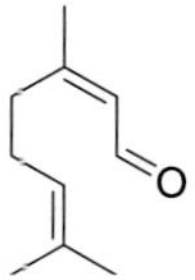	73(83)	1[c]	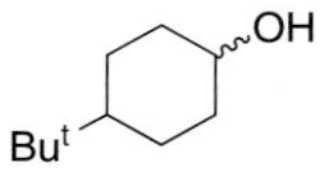	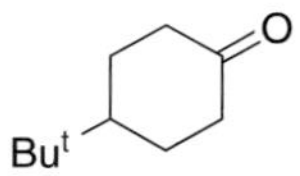	84(87)	2[f]
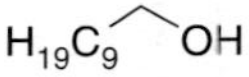	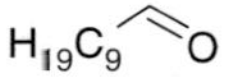	73(87)	0.75[d]	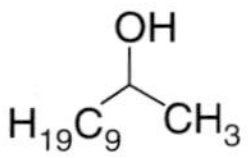	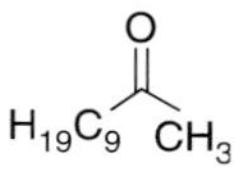	88(90)	2[e]

[a] For conditions see Fig. 24.
[b] > 95% (*E*)-geranial.
[c] > 95% (*Z*)-neral.
[d] 10 mol% [CuCl-DBADH$_2$] used.
[e] 5 mol% DBAD instead of DBADH$_2$.
[f] 10 mol% [CuCl-DBAD] used.

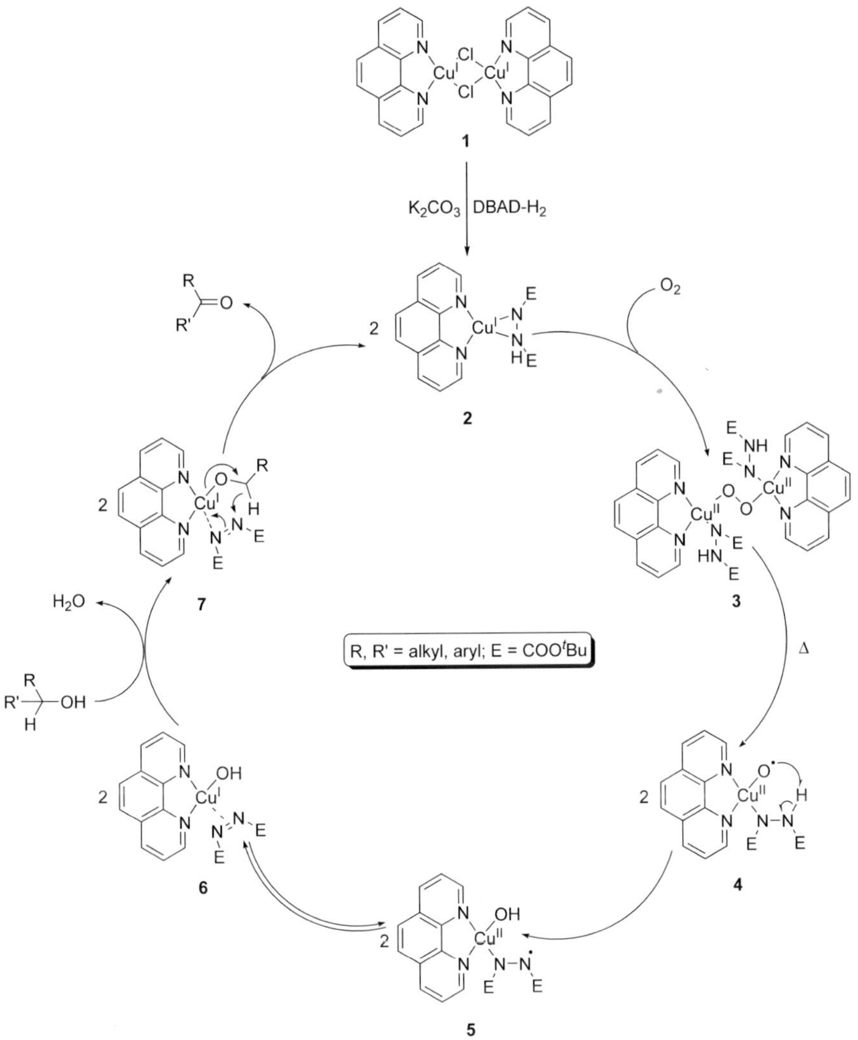

FIG. 25. Proposed catalytic cycle for the copper-DBAD aerobic oxidation of alcohols (*88,91*).

In 2001, Markó *et al.* reported a neutral variant of their $Cu^{I}Cl$(phen)-$DBADH_2$-base system (*91*). A catalytic amount of base was used, i.e., 5 mol% potassium *tert*-butoxide, which was advantageous for pH-sensitive substrates and products. The order of addition of the different reactants turned out to be crucial for the reactivity. The best catalytic procedure was obtained when the base was added to the $[Cu^{I}Cl(Phen)]_2$ complex in the presence of the alcohol, followed by the addition of DBAD. A number of sensitive and/or sterically hindered alcohols were oxidized at 80°C using this optimized protocol and the results are shown in Table IX (*91*). (1*S*,2*S*,5*R*)-Neomenthol was converted to menthone without any epimerization to isomenthone. A rather hindered decaline derivative could be smoothly oxidized as well as

TABLE VII

[CuCl-DBAD]-CATALYZED AEROBIC OXIDATION OF ALCOHOLS WITH 0.25 EQUIVALENT K_2CO_3 IN FLUOROBENZENE[a] (*89*)

Alcohol	Product	Yield %[a]	Alcohol	Product	Yield %[b]
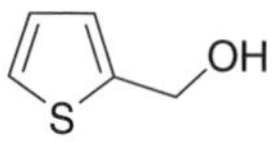	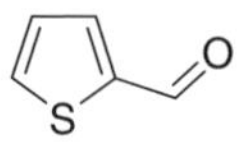	85	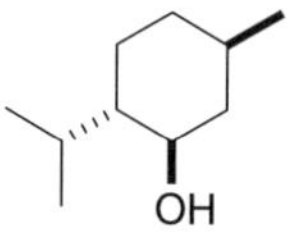	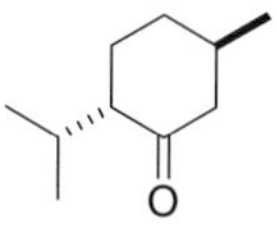	75
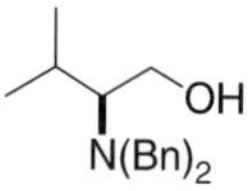	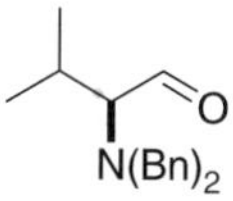	85[b]	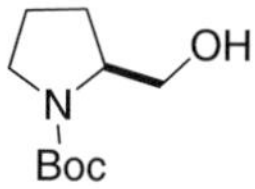	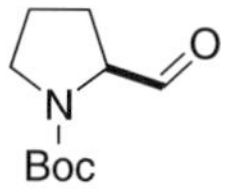	80[c]
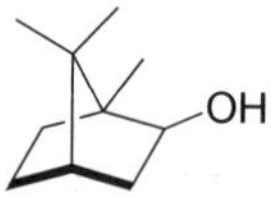	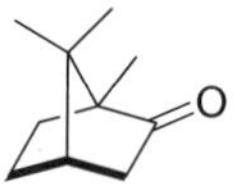	93		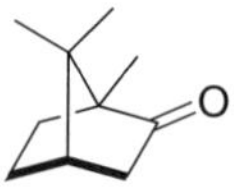	90

[a]For conditions, see Fig. 24.
[b]Yields of pure isolated aldehydes.
[c]No racemization observed in this oxidation reaction.

TABLE VIII

[CuCl-$DEADH_2$]-CATALYZED AEROBIC OXIDATION OF SECONDARY ALCOHOLS WITH 2 EQUIVALENTS K_2CO_3[a] (*90*)

Alcohol	Product	Conversion %[b]
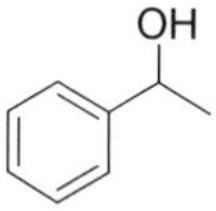	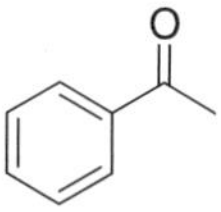	100
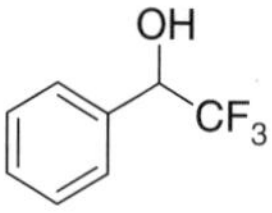	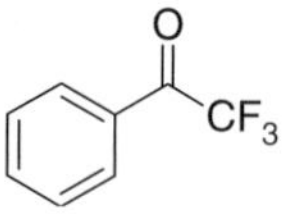	100
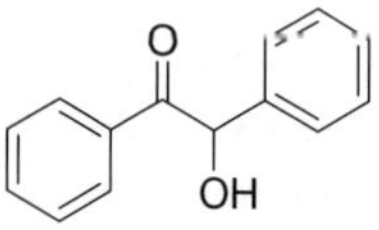	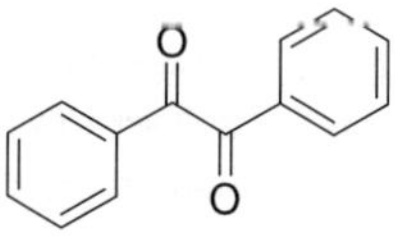	100

[a]For conditions, see Fig. 24.
[b]Determined by ^1H NMR or by GC.

TABLE IX

[CuCl-DBAD]-Catalyzed Aerobic Oxidation of Alcohols in Fluorobenzene with 5 mol % ^{t}BuOK[a] (*91*)

Alcohol	Product	Yield %[a]	Alcohol	Product	Yield %[b]
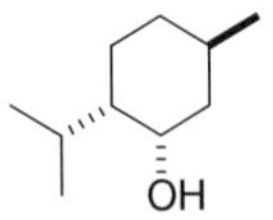	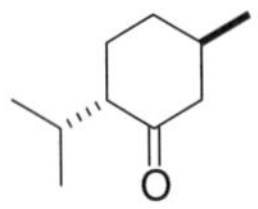	92	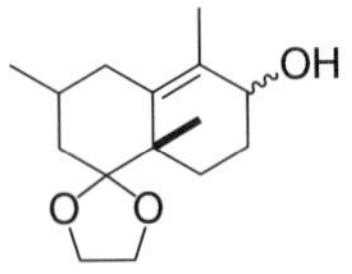	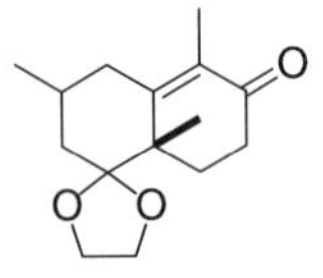	84
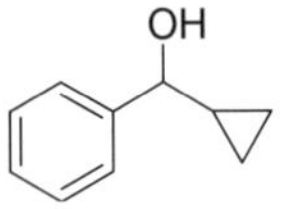	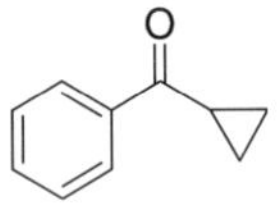	87	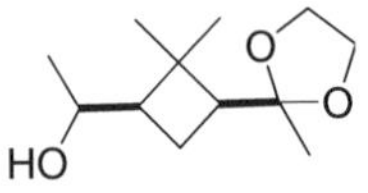	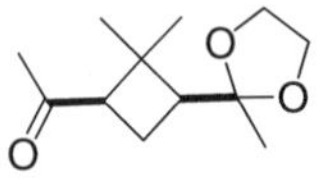	78

[a]For conditions, see Fig. 24.
[b]Yields of pure isolated aldehydes.

cyclopropyl(phenyl)methanol or (*S*)-1-((1*R*,3*S*)-2,2-dimethyl-3-(2-methyl-1,3-dioxolan-2-yl)cyclobutyl)ethanol.

More recently, a further significant advance has been described by the group of Markó (*8*). An additive, i.e., 1-methylimidazole, has been adjoined to the catalytic system involving potassium *tert*-butoxide as basic co-catalyst (*91*), which dramatically enhanced the activity. The remarkable role of 1-methylimidazole would be to avoid the formation of inactive species during the catalytic cycle by stabilization of the copper(I) complex **2** (Fig. 25). Thus, decan-1-ol was transformed to 100 % decan-1-al with this new catalyst while only 60 % conversion was reached in the absence of additive (*8*). This new catalytic procedure was used to oxidize a variety of functionalized primary alcohols (Table X). All the substrates employed were selectively converted to the aldehydes with high yields, ranging from 83 to 97 %. This system oxidizes aliphatic and hindered alcohols and tolerates functional groups such as benzyl or *tert*-butyldimethylsilyl. However, unsaturated alcohols like 2-(4-methylcyclohex-3-enyl)propan-1-ol, are not affected, but base-sensitive substrates like 4-((2*S*,4*R*,6*R*)-6-benzyl-4-chloro-tetrahydro-2H-pyran-2-yl)butan-1-ol, are easily oxidized.

F. Copper-TEMPO Catalytic Systems

2,2,6,6-Tetramethyl-1-piperidinyloxy (TEMPO) is a secondary amine nitrogen oxide radical for which the unpaired electron is delocalized between the nitrogen and the oxygen atom (*92*). This nitroxyl radical is extremely stable towards dimerization or decomposition and inert to usual organic molecules (*93*). In the last decade, catalytic oxidation using TEMPO has become one of the most promising procedures to convert alcohols into the corresponding

TABLE X

[CuCl-DBAD]-CATALYZED AEROBIC OXIDATION OF ALCOHOLS IN FLUOROBENZENE WITH 5 mol% ^{t}BuOK AND 1-METHYLIMIDAZOLE AS AN ADDITIVE[a] (*8*)

Alcohol	Product	Yield %[a]	Alcohol	Product	Yield %[b]
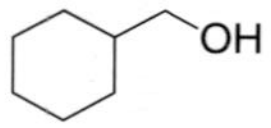	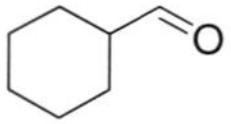	93	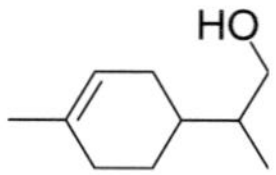	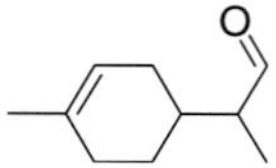	94
	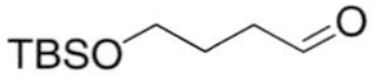	94			97
		95	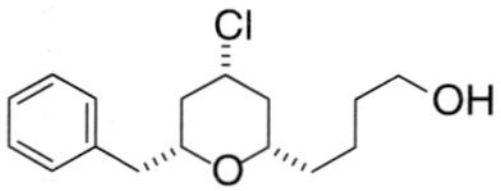	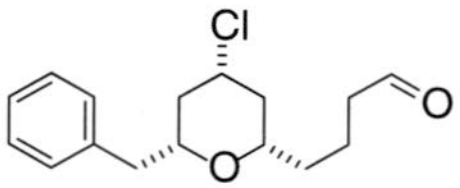	83

[a]For typical conditions, see Fig. 24 and text.
[b]Yields of pure isolated aldehydes.

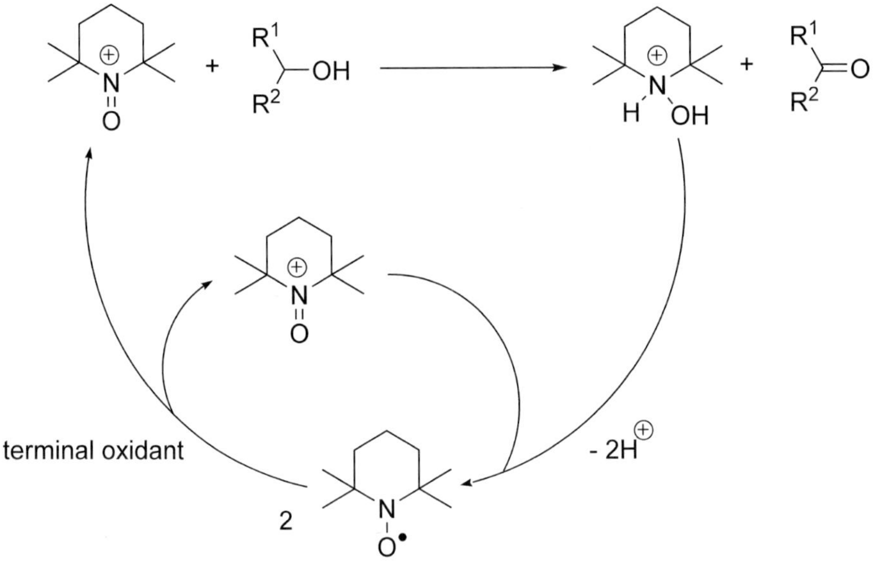

FIG. 26. TEMPO-mediated oxidation of activated alcohols (*101*).

carbonyl compounds (*59*). Despite the promising results the method has some drawbacks with respect to the environmental impact and cost issues, which have to be addressed in order to meet the requirements of the 21st century (*94*). The most serious concern regarding the current procedures arises from the fact that halide-based reagents, like sodium hypochlorite, sodium chlorite, sodium bromite, often in combination with sodium bromide, are required. In these systems, an oxoammonium cation is generated and acts as the active oxidant (Fig. 26) (*95*). The hydroxylamine which is formed is reoxidized by the terminal oxidant to regenerate TEMPO (Fig. 26).

Alternatively, dioxygen can be used as the primary oxidant in combination with TEMPO, using either copper salts, i.e., $Cu^{I}Cl$ as reported by Semmelhack *et al.* (*96*), or laccase (see above) as co-catalysts. The $Cu^{I}Cl$ catalytic system, however, is only effective with activated benzylic and allylic alcohols (*96*).

F.1. [$Cu^{I}Br \cdot Me_2S$-bipyridine] under Fluorous Biphasic Conditions (97,98)

Recently, Betzemeier *et al.* showed that the scope of the copper-catalyzed oxidation reaction could be improved by using $Cu^{I}Br \cdot Me_2S$ instead of $Cu^{I}Cl$ (*97*). The coordination of a perfluoroalkylated bipyridine to the metal centre produced a catalyst which could be used under biphasic fluorous conditions. Typically, 40 μmol of perfluoroalkylated bipyridine was dissolved in 2 mL of perfluorooctane. Next, 40 μmol of $Cu^{I}Br \cdot Me_2S$ was added, followed by a solution of 2 mmol alcohol and 64 μmol TEMPO in 2 mL chlorobenzene. The resulting biphasic reaction mixture was heated to 90°C under dioxygen (Fig. 27).

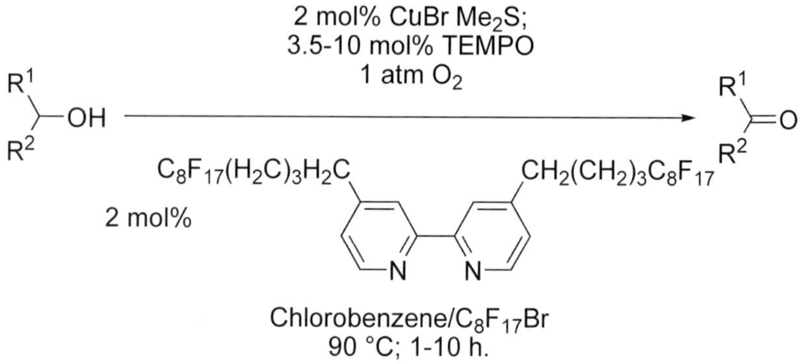

FIG. 27. [Cu^I-TEMPO]-catalyzed aerobic oxidation of alcohols using fluorous biphasic conditions (*97,98*).

A number of primary and secondary aliphatic, allylic, and benzylic alcohols could be selectively and smoothly converted to the corresponding carbonyl compounds within 1 to 10 hours reaction time, with excellent isolated yields ranging from 71 to 96% (Table XI). Benzylic alcohols reacted faster (2–7 h) compared to aliphatic alcohols (7–10 h) (*98*).

The recyclability of the catalyst was demonstrated by using the fluorous phase containing the catalyst several times for further reaction runs. No signifcant decrease in yield and in rate could be observed. For example, even after eight cycles, 86% of analytically pure 4-nitrobenzaldehyde (93% conversion was achieved during the first run) could be obtained from 4-nitrobenzyl alcohol (*97*).

F.2. [Cu^ICl-TEMPO] Catalyst in Ionic Liquid (99)

In 2002, Ansari and Gree developed a simple and mild TEMPO-CuCl catalyzed aerobic oxidation of primary and secondary alcohols to the corresponding aldehydes and ketones in 1-butyl-3-methylimidazolium hexafluorophosphate [bmim][PF_6] (Fig. 28) (*99*). Thus, a mixture of alcohol (1 mmol) and TEMPO (0.05 mmol) was reacted with 1 atm O_2 at 65°C in the ionic liquid. The products were simply isolated by extraction with diethyl ether. We note, however, that it is a pity that a more environmentally acceptable solvent was not used for product extraction.

Various primary and secondary benzylic, allylic, and aliphatic alcohols were successfully used and the results are depicted in Table XII (*99*). Analogous to the system of Betzemeier *et al.* (*97*), it is clear that the conversion of benzylic and allylic alcohols to the corresponding carbonyl compounds is faster and more efficient.

One of the main objectives of using ionic liquid as reaction medium is the possibility to recycle the catalyst. The catalytic solution could indeed be reused eight times for the oxidation of benzyl alcohol to benzaldehyde, with only a slight loss of activity. In addition, the recovered ionic liquid solvent could be recovered and reused for new oxidations of other substrates.

TABLE XI

COPPER-CATALYZED AEROBIC OXIDATION OF ALCOHOLS UNDER FLUOROUS BIPHASIC CONDITIONS[a] (*97,98*)

Alcohol	Product	Yield %[b]	Alcohol	Product	Yield %[b]
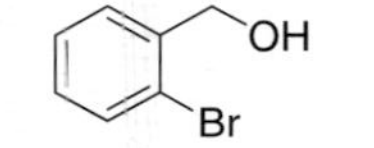	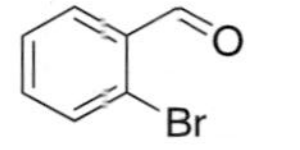	96	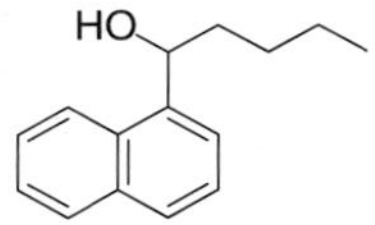	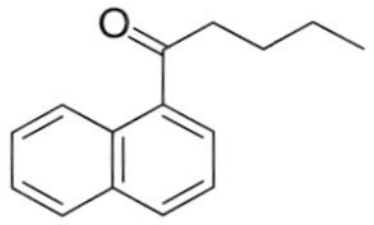	91
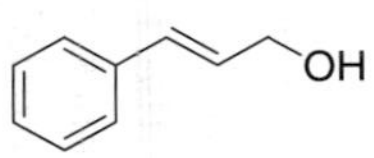	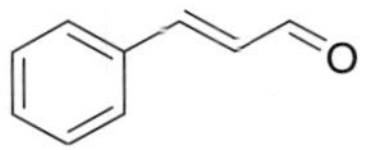	79	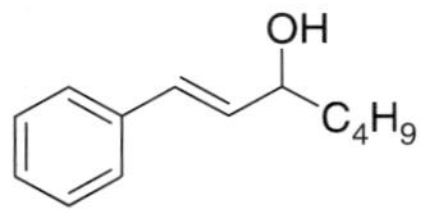	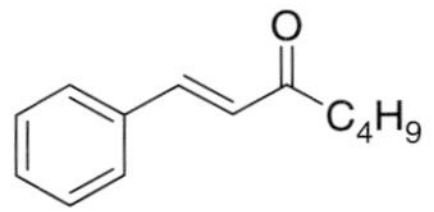	84
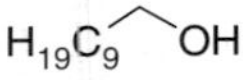	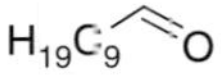	73	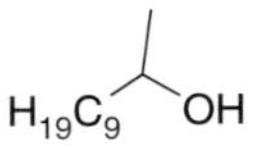	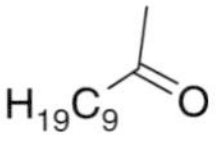	71
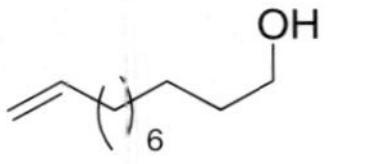	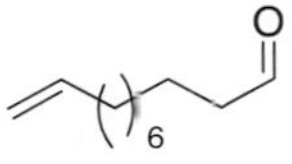	78	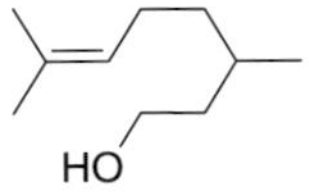	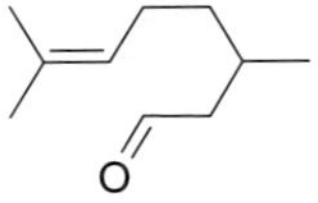	92
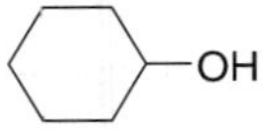	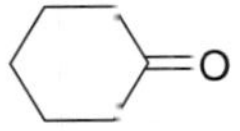	74	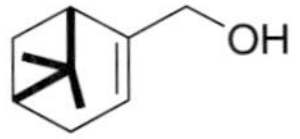	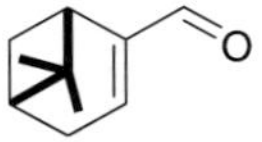	76
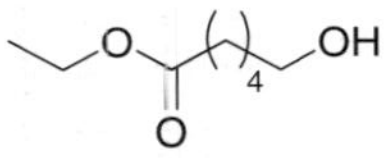	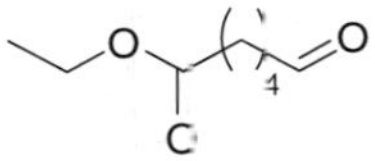	89		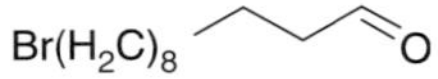	81

[a]For conditions, see Fig. 27.
[b]Isolated yield.

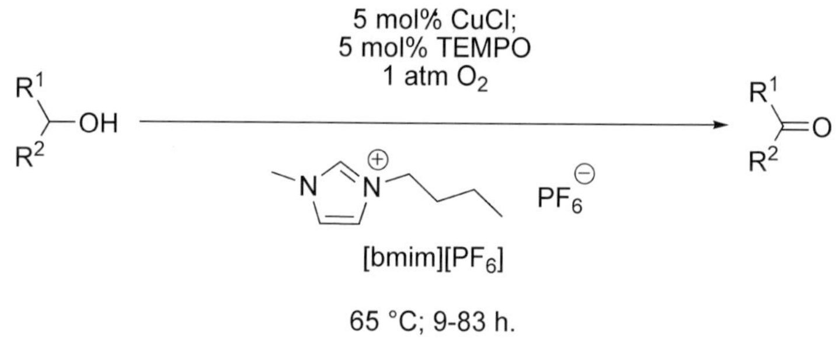

FIG. 28. [Cu^{I}-TEMPO]-catalyzed aerobic oxidation of alcohols in ionic liquid (*99*).

TABLE XII

TEMPO-CATALYZED AEROBIC OXIDATION OF ALCOHOLS IN IONIC LIQUID [BMIM][PF$_6$][a] (*99*)

Alcohol	Product	Yield %[b]	Alcohol	Product	Yield %[b]
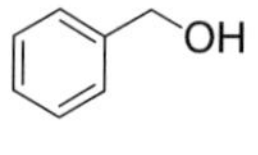	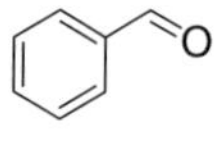	72	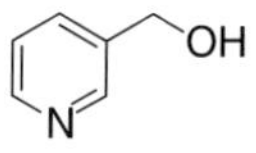	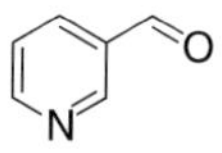	75
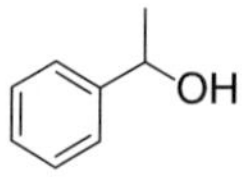	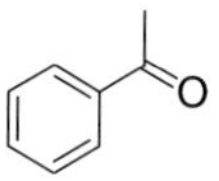	73	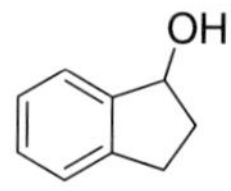	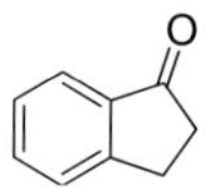	91
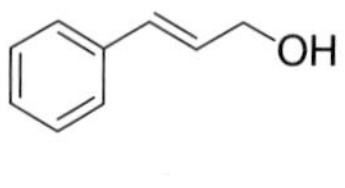	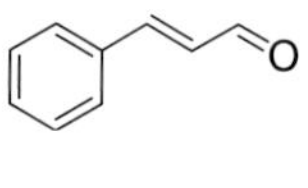	75	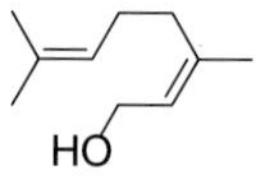	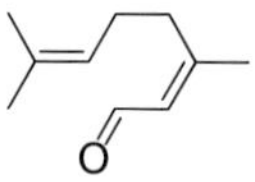	85
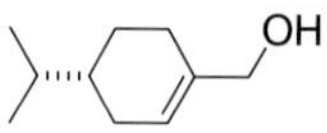	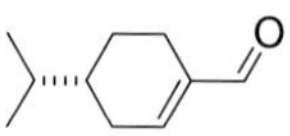	90	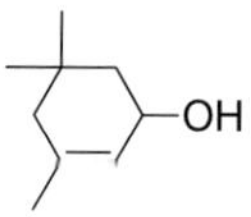	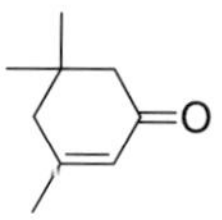	60
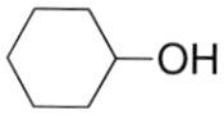	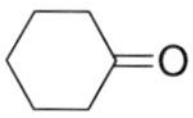	50	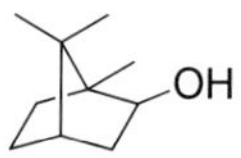	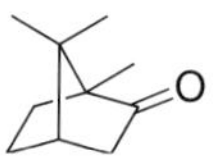	30
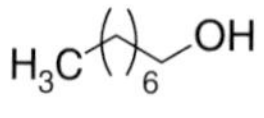	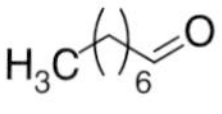	70[c]	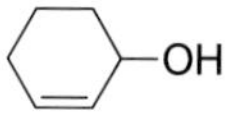	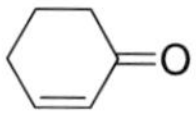	75

[a]For conditions, see Fig. 28.
[b]Isolated yield.
[c]% Conversion.

F.3. [Cu^{I}Cl-TEMPO] Catalyst in DMF (100)

Dijksman *et al.* studied the mechanism of the CuCl-TEMPO catalyzed oxidation of alcohol first reported by Semmelhack *et al.* (*96*). Thus, various alcohol substrates were reacted with dioxygen in DMF at room temperature.

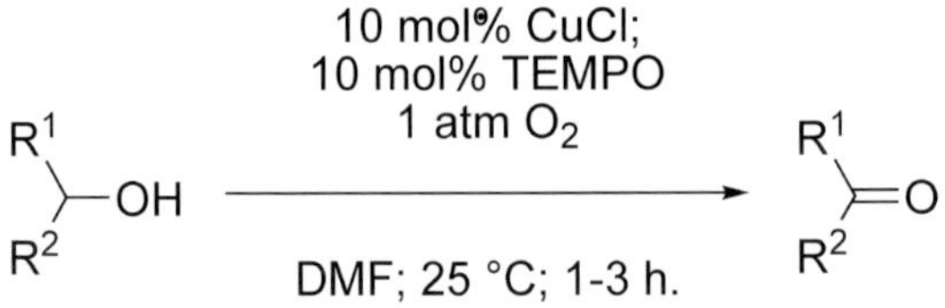

FIG. 29. [Cu^{I}Cl-TEMPO]-catalyzed oxidation of activated alcohols (*100*).

TABLE XIII

CuCl-TEMPO CATALYZED AEROBIC OXIDATION OF ALCOHOLS[a] (*100*)

Alcohol[b]	Product	Time (h)	Conv. %
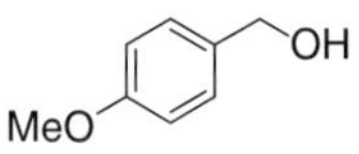	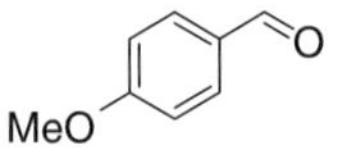	1	96
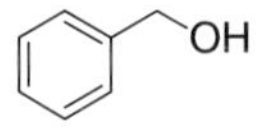	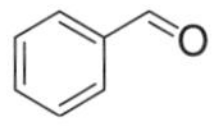	24[c] 4[d]	97 19
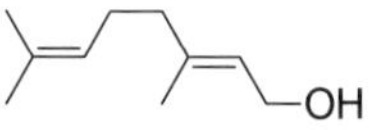	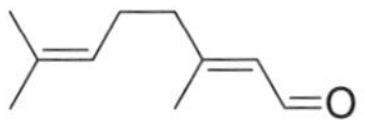	1.75	92
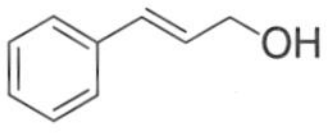	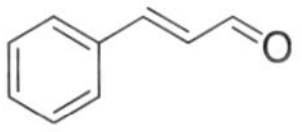	1.75	92

[a]For conditions, see Fig. 29
[b]No reactivity was achieved for 1- and 2-octanol.
[c]1 mol % [CuCl-TEMPO] instead of 10 mol %.
[d]5 mol % [CuCl-TEMPO] instead of 10 mol %.

The reaction was catalyzed by 10 mol% CuCl and 5 mol% TEMPO as co-catalyst (Fig. 29 and Table XIII) (*100*).

As shown in Table XIII, benzylic and allylic alcohols were smoothly oxidized under these conditions. Surprisingly, simple aliphatic alcohols like 1-octanol were unreactive, which is not consistent with a mechanism involving an 'oxoammonium' species (see Fig. 26) (*96*). Indeed, oxoammonium cations are known to oxidize a broad range of alcohols (*101,102*). The results from Hammett correlation investigations and primary kinetic isotope effects for the CuCl-TEMPO catalytic system confirmed the discrepancy regarding an oxoammonium based mechanism. A catalytic cycle was thus proposed, where TEMPO was believed to undergo a one-electron oxidation of Cu^{I} to Cu^{II} (Fig. 30) (*100*).

The active piperidinyloxy copper(II) species is produced *via* TEMPO-mediated oxidation of Cu^{I}Cl. Replacement by a substrate molecule, followed by coordination of a second TEMPO molecule and intramolecular β-hydrogen

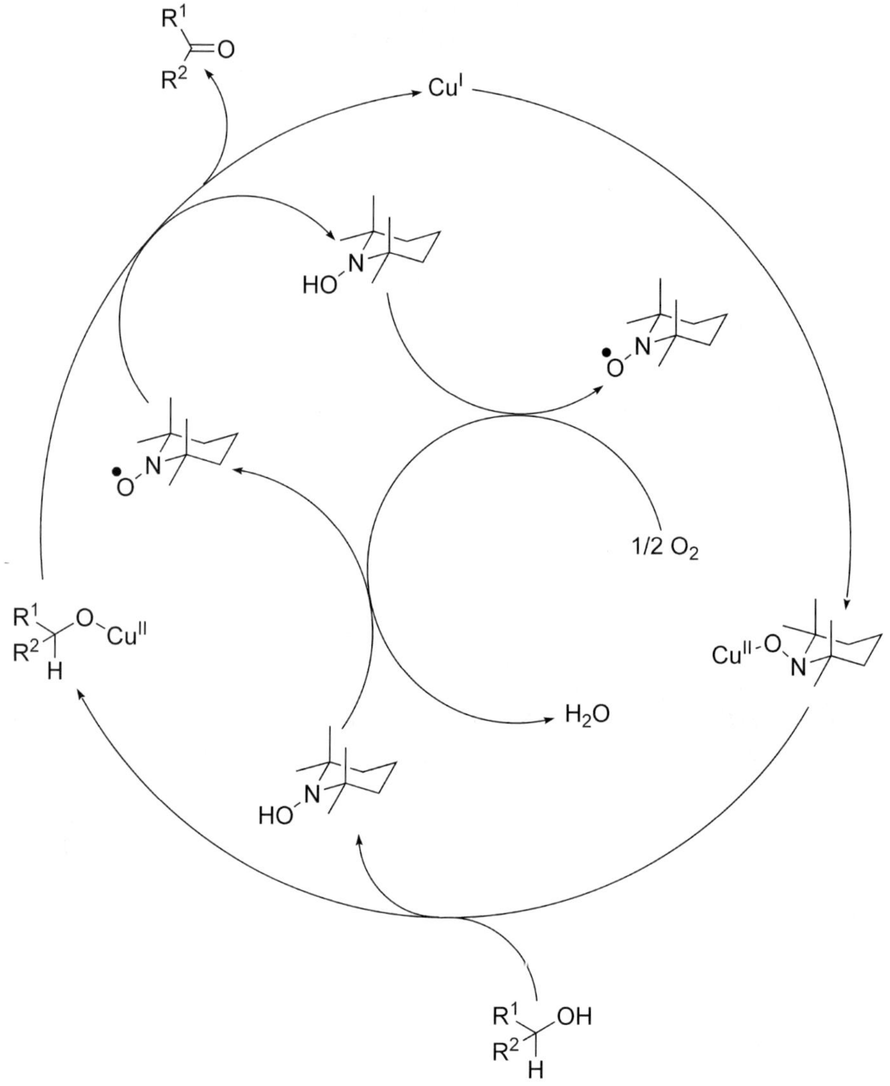

FIG. 30. Proposed mechanism for the [CuCl-TEMPO]-catalyzed oxidation of benzylic and allylic alcohols (*100*).

abstraction gives the carbonyl product, copper(I) species and TEMPOH. Finally, TEMPO is regenerated by rapid air oxidation of TEMPOH with release of a water molecule.

F.4. [$Cu^{II}Br_2$(bpy)-TEMPO] Catalyst in Acetonitrile/Water (2:1) (9,103)

Recently, an alternative to the catalytic system described above was reported by Gamez *et al.* (*9,103*) The new catalytic procedure for the selective aerobic oxidation of primary alcohols to aldehydes was based on

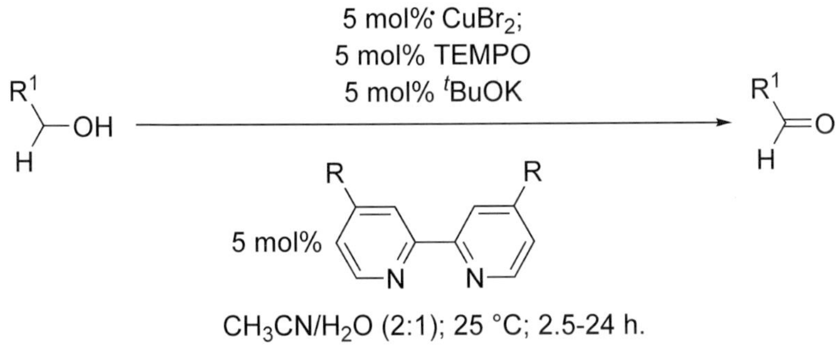

FIG. 31. Selective copper-catalyzed aerobic oxidation of primary alcohols (*9,103*).

a $Cu^{II}Br_2$(Bpy)–TEMPO system (Bpy = 2,2′-bipyridine). The reactions were carried out under air at room temperature and were catalyzed by a [copper^{II}(bipyridine ligand)] complex and TEMPO and base (KOtBu) as co-catalysts (Fig. 31).

Several primary benzylic, allylic and aliphatic alcohols were successfully oxidized with excellent conversions (61–100 %) and high selectivities (Table XIV). As previously mentioned, TEMPO most likely acts as a one-electron oxidant but also as a hydrogen acceptor during the catalytic cycle (Fig. 33). The alcoholate coordinates to the copper(II) complex **I** leading to alkoxy species **II**. TEMPO most likely coordinates to the copper(II) ion in an η^2 manner **III** as reported by Caneschi *et al.* (Fig. 32) (*104*). The β-hydrogen is then transferred to TEMPO resulting in a radical-TEMPOH copper species **IV**. Intramolecular one-electron transfer leads to the aldehyde, TEMPOH and Cu(I) species **V**. Finally, species **I** is regenerated by the TEMPO-mediated oxidation of Cu(I). The catalytic cycle is completed by the aerobic oxidation of the TEMPOH to TEMPO.

Experiments (Table XIV) showed that the oxidation of activated alcohols is faster than of aliphatic ones. It is likely that hydrogen abstraction from the α-carbon atom by TEMPO (Fig. 33, species **III** to **IV**) is involved in the rate-determining step. No reaction, however, was observed for 1-phenylethanol, an activated secondary benzylic alcohol, or for octan-2-ol. This selectivity for the oxidation of primary alcohols, resembles the activity and selectivity exhibited by GOase (see above). Indeed, the proposed active species for this catalytic oxidation can be considered as a model of the active site of GOase (Fig. 7). The bipyridine N atoms mimic the two histidine ligands, while TEMPO plays the role of both the phenolate and the phenoxyl groups.

Competitive oxidation experiments were carried out in order to confirm this high selectivity towards primary alcohols. Thus, mixtures of benzyl alcohol with octan-2-ol or 1-phenylethanol were reacted and no carbonyl products resulting from the oxidation of the secondary alcohols were observed (Table XIV). This lack of reactivity of secondary alcohols may be explained by two factors affecting the catalytic cycle. (1) Steric effects between the methyl groups of TEMPO and the alkyl group of the secondary alcohol can hinder the formation of the active species **III** (Fig. 34a), crucial for the C–H abstraction. (2) In addition, in the case of primary alcohols, the

TABLE XIV

CuBr$_2$(Bpy)-TEMPO-CATALYZED OXIDATION OF ALCOHOLS TO ALDEHYDES UNDER AIR[a] (*103*)

Alcohol	Product	Time (h.)	Yield %
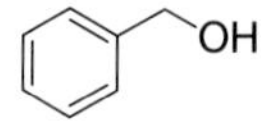	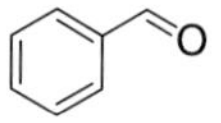	2.5	100
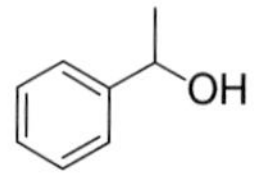	–	5	No reaction
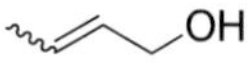	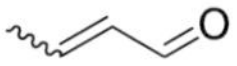	5	91
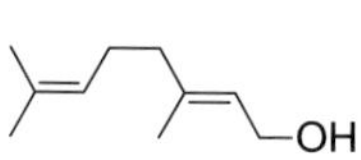	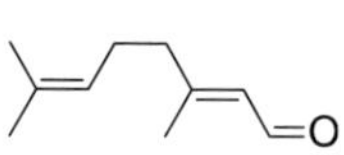	5	100
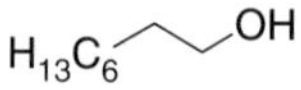	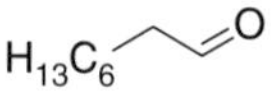	24	61[b]
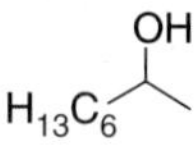	–	5	No reaction
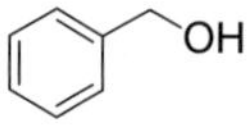 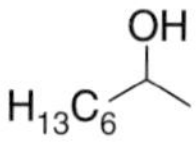	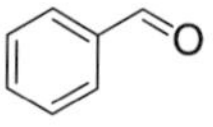	1.5	67/0[c]
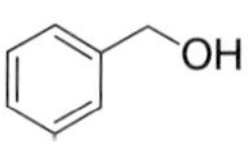 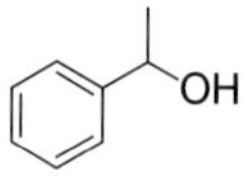	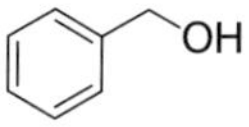	1.5	63/0[d]

[a]For conditions, see Fig. 31.
[b]95 % conversion at 40°C.
[c]67 % Benzaldehyde and no octan-2-one detected.
[d]63 % Benzaldehyde and no acetophenone detected.

second β-hydrogen atom can be H-bonded to the oxygen atom of TEMPOH, stabilizing the radical intermediate **IV** (Fig. 34b).

This CuBr$_2$-bpy/TEMPO-based catalytic system can be considered as the first synthetic functional model of galactose oxidase, as both the achieved chemoselectivity, and the proposed reaction mechanism, resemble that of the biological copper enzyme. Nevertheless, this functional model is not able to compete with the natural enzyme in terms of catalytic efficiency. Indeed, the rate of turnover is only 0.006 s^{-1} while a TOF of 800 s^{-1} is reached by GOase for its native substrate. The objective of future research investigations is therefore to enhance the proficiency of the catalyst to obtain an economically interesting system for industrial applications.

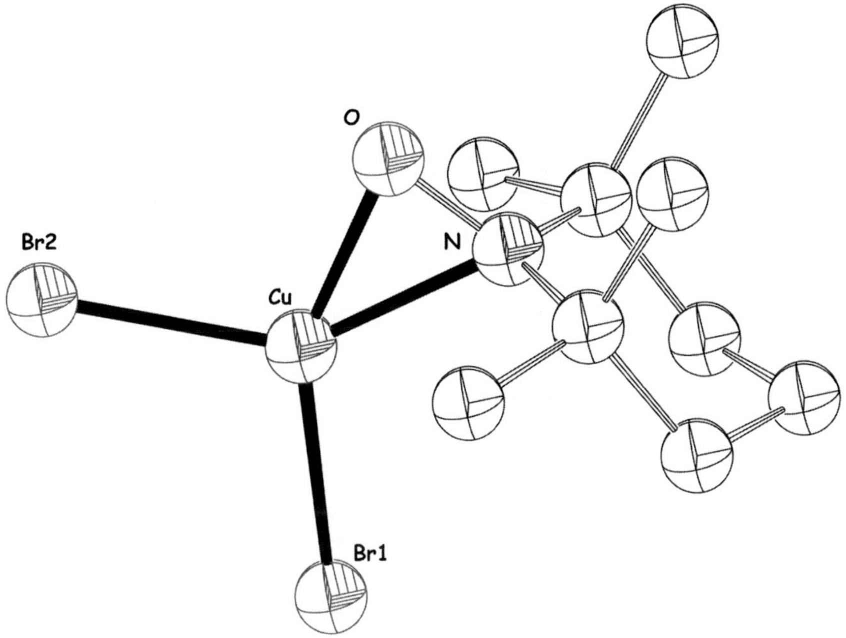

FIG. 32. Molecular structure of $CuBr_2$(Tempo), showing the atom label scheme. Thermal ellipsoids are drawn at the 50 % probability level. Selected distances (Å) and angles (°) are as follows: Cu–Br(av) = 2.270 (2), Cu–O = 1.860 (5), Cu–N = 1.998 (7), O–N = 1.304 (8); Br1–Cu–Br2 = 106.5 (1), O–Cu–N = 39.3 (2) (*104,107*).

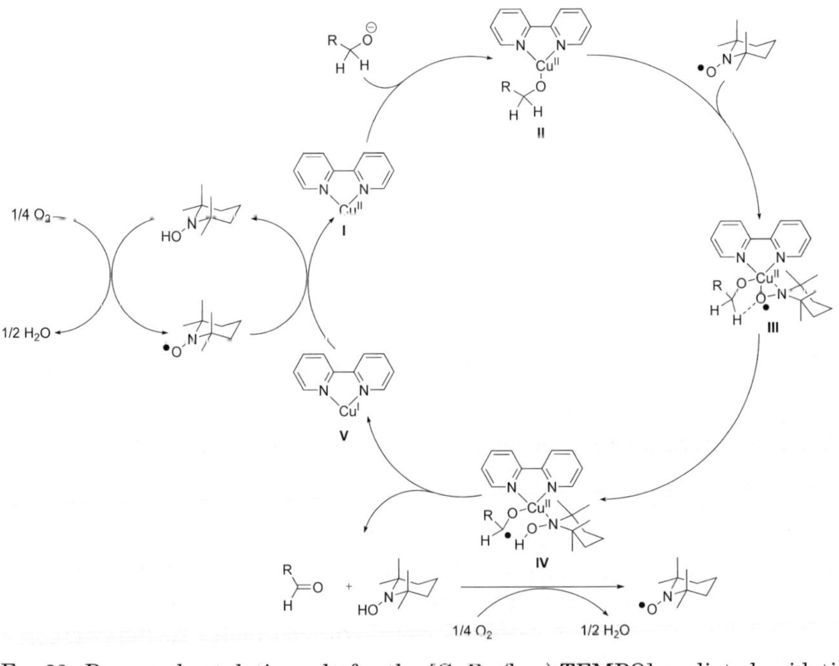

FIG. 33. Proposed catalytic cycle for the [$CuBr_2$(bpy)-TEMPO]-mediated oxidation of primary alcohols to the corresponding aldehydes (*9,103*).

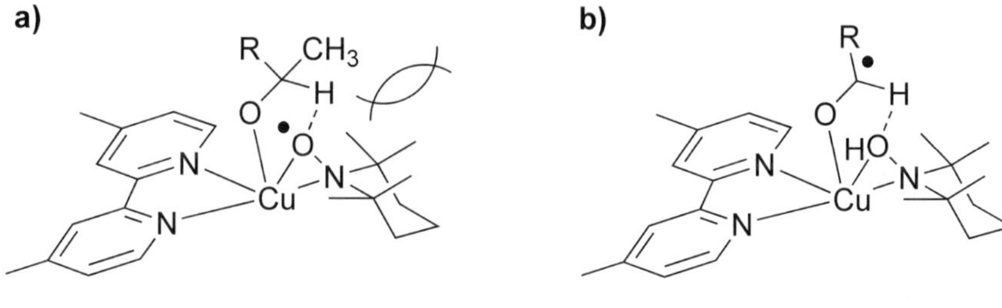

FIG. 34. (a) Steric hindrance between the alkyl group of the secondary alcohol and the methyl groups of TEMPO preventing the formation of the active species in the catalytic cycle; (b) Stabilization of the intermediate radical species by the second β-hydrogen of the primary alcohol (*103*).

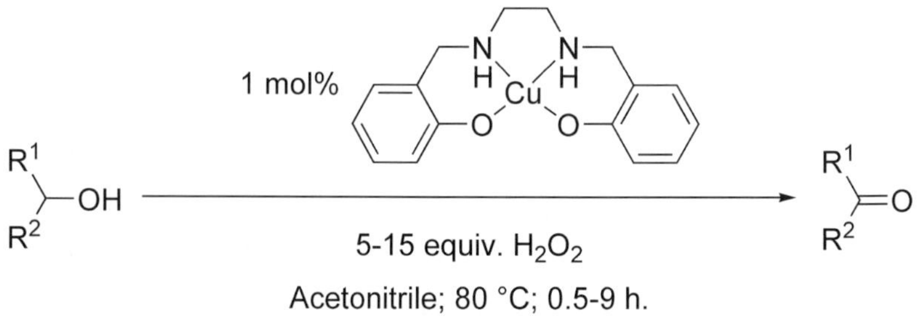

FIG. 35. Copper-salen catalyzed oxidation of alcohols using H_2O_2 as the oxidant (*105*).

G. Cu-salen Using Dihydrogen Peroxide as Oxidant

Velusamy and Punniyamurthy reported the first example of a clean and effective copper-catalyzed oxidation of primary alcohols to the corresponding carboxylic acids and of secondary alcohols to ketones using H_2O_2 as the oxygen source (*105*). Typically, the reaction was carried out in acetonitrile at 80°C with 5 to 15 equivalents of dihydrogen peroxide (Fig. 35). The conversion of the alcohols to the corresponding carbonyl products was catalyzed by 1 mol% of a salen-type copper(II) complex.

Various aliphatic and aromatic alcohols were oxidized with excellent yields, ranging from 82 to 99%, using this procedure and the results are shown in Table XV.

As mentioned above, primary alcohols like benzyl alcohol or 1-decanol were converted to the corresponding carboxylic acid derivatives in the presence of 10 to 15 equivalents of dihydrogen peroxide. Secondary alcohols like cyclohexanol or 1-phenylethanol were oxidized faster to the corresponding ketones using only 5 equivalents of dihydrogen peroxide. Substrates having both electron-donating, e.g., 4′-methoxybenzyl alcohol, as well as

TABLE XV

COPPER(II)-CATALYZED OXIDATION OF ALCOHOLS WITH DIHYDROGEN PEROXIDE[a] (*105*)

Alcohol	Product	Yield %[b]	Alcohol	Product	Yield %[b]
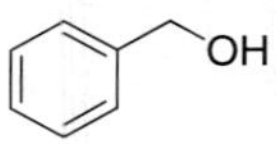	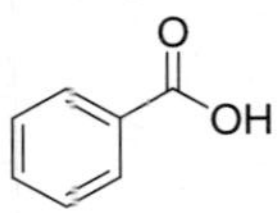	92	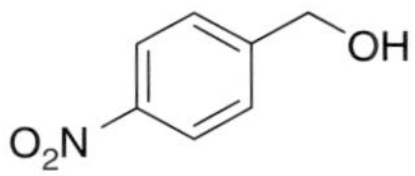	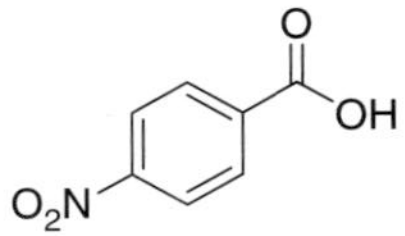	82
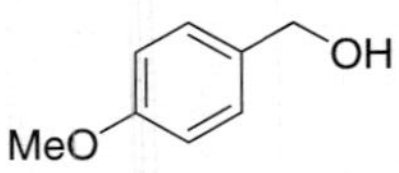	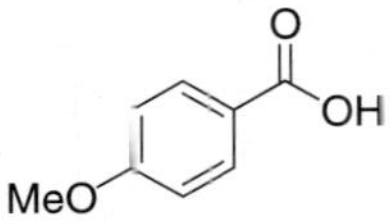	93	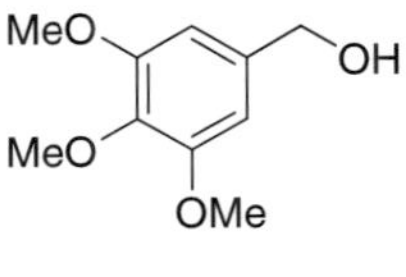	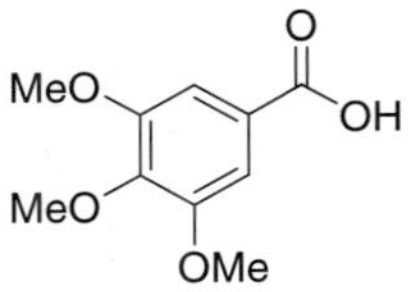	93
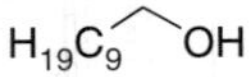	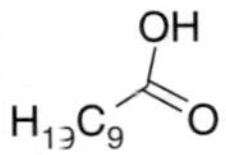	93	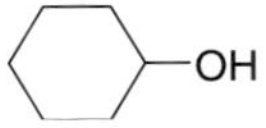	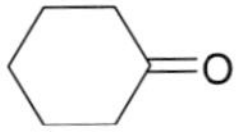	99[c]
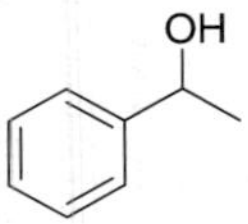	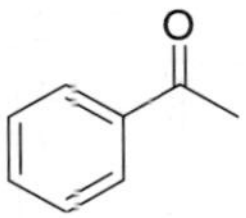	99[c]	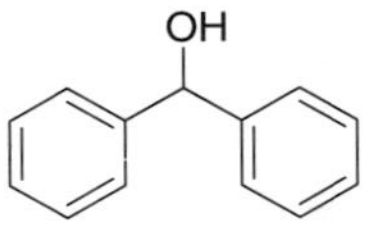	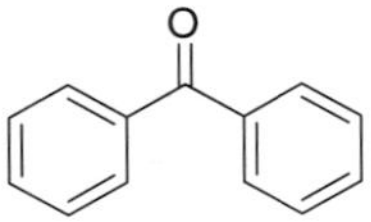	98

[a]For conditions, see Fig. 35
[b]Isolated yield.
[c]GC yield.

electron-withdrawing substituents, e.g., 4′-nitrobenzyl alcohol, are compatible with this catalytic system. The requirement for a large excess of dihydrogen peroxide (5–15 equivalents) suggests that substantial decomposition of the dihydrogen peroxide takes place (catalase reaction) and severely restricts the synthetic utility of this method.

IV. Future Outlook

In the last decade significant advances have been booked regarding the understanding of the mechanistic details of oxidations mediated by copper-dependent oxidases. Amongst these galactose oxidase, GOase, occupies a special place; it catalyzes the aerobic oxidation of alcohols *via* a unique mechanism involving electron transfer between the coordinated substrate, the copper centre and a coordinated tyrosyl radical in the active site (*63,64*). In addition to its natural substrate galactose, GOase is able to mediate the oxidation, albeit with lower rates, of several primary alcohols to the corresponding aldehydes. Nonetheless, the substrate specificity of GOase is narrow. The application of protein engineering techniques, such as directed evolution *via* gene shuffling, could pave the way for a broadening of the substrate specificity of GOase to include simple aliphatic and benzylic alcohols. The recently reported modification of GOase, by directed evolution, to afford an enzyme capable of oxidizing glucose provides a basis for optimism in this respect (*24*).

Laccase, in contrast to GOase, does not mediate the aerobic oxidation of alcohols *in vivo*, but in the presence of a variety of electron mediators, notably TEMPO, is able to catalyze alcohol oxidations *in vitro* (*47,50*). Recent kinetic and EPR studies (*58*) have established the mechanistic details of TEMPO/laccase-catalyzed oxidations of alcohols and provided insights into reasons for the relatively low rates of conversion and the need for high loadings of TEMPO. This provides a sound basis for further improvement of the synthetic utility of this system, e.g., by stabilizing the laccase *via* immobilization and/or the use of more stable TEMPO derivatives.

In parallel with advances in the mechanistic understanding of copper-dependent oxidases, significant improvements have been achieved in copper-based biomimetic systems for the aerobic oxidation of alcohols. Various GOase models have been described (see earlier), of increasing simplicity, which are capable of oxidizing alcohols with catalytic turnovers (*69*). Nonetheless, there is still room for further improvement towards catalytic systems that couple high activities with broad substrate specificities, i.e., with real synthetic utility. From the viewpoint of synthetic utility the systems based on CuCl/phen/DBAD (*8*) and the $CuBr_2$/Bpy/TEMPO (*103*) constitute significant advances. The former system is effective with a broad range of primary and secondary alcohols while the latter is highly effective for the selective oxidation of primary alcohol moieties. We expect that the mechanistic understanding will provide a sound basis for the further improvement of the activity and synthetic scope of these biomimetic copper catalysts.

In short, we expect that the synthetic utility of both biocatalytic and biomimetic copper-based systems will be further improved in the future. This will provide economically viable catalytic methodologies for the green, atom efficient oxidations of alcohols employing molecular oxygen as the primary oxidant, and affording water as the sole byproduct.

References

1. Cainelli, G.; Cardillo, G. "*Chromium Oxidants in Organic Chemistry*"; Springer: Berlin, **1984**.
2. Sheldon, R. A.; Kochi, J. K. "*Metal-Catalysed Oxidations of Organic Compounds*"; Academic Press: New York, **1981**.
3. Tidwell, T. T. "*Organic Reactions*"; Wiley: New York, **1990**; p. 297.
4. Yamaguchi, K.; Mizuno, N. *Chem. Eur. J.* **2003**, *9*, 4353–4361.
5. Anderson, R.; Griffin, K.; Johnston, P.; Alsters, P. L. *Adv. Synth. Catal.* **2003**, *345*, 517–523.
6. Besson, M.; Gallezot, P. *Catal. Today* **2000**, *57*, 127–141.
7. Sato, K.; Aoki, M.; Takagi, J.; Noyori, R. *J. Am. Chem. Soc.* **1997**, *119*, 12386.
8. Marko, I. E.; Gautier, A.; Dumeunier, R. L.; Doda, K.; Philippart, F.; Brown, S. M.; Urch, C. J. *Angew. Chem. Int. Ed.* **2004**, *43*, 1588–1591.
9. Gamez, P.; Arends, I. W. C. E.; Reedijk, J.; Sheldon, R. A. *Chem. Commun.* **2003**, 2414–2415.
10. McGuirl, M. A.; Dooley, D. M. *Curr. Opin. Chem. Biol.* **1999**, *3*, 138–144.
11. Ito, N.; Phillips, S. E. V.; Stevens, C.; Ogel, Z. B.; Mcpherson, M. J.; Keen, J. N.; Yadav, K. D. S.; Knowles, P. F. *Nature* **1991**, *350*, 87–90.
12. Klabunde, T.; Eicken, C.; Sacchettini, J. C.; Krebs, B. *Nat. Struct. Biol.* **1998**, *5*, 1084–1090.
13. Ducros, V.; Brzozowski, A. M.; Wilson, K. S.; Brown, S. H.; Ostergaard, P.; Schneider, P.; Yaver, D. S.; Pedersen, A. H.; Davies, G. J. *Nat. Struct. Biol.* **1998**, *5*, 310–316.
14. Whittaker, J. W. In: "*Metal Ions in Biological Systems*"; Ed. Sigel, H.; vol. 30, Marcel Dekker: New York, **1994**; pp. 315–360.
15. Messerschmidt, A.; (Ed.) "*Multi-Copper Oxidases*"; World Scientific: Singapore, **1997**.
16. Whittaker, M. M.; Whittaker, J. W. *Protein Expr. Purif.* **2000**, *20*, 105–111.
17. Sun, L. H.; Petrounia, I. P.; Yagasaki, M.; Bandara, G.; Arnold, F. H. *Protein Eng.* **2001**, *14*, 699–704.
18. Cooper, J. A. D.; Smith, W.; Bacila, M.; Medina, H. *J. Biol. Chem.* **1959**, *234*, 445–448.
19. Avigad, G.; Amaral, D.; Asensio, C.; Horecker, B. L. *J. Biol. Chem.* **1962**, *237*, 2736.
20. Nobles, M. K.; Madhosingh, C. *Biochem. Biophys. Res. Commun.* **1963**, *12*, 146.
21. Amaral, D.; Kelly-Falcoz, F.; Horecker, B. L. *Methods Enzymol.* **1966**, *9*, 87.
22. Schlegel, R. A.; Gerbeck, C. M.; Montgomery, R. *Carbohydr. Res.* **1968**, *32*, 43.
23. Maradfu, A.; Perlin, A. S. *Carbohydr. Res.* **1974**, *32*, 43.
24. Sun, L. H.; Bulter, T.; Alcalde, M.; Petrounia, I. P.; Arnold, F. H. *Chembiochem.* **2002**, *3*, 781–783.
25. Zancan, G. T.; Amaral, D. *Biochim. Biophys. Acta* **1970**, *198*, 146.
26. Hamilton, G. A.; deJersey, J.; Adolf, P. K. In: "*Oxidases and Related Systems*"; Eds. King, T.; Mason, H. S.; Morrison, M.; University Park Press: Baltimore, Md., **1973**, p. 103.
27. Cleveland, L.; Coffman, R. E.; Coon, P.; Davis, L. *Biochemistry* **1975**, *14*, 1108.
28. Maradfu, A.; Cree, G. M.; Perlin, A. S. *Can. J. Chem.* **1971**, *49*, 3429.
29. Klibanov, A. M.; Alberti, B. N.; Marletta, M. A. *Biochem. Biophys. Res. Commun.* **1982**, *108*, 804–808.
30. Drueckhammer, D. G.; Hennen, W. J.; Pederson, R. L.; Barbas, C. F.; Gautheron, C. M.; Krach, T.; Wong, C. H. *Synthesis* **1991**, 499–525.
31. Schoevaart, R.; Kieboom, T. *Top. Catal.* **2004**, *27*, 3–9.
32. Franke, D.; Machajewski, T.; Hsu, C. C.; Wong, C. H. *J. Org. Chem.* **2003**, *68*, 6828–6831.
33. Hakulinen, N.; Kiiskinen, L. L.; Kruus, K.; Saloheimo, M.; Paananen, A.; Koivula, A.; Rouvinen, J. *Nat. Struct. Biol.* **2002**, *9*, 601–605.

34. Bertrand, T.; Jolivalt, C.; Briozzo, P.; Caminade, E.; Joly, N.; Madzak, C.; Mougin, C. *Biochemistry* **2002**, *41*, 7325–7333.
35. Piontek, K.; Antorini, M.; Choinowski, T. *J. Biol. Chem.* **2002**, *277*, 37663–37669.
36. Solomon, E. I.; Sundaram, U. M.; Machonkin, T. E. *Chem. Rev.* **1996**, *96*, 2563–2605.
37. ten Have, R.; Teunissen, P. J. M. *Chem. Rev.* **2001**, *101*, 3397–3413.
38. Bourbonnais, R.; Paice, M. G. *FEBS Lett.* **1990**, *267*, 99–102.
39. Rochefort, D.; Leech, D.; Bourbonnais, R. *Green. Chem.* **2004**, *6*, 14–24.
40. Call, H. P.; Mucke, I. *J. Biotechnol.* **1997**, *53*, 163–202.
41. Gianfreda, L.; Xu, F.; Bollag, J. M. *Bioremediation J.* **1999**, *3*, 1–25.
42. Eggert, C.; Temp, U.; Dean, J. F. D.; Eriksson, K. E. L. *FEBS Lett.* **1996**, *391*, 144–148.
43. Srebotnik, E.; Hammel, K. E. *J. Biotechnol.* **2000**, *81*, 179–188.
44. Sealey, J.; Ragauskas, A. J. *J. Wood Chem. Technol.* **1998**, *18*, 403–416.
45. Xu, F.; Kulys, J. J.; Duke, K.; Li, K. C.; Krikstopaitis, K.; Deussen, H. J. W.; Abbate, E.; Galinyte, V.; Schneider, P. *Appl. Environ. Microbiol.* **2000**, *66*, 2052–2056.
46. Potthast, A.; Rosenau, T.; Chen, C. L.; Gratzl, J. S. *J. Mol. Catal. A-Chem.* **1996**, *108*, 5–9.
47. Fabbrini, M.; Galli, C.; Gentili, P.; Macchitella, D. *Tetrahedron Lett.* **2001**, *42*, 7551–7553.
48. Viikari, L.; Niku-Paavola, M. L.; Buchert, J.; Forssell, P.; Teleman, A.; Kruus, K. (WO 9923240). **1999**.
49. Jetten, J. M.; van den Dool, R. T. M.; van Hartingsveldt, W.; van Wandelen, M. T. R. (WO 00/50621). **2000**.
50. Fabbrini, M.; Galli, C.; Gentili, P. *J. Mol. Catal. B-Enzym.* **2002**, *16*, 231–240.
51. d'Acunzo, F.; Baiocco, P.; Fabbrini, M.; Galli, C.; Gentili, P. *Eur. J. Org. Chem.* **2002**, 4195–4201.
52. Cantarella, G.; Galli, C.; Gentili, P. *J. Mol. Catal. B-Enzym.* **2003**, *22*, 135–144.
53. Crestini, C.; Jurasek, L.; Argyropoulos, D. S. *Chem. Eur. J.* **2003**, *9*, 5371–5378.
54. Marjasvaara, A.; Torvinen, M.; Vainiotalo, P. *J. Mass Spectrometry* **2004**, *39*, 1139–1146.
55. Baiocco, P.; Barreca, A. M.; Fabbrini, M.; Galli, C.; Gentili, P. *Org. Biomol. Chem.* **2003**, *1*, 191–197.
56. Golubev, V. A.; Rudyk, T. S.; Sen, V. D.; Aleksandrov, A. L. *Izv. Ak. Nauk. SSSR Ser. Khim.* **1976**, *4*, 763–771.
57. Golubev, V. A.; Kozlov, Y. N.; Petrov, A. N.; Purmal, A. P. In: "*Bioactive Spin Labels*"; Ed. Zhdanov, R. I.; Springer: Berlin, **1992**, pp. 119–140.
58. Li, Y. X.; Arends, I. W. C. E.; Sheldon, R. A. *submitted for publication* **2005**.
59. deNooy, A. E. J.; Besemer, A. C.; vanBekkum, H. *Synthesis* **1996**, 1153–1174.
60. Whittaker, J. W. *Chem. Rev.* **2003**, *103*, 2347–2363.
61. Ito, N.; Phillips, S. E. V.; Yadav, K. D. S.; Knowles, P. F. *J. Mol. Biol.* **1994**, *238*, 794–814.
62. Whittaker, M. M.; Ballou, D. P.; Whittaker, J. W. *Biochemistry* **1998**, *37*, 8426–8436.
63. Knowles, P. F.; Ito, N. Perspectives in Bioinorganic Chemistry. In: "*Perspectives in Bioinorganic Chemistry*"; Ed. Hay, R. W.; JAI Press: London, **1993**, pp. 207–244.
64. Baron, A. J.; Stevens, C.; Wilmot, C.; Seneviratne, K. D.; Blakeley, V.; Dooley, D. M.; Phillips, S. E. V.; Knowles, P. F.; Mcpherson, M. J. *J. Biol. Chem.* **1994**, *269*, 25095–25105.
65. Michel, F.; Thomas, F.; Hamman, S.; Saint-Aman, E.; Bucher, C.; Pierre, J. L. *Chem. Eur. J.* **2004**, *10*, 4115–4125.
66. Jazdzewski, B. A.; Tolman, W. B. *Coord. Chem. Rev.* **2000**, *200*, 633–685.
67. Pierre, J. L. *Chem. Soc. Rev.* **2000**, *29*, 251–257.
68. Zurita, D.; GautierLuneau, I.; Menage, S.; Pierre, J. L.; Saint-Aman, E. *J. Biol. Inorg. Chem.* **1997**, *2*, 46–55.
69. Mahadevan, V.; Gebbink, R. J. M. K.; Stack, T. D. P. *Curr. Opin. Chem. Biol.* **2000**, *4*, 228–234.
70. Gebbink, R. J. M. K.; Watanabe, M.; Pratt, R. C.; Stack, T. D. P. *Chem. Commun.* **2003**, 630–631.
71. Barooah, N.; Sharma, S.; Sarma, B. C.; Baruah, J. B. *Appl. Organometal. Chem.* **2004**, *18*, 440–445.
72. Chaudhuri, P.; Hess, M.; Muller, J.; Hildenbrand, K.; Bill, E.; Weyhermuller, T.; Wieghardt, K. *J. Am. Chem. Soc.* **1999**, *121*, 9599–9610.

73. Wang, Y. D.; DuBois, J. L.; Hedman, B.; Hodgson, K. O.; Stack, T. D. P. *Science* **1998,** *279,* 537–540.
74. Chaudhuri, P.; Hess, M.; Florke, U.; Wieghardt, K. *Angew. Chem. Int. Ed.* **1998,** *37,* 2217–2220.
75. Lions, F.; Martin, K. V. *J. Am. Chem. Soc.* **1957,** *79,* 1273–1275.
76. Wang, Y. D.; Stack, T. D. P. *J. Am. Chem. Soc.* **1996,** *118,* 13097–13098.
77. Pratt, R. C.; Stack, T. D. P. *J. Am. Chem. Soc.* **2003,** *125,* 8716–8717.
78. Whittaker, M. M.; Whittaker, J. W. *Biophys. J.* **1993,** *64,* 762–772.
79. Chaudhuri, P.; Hess, M.; Weyhermuller, T.; Wieghardt, K. *Angew. Chem. Int. Ed.* **1999,** *38,* 1095–1098.
80. Speier, G.; Csihony, J.; Whalen, A. M.; Pierpont, C. G. *Inorg. Chem.* **1996,** *35,* 3519–3524.
81. Bruni, S.; Caneschi, A.; Cariati, F.; Delfs, C.; Dei, A.; Gatteschi, D. *J. Am. Chem. Soc.* **1994,** *116,* 1388–1394.
82. Paine, T. K.; Weyhermuller, T.; Wieghardt, K.; Chaudhuri, P. *Dalton Trans.* **2004,** 2092–2101.
83. Itoh, S.; Takayama, S.; Arakawa, R.; Furuta, A.; Komatsu, M.; Ishida, A.; Takamuku, S.; Fukuzumi, S. *Inorg. Chem.* **1997,** *36,* 1407–1416.
84. Itoh, S.; Taki, M.; Takayama, S.; Nagatomo, S.; Kitagawa, T.; Sakurada, N.; Arakawa, R.; Fukuzumi, S. *Angew. Chem. Int. Ed.* **1999,** *38,* 2774–2776.
85. Itoh, S.; Taki, M.; Fukuzumi, S. *Coord. Chem. Rev.* **2000,** *198,* 3–20.
86. Saint-Aman, E.; Menage, S.; Pierre, J. L.; Defrancq, E.; Gellon, G. *New. J. Chem.* **1998,** *22,* 393–394.
87. Sanzenbacher, R.; Bottcher, A.; Elias, H.; Huber, M.; Haase, W.; Glerup, J.; Jensen, T. B.; Neuburger, M.; Zehnder, M.; Springborg, J.; Olsen, C. E. *Inorg. Chem.* **1996,** *35,* 7493–7499.
88. Marko, I. E.; Giles, P. R.; Tsukazaki, M.; Brown, S. M.; Urch, C. J. *Science* **1996,** *274,* 2044–2046.
89. Marko, I.; Gautier, A.; Chelle-Regnaut, I.; Giles, P. R.; Tsukazaki, M.; Urch, C. J.; Brown, S. M. *J. Org. Chem.* **1998,** *63,* 7576–7577.
90. Marko, I. E.; Giles, P. R.; Tsukazaki, M.; Chelle-Regnaut, I.; Gautier, A.; Brown, S. M.; Urch, C. J. *J. Org. Chem.* **1999,** *64,* 2433–2439.
91. Marko, I. E.; Gautier, A.; Mutonkole, J. L.; Dumeunier, R.; Ates, A.; Urch, C. J.; Brown, S. M. *J. Organomet. Chem.* **2001,** *624,* 344–347.
92. Rozantsew, E. G.; Ulrich, H. "*Free Nitroxyl Radicals*"; Plenum Press: NJ-London, **1970.**
93. Bowman, D. F.; Gillan, T.; Ingold, K. U. *J. Am. Chem. Soc.* **1971,** *93,* 6555–6561.
94. Adam, W.; Saha-Moller, C. R.; Ganeshpure, P. A. *Chem. Rev.* **2001,** *101,* 3499–3548.
95. Bragd, P. L.; van Bekkum, H.; Besemer, A. C. *Top. Catal.* **2004,** *27,* 49–66.
96. Semmelhack, M. F.; Schmid, C. R.; Cortes, D. A.; Chou, C. S. *J. Am. Chem. Soc.* **1984,** *106,* 3374–3376.
97. Betzemeier, B.; Cavazzini, M.; Quici, S.; Knochel, P. *Tetrahedron Lett.* **2000,** *41,* 4343–4346.
98. Ragagnin, G.; Betzemeier, B.; Quici, S.; Knochel, P. *Tetrahedron* **2002,** *58,* 3985–3991.
99. Ansari, I. A.; Gree, R. *Org. Lett.* **2002,** *4,* 1507–1509.
100. Dijksman, A.; Arends, I. W. C. E.; Sheldon, R. A. *Org. Biomol. Chem.* **2003,** *1,* 3232–3237.
101. Miyazawa, T.; Endo, T., Shiihashi, S.; Okawara, M. *J. Org. Chem.* **1985,** *50,* 1332–1334.
102. Bobbitt, J. M.; Flores, M. C. L. *Heterocycles* **1988,** *27,* 509–533.
103. Gamez, P.; Arends, I. W. C. E.; Sheldon, R. A.; Reedijk, J. *Adv. Synth. Catal.* **2004,** *346,* 805–811.
104. Caneschi, A.; Grand, A.; Laugier, J.; Rey, P.; Subra, R. *J. Am. Chem. Soc.* **1988,** *110,* 2307–2309.
105. Velusamy, S.; Punniyamurthy, T. *Eur. J. Org. Chem.* **2003,** 3913–3915.
106. Bencini, A.; Ciofini, I.; Giannasi, E.; Daul, C. A.; Doclo, K. *Inorg. Chem.* **1998,** *37,* 3719–3725.
107. Laugier, J.; Latour, J. M.; Caneschi, A.; Rey, P. *Inorg. Chem.* **1991,** *30,* 4474–4477.
108. Ettinger, T. M. J.; Kosman, D. J. In: "*Copper Proteins*"; Ed. Spiro, T. G.; Wiley&Sons: New York, **1981,** pp. 219–261.
109. Whittaker, M. M.; Whittaker, J. W. *Biochemistry* **2001,** *40,* 7140–7148.

INDEX

CONTENTS OF PREVIOUS VOLUMES